AF324605

CLIMATE CHANGE

CLIMATE CHANGE

AN ENCYCLOPEDIA OF SCIENCE, SOCIETY, AND SOLUTIONS

Volume 2
Weather and Global Warming

Bruce E. Johansen

An Imprint of ABC-CLIO, LLC
Santa Barbara, California • Denver, Colorado

Copyright © 2017 by ABC-CLIO, LLC

All rights reserved. No part of this publication may be reproduced, stored in
a retrieval system, or transmitted, in any form or by any means, electronic,
mechanical, photocopying, recording, or otherwise, except for the inclusion of brief
quotations in a review, without prior permission in writing from the publisher.

Library of Congress Cataloging-in-Publication Data
Names: Johansen, Bruce E. (Bruce Elliott), 1950–
Title: Climate change : an encyclopedia of science, society, and solutions /
    [edited by] Bruce E. Johansen.
Description: Santa Barbara, California : ABC-CLIO, LLC, 2017. |
    Includes bibliographical references and index.
Identifiers: LCCN 2016059526 (print) | LCCN 2016059775 (ebook) |
    ISBN 9781440840852 (hardcopy : alk. paper : set) | ISBN 9781440848698
    (vol. 1) | ISBN 9781440848704 (vol. 2) | ISBN 9781440848711 (vol. 3) |
    ISBN 9781440840869 (ebook)
Subjects: LCSH: Climatic changes—Encyclopedias.
Classification: LCC QC902.92 .C54 2017 (print) | LCC QC902.92 (ebook) |
    DDC 577.2/2—dc23
LC record available at https://lccn.loc.gov/2016059526

ISBN: 978-1-4408-4085-2 (set)
        978-1-4408-4869-8 (vol. 1)
        978-1-4408-4870-4 (vol. 2)
        978-1-4408-4871-1 (vol. 3)
EISBN: 978-1-4408-4086-9

21 20 19 18 17   1 2 3 4 5

This book is also available as an eBook.

ABC-CLIO
An Imprint of ABC-CLIO, LLC

ABC-CLIO, LLC
130 Cremona Drive, P.O. Box 1911
Santa Barbara, California 93116-1911
www.abc-clio.com

This book is printed on acid-free paper ∞

Manufactured in the United States of America

Contents

Oceans

VOLUME II: WEATHER AND GLOBAL WARMING

Weather

Global Warming

VOLUME III: HUMAN IMPACT AND PRIMARY DOCUMENTS

Agriculture and Food

Human Health

Indigenous Peoples and Cultures

WEATHER

OVERVIEW

Weather is the story and climate is the plot. Both are changing, although not always in expected ways. Who might have expected, for example, that a frigid Arctic winter in the eastern United States (or a series of "snow hurricanes" in Boston) could be a signal of a warming planet? Yet, in February 2015, a jet stream arching northward over Alaska brought unseasonal warmth there and in the western United States, while conveying record cold east of the Mississippi River. California had its warmest winter on record for the second year in a row, and San Francisco had its first January without any rain. Flowers bloomed in Oregon and ski slopes went bare, while Boston was paralyzed by eight feet of snow in roughly six weeks. Scientists theorized that the contortion of the jet stream was related to a receding Arctic ice cap.

The story featured some amazing contradictions, but the planetwide plot was record warmth. During 2014, the only populated area on Earth that had below-average temperatures was the eastern half of the United States, where a kink in the jet stream pulled air from the Arctic. The same jet stream flowed abnormally northward into Alaska, aiding record heat there and in the U.S. West. Much of Siberia and Europe also set record highs, as did Australia and Argentina with sizzling summers. In 2015, the only areas with below-average temperatures were adjacent to Greenland and Antarctica, where melting glaciers cooled nearby ocean waters.

"It is exceptionally unlikely that we would be witnessing a record year of warmth, during a record-warm decade, during a several decades-long period of warmth that appears to be unrivaled for more than a thousand years, were it not for the rising levels of planet-warming gases produced by the burning of fossil fuels," Michael E. Mann, a climate scientist at Pennsylvania State University, told Justin Gillis of *The New York Times* (2015, January 16). "We can anticipate that global warming will continue on decadal time scales, because Earth is out of energy balance—more energy coming in than going out—as a result of increased atmospheric greenhouse gases, especially carbon dioxide," asserted James E. Hansen and colleagues (2015).

A World Bank report issued in 2016 emphatically states that global warming's most severe impact worldwide will be on water supplies because of changes in the hydrological cycle. The report suggests that by 2050 inadequate water supplies could reduce gross domestic product by as much as 6 percent in some parts of the world, "sending them into sustained negative growth" (Mooney 2016). "When we look at any of the major impacts of climate change, they one way or another come through water," said Richard Damania, an economist at the bank and lead author of the report. "So it will be no exaggeration to claim that climate change is really in fact about hydrological change. Growing populations, rising incomes, and expanding cities will converge upon a world where the demand for water rises exponentially, while supply becomes more erratic and uncertain," the report said. It projects large increases in demand for water as climate constricts supplies. Within three decades, the "global food system will require between 40 to 50 percent more water; municipal and industrial water demand will increase by 50 to 70 percent; the energy sector will see water demand increase by 85 percent; and the environment, already the residual claimant, may receive even less" (Mooney 2016).

Droughts and deluges can alternate in this new weather regime. As temperatures have risen in a linear fashion, the hydrological cycle has been changing exponentially, with some wild swings in the same areas. In the midst of a years-long drought, on June 4 and 5, 2015, the Denver area was pounded by as much as four feet of hail. People dug out with snow shovels. A crippling drought had lasted five years (as of this writing) in California by 2015 as the United States as a whole recorded its wettest May on record—that was, until 2015 and 2016, when an El Niño weather pattern in the equatorial Pacific Ocean doused Southern California in occasional flooding rains.

A relatively modest change in average temperatures can produce large increases in extremes. In the case of climate change, heat waves, deluges, and droughts increase several-fold even with a global average temperature rise of a degree or two. Even a temperature rise of 2°C, considered the maximum safe level to avoid climatic chaos, may produce a 14-fold increase in the chances of a major heat wave at a given location. If the average temperature rises by more than roughly the same amount according to some models (Fischer and Knutti 2015), the probability of heat extremes at a given location rises 62-fold.

Fall Colors and Warming

In New England, which has long been noted for sharp and brilliant fall colors, observers have found those colors muted in recent years because of rising temperatures. Falling temperatures in the fall are part of the natural cycle that halts production of chlorophyll in plants and causes leaves to change color (sustained freezing early in the season kills leaves before they can change color, however). Shorter days cause plants to form a layer at the base of their stems that cuts off water and nutrients, and cold nights (close to or just above freezing) accelerate this process.

Warmer temperatures also encourage fungi that kill the leaves before they change color, especially among the red and sugar maples that are the most colorful fall trees in New England. The turning of leaves is such a tourist draw in Vermont ($3.4 million spent during 2005) that state tourism officials have been denying that any problem exists. In the meantime, the Associated Press has quoted other Vermonters as saying that the leaves not only are blander most years but also that the peak season now occurs 10 days to two weeks later than prior years ("Climate Blamed" 2007).

Many trees go from the deep green of late summer to the dull brown of late fall without the intervening bright yellows and oranges that used to occur in October. "It's nothing like it used to be," said University of Vermont plant biologist Tom Vogelmann ("Climate Blamed" 2007).

Further Reading

"Climate Blamed for Blah Foliage." Associated Press in *Omaha World-Herald*, October 29, 2007, 3A.

"People can argue that we had these kinds of extremes well before human influence on the climate   we had them centuries ago," said Erich M. Fischer, lead author of a study published by *Nature Climate Change*. "And that's correct. But the odds have changed, and we get more of them" (Gillis 2015, April 27). "Climate change doesn't 'cause' any single weather event in a deterministic sense," said Fischer, a climate scientist at the Swiss Federal Institute of Technology in Zurich, Switzerland, the study's lead author. "But a warmer and moister atmosphere does clearly favor more frequent hot and wet extremes" (Schiermeier 2015).

This section focuses on how weather and climate work and how rising greenhouse gas levels in the atmosphere are changing weather and climate. We begin with a look at past climates in which high greenhouse gas levels played a role and then move on to peat's large role in carbon dioxide and methane emissions, as well as changes in atmospheric circulation, heat wave intensity and frequency, the role of water vapor in global warming, and the relationship between ozone depletion and atmospheric warming in the Arctic and Antarctic.

The next area of emphasis involves anecdotal observations and scientific context for rising temperatures around the world and the role played by rising levels of carbon dioxide, methane, and other heat-trapping gases. As previously observed, the relationship is not linear and not everyone receives the same dollop of heat at the same time, something residents of eastern North America have learned in recent years courtesy of the Arctic Oscillation. This section concludes with entries that describe carbon cycle feedbacks and how they compound each other as well as solar influences on climate.

Attention next focuses on storms, including observations and scientific debates regarding the intensity of tropical cyclones around the world, as well as thunderstorms and tornadoes and the expansion of tropical zones northward and southward of the equator. This section concludes with an entry on intensification of "lake-effect" blizzards on the shores of North America's Great Lakes, as well as the previously mentioned snow hurricanes that have devastated coastal New England in recent years.

This initial section continues with work on changes in the hydrological cycle, including the marked tendency in a warming world toward increasing intensity of droughts and deluges, with special emphasis on episodic droughts in the U.S. West and Great Plains. Research from the National Aeronautics and Space Administration (NASA) indicates that global warming plays a role in a long-term "megadrought" that has developed in these areas. Attention is also paid here to models of extreme weather and the role of El Niño and La Niña in changes in the hydrological cycle. Additional work details the effects of drought on forests, including in the U.S. West and worldwide, and the influence of deforestation on the carbon cycle, with special attention to the Amazon River valley.

The section on climate and weather concludes with an eye to the future, further developing the propensity of carbon cycle feedbacks to compound each other and provoke accelerating temperature changes and climatic changes in a destructive manner. Subjects include tar sands and tipping points, the abrupt nature of climate change, thermal inertia, climatic equilibrium, feedback loops, and tipping points.

Further Reading

Fischer, Erich M., and Reto Knutti. "Anthropogenic Contribution to Global Occurrence of Heavy-Precipitation and High-Temperature Extremes." *Nature Climate Change*, April 27, 2015. http://www.nature.com/nclimate/journal/vaop/ncurrent/full/nclimate2617.html.

Gillis, Justin. "2014 Breaks Heat Record, Challenging Global Warming Skeptics." *The New York Times*, January 1, 2015. http://www.nytimes.com/2015/01/17/science/earth/2014 -was-hottest-year-on-record-surpassing-2010.html.

Gillis, Justin. "New Study Links Weather Extremes to Global Warming." *The New York Times*, April 27, 2015. http://www.nytimes.com/2015/04/28/science/new-study-links-weather -extremes-to-global-warming.html.

Hansen, James E., et al. "Global Temperature in 2014 and 2015." January 16, 2015. http:// csas.ei.columbia.edu/2015/01/16/global-temperature-in-2014-and-2015/.

Mooney, Chris. "World Bank: The Way Climate Change Is Really Going to Hurt Us Is through Water." *Washington Post*, May 3, 2016. https://www.washingtonpost.com/news /energy-environment/wp/2016/05/03/world-bank-the-way-climate-change-is-really -going-to-hurt-us-is-through-water/?wpmm=1&wpisrc=nl_evening.

Schiermeier, Quirin. "Global Warming Brews Weird Weather." *Nature* Online, April 27, 2015. http://www.nature.com/news/global-warming-brews-weird-weather-1.17407.

DROUGHT, UNITED STATES
California

In 2015, scientists reported that a five-year drought in California had been intensified 15 percent to 20 percent by rising global temperatures. They projected that drought is almost certain to get worse as temperatures continue to rise. Melting ice in the Arctic may play a role in anchoring high pressure off the North American west coast and diverting storms in winter (as well as the usual summer pattern), a matter that is open to scientific debate. There is little question, however, that warmer air accelerates evaporation (Gillis 2015). A. Park Williams and colleagues wrote in *Geophysical Research Letters* (2015),

Regionally, the 2012–2014 drought was record breaking in the agricultur-ally important southern Central Valley and highly populated coastal areas. Contributions of individual climate variables to recent drought are also exam-ined, including the temperature component associated with anthropogenic warming. Precipitation is the primary driver of drought variability but anthro-pogenic warming is estimated to have accounted for 8 to 27 percent of the observed drought anomaly in 2012 to 2014 and 5 to 18 percent in 2014. Although natural variability dominates, anthropogenic warming has substan-tially increased the overall likelihood of extreme California droughts.

"California is facing a new climate reality," wrote Noah S. Diffenbaugh and Chris-topher B. Field in *The New York Times* (2015). "Extreme drought is more likely. The state's water rights, infrastructure and management were designed for an old climate, one that no longer exists." Their research, they said,

has shown that global warming has doubled the odds of the warm, dry con-ditions that are intensifying and prolonging this drought, which now holds records not only for lowest precipitation and highest temperature, but also for lowest spring snowpack in the Sierra Nevada in at least 500 years. These changing odds make it much more likely that similar conditions will occur again, exacerbating other stresses on agriculture, ecosystems, and people.

At the same time, the intensity of extreme wet periods will increase even during droughts because a warming atmosphere can carry a larger load of water vapor. El Niño conditions during 2015 and 2016 forced Californians to face both flooding and drought simultaneously. The more rainfall there is, the more water will be lost as runoff or river flow, increasing the risk of flooding and landslides. Add in the fact that the drought and wildfires have hardened the ground, and a paradox arises wherein the closer El Niño comes to deliv-ering enough precipitation to break the drought this year, the greater the potential for those hazards. (Diffenbaugh and Field 2015)

An August 2015 analysis of tree rings from blue oaks in California's Central Val-ley indicated that snowfall in nearby mountains was lower by 2015 than it had been in 500 years. By that time, the Sierra Nevada was entirely bereft of snow for the first time in 75 years of recorded history. "The results were astonishing," said Val-erie Trouet, an associate professor at the University of Arizona, an author of the analysis in *Nature Climate Change* (Belmecheri et al. 2015). "We knew it was an all-time low over a historical period, but to see this as a low for the last 500 years, we didn't expect that. There's very little doubt about it," she said (Fears 2015).

California's drought caused $2.2 billion in agricultural loses as fields were left fallow in just 2014 alone. More than 12 million trees had died in the five-year drought by 2015. This may be a foretaste: "Future droughts will be compounded by more-intense heat waves and more wildfires. Soaring temperatures will increase demand for energy just when water for power generation and cooling is in short supply. Such changes will increase the tension between human priorities and nature's share" (AghaKouchak et al. 2015).

During the winter of 2014–2015, rain teased California for a few December days but evaporated in January with record heat and low rainfall. By March, temperatures in Southern California had risen to nearly 100°F in some places as mountain snowpacks eroded to the lowest on record. On March 3, statewide snowpack was only one-fifth of usual. Rising temperatures were shortening the period in which snow could accumulate and were melting snow at lower altitudes. What little precipitation did fall came more often as rain, which melted existing snow and then soaked into the ground before reaching reservoirs. Winter 2013–2014 was California's warmest on record, at 45.6°F, until the winter of 2014–2015 averaged 47.4°F. Higher temperatures increase evaporation and human demand for water. The winter of 2016–2017 brought record rainfall.

"The normal cyclical conditions in California are different now from what they used to be, and that's not because the long-term annual precipitation changed," said Noah Diffenbaugh, a senior fellow at the Stanford Woods Institute for the Environment. "What is really different is there has been a long-term warming in California," he said. "And we know from looking at the historical record that low precipitation years are much more likely to result in drought conditions if they occur with high temperatures" (Nagourney 2015).

Drought: U.S. West

Since the 1990s, an epic drought has seized the western United States from the Navajo reservation (which has been inundated by sand dunes) to California, where the continent's most productive agricultural valleys have been parched. With a few exceptions, the dominant jet stream pattern, locked in place by advancing Hadley cells in the upper troposphere, has moved rain-bearing storms to the north, even during the usual winter rainy season. The winter of 2010–2011 was an exception. Snow was so plentiful that the Missouri flooded for months all summer long. However, the drought soon locked in once again.

"Widespread annual droughts, once a rare calamity, have become more frequent and are set to become the 'new normal,'" wrote Christopher Schwalm, Christopher A. Williams, and Kevin Schaeffer in *The New York Times* (2012). Schwalm is a research assistant professor of Earth sciences at Northern Arizona University, Williams is an assistant professor of geography at Clark University, and Schaefer is a research scientist at the National Snow and Ice Data Center.

"It sounds like the plot of a disaster movie that somehow fuses the dangers of drought, earthquakes, and overpopulation—'Mad Max' meets 'Soylent Green'," wrote Justin Moyer in the *Washington Post* in 2015.

First, we burned too many fossil fuels. Planet Earth heated up. Meanwhile, in a beautiful state known for its endless sunshine, legislators made it too easy to farm the desert. With land overdeveloped and water resources overtaxed, and inevitable drought fused with water mismanagement and global warming to bring about a slowly unrolling apocalypse west of the Rockies.

With snow scarce in western North America and temperatures abnormally warm, intense wildfires broke out early in May 2015 as far north as the Canada's Northwest Territories. The fires were so numerous that the NASA Earth Observatory described "[heavy] smoke from the growing blazes merged with lingering smoke from previous days, making air quality a major concern." "Fires this early are very unusual," said Mike Flannigan of the University of Alberta. He noted that the Northwest Territories and northern Alberta usually see the most active part of the fire season in July, and British Columbia's fire season typically runs from the end of July through August. Some of the fires were so large and intense that they generated their own pyrocumulonimbus clouds and thunderstorms. In Alberta, NASA said, more than 60 fires displaced thousands of residents and led some energy companies to suspend oil operations and evacuate staff, which reduced tar-sands production ("Intense Fires" 2015; "Alberta Wildfires" 2015).

Wildfires across Canada during the summer of 2015 exported smoke to urban areas. Across the western half of North America, from Alaska to California, record heat and drought provoked an unprecedented fire season. During the first week of July, smoke that had risen into the stratosphere from fires in Alberta and Saskatchewan obscured the sun across the U.S. upper Midwest, including in Omaha, Des Moines, and other cities. Smoke was so thick some days in Denver that flights were delayed at its international airport. Colorado state health officials warned people with asthma to stay indoors on days when the smoke was thickest.

Wildfires in Alaska and Canada

By the end of July 2015, as many as 300 wildfires, most started by lightning, were burning at one time across Alaska as record temperatures prompted early snowmelt and drought turned forests to tinder. Nearly 5 million acres had already burned, making the season among Alaska's worst, with several weeks to go until the first snows of early fall snuff residual cinders. During the record year 2004, 6.59 million acres burned (Mooney 2015).

More land burned in Alaska in 2015 than in the rest of the United States combined, producing smoke so dense on some days that many outdoor events were canceled in Fairbanks. The intense burning was adding carbon dioxide to the atmosphere in two ways: the torched trees themselves, as well as the smoldering of what was once frozen ground (permafrost), much of which is laced with carbon-rich organic peat. "The more severe the fire, the deeper that it burns through the organic layer, the higher the chance it will go through this complete conversion," said Ted Schuur, a Northern Arizona University ecologist and permafrost specialist. "What happens in the summer of 2015 has the potential to change the whole trajectory of [the burned area] for the next 100 years or more" (Mooney 2015). "Across the Arctic, permafrost in North America, Northern Europe, and Siberia contains twice as much carbon as Earth's atmosphere. The permafrost that we degrade now in these forest fires might never return in our lifetimes," Schuur said (Mooney 2015).

Smoke from hundreds of fires burning across British Columbia swept into Vancouver on the coast and extended into nearby interior areas. NASA detected a thick pall of smoke July 5 and 6 that "settled over Vancouver and adjacent areas of British Columbia, leading some residents to wear face masks and health officials to warn residents and World Cup tourists against outdoor activities" ("Metro Vancouver" 2015). "Tan and gray smoke almost completely obscured the Strait of Georgia and southern Vancouver Island. Winds shifted abruptly between July 5 and 6, driving the smoke plume toward the east, dispersing it in some places while fouling the air in areas to the east, such as the Fraser Valley" ("Smoke Blankets" 2015). For a few days, officials in Vancouver compared its air quality to that of Beijing.

"When it comes to the skies overhead, it appears that orange is the new blue," reported Yvonne Zacharias in the *Vancouver Sun* (2015).

> Get ready for more of the apocalyptic haze as scientists warn that climate change is blazing a whole new trail for the Earth. He predicted global warning will not only increase the length and intensity of the forest fire season, but will also affect sockeye salmon runs, ski resorts like Whistler, which is changing to an all-season resort, and the inaccessibility of remote areas in the north because of an early breakup that makes transportation routes impassible. And as for forestry, "we can no longer rely on the trees we are planting today to be there in 80 or 100 years' time," he said. "I hope this is a wake-up call," said UBC forestry professor Lori Daniels. In other words, by the time today's youth grow into adulthood, "this is what is going to become the new norm." (Zacharias 2015)

As smoke from at least 110 fires in the province of Saskatchewan poured southward into large areas of the United States, some 13,000 indigenous people of the Lac La Ronge and other bands were forced to evacuate their homes during the first week of July 2015. One hundred miles to the south of this evacuation, the Montreal Lake Cree Nation declared a state of emergency as its community was surrounded by a wall of fire so fierce that several hundred firefighters were forced to withdraw. Residents were forced out of their homes as several structures burned to the ground.

"There was no stopping it—four water bombers, two helicopters bucketing, crews on the ground doing whatever they could, but we just couldn't do anything when the wind picked up," said Chief Edward Henderson, describing the wall of flames in La Ronge. "I've never seen anything like this," Lac La Ronge Indian Band Chief Tammy Cook-Searson told CBC News. "We've had to evacuate all six of our communities" ("Saskatchewan" 2015). At the height of the fires, 2,500 firefighters and Canadian armed forces troops from Ontario, New Brunswick, Newfoundland, Alberta, and Saskatchewan battled the flames along with reinforcements from South Dakota.

As many as 80 million piñon trees (the state tree of New Mexico) died in Arizona and New Mexico between 2001 and 2005 because of intense drought. That represents some 90 percent of the area's piñon trees (Carlton 2006, A1). Four million have died in Santa Fe alone. The trees also have fallen victim to a bark beetle the size of a grain of rice that feasts on dying and diseased trees.

Eroding Snowpacks and Urban Water Shortages

Confirming many anecdotal observations, a study of data collected over the past 50 years and published online on January 31, 2008, by the journal *Science* said that sharp declines in mountain snowpacks are being caused primarily by human-induced global warming, As temperatures warm, mountains receive more rain and less snow, which alters the hydrological cycle in many ways that are detrimental to nature and human activity. Snows have been melting earlier and more rivers have been running dry in the summer.

The study examined natural variability of temperatures and precipitation and the effect of sun-spot cycles as well as worldwide volcanic activity and human-induced greenhouse gas releases. "We've known for decades that the hydrology of the West is changing, but for much of that time people said it was because of Mother Nature and that she would return to the old patterns in the future," said lead author Tim Barnett of the Scripps Institution of Oceanography at the University of California–San Diego. "But we have found very clearly that global warming has done it, that it is the mechanism that explains the change and that things will be getting worse" (Kaufman 2008).

Gradual annual melting of snowpacks is vital for generating electricity, agriculture, drinking water, and to provide enough water for salmon to reach their breeding grounds. More abundant rain tends to run off too quickly to be useful, especially during dry summers, and may cause floods. "Our results are not good news for those living in the western United States," the researchers wrote, adding that the changes may make "modifications to the water infrastructure of the western U.S. a virtual necessity" (Kaufman 2008).

Water Supplies and Snowpack Erosion in the Pacific Northwest

What harm could global warming do in Seattle, a place where a popular joke is that when summer comes, everyone hopes it arrives on a Saturday? Mark Twain visited that area and allegedly said that the mildest winter he ever spent was a summer on Puget Sound. Seattle residents should not hold out hope that global warming will improve their chilly, soggy climate because its most obvious evidence probably will not arrive there in the summer. Residents of the Pacific Northwest probably will feel the brunt of climate change in the winter when snow levels will rise and wash away the next summer's irrigation and power-generating snowpack.

To forecast the severity of snowpack loss from global warming in the Washington Cascades, scientists first took a step back in time. They examined a half-century of temperature and snowfall data at weather stations from Arizona to British Columbia. "The results were striking; I was shocked by the magnitude of the (snow-moisture) declines," said University of Washington (UW) climate scientist Phil Mote, who conducted much of the survey. "In some places, particularly in Oregon, we saw declines of 100 percent. It had gotten so warm there was no snow left in April at all" (Welch 2004). Mote studied snowpack records for April 1 in four Northwest states and British Columbia at 145 sites from 1950 through 1992. He found that the amount of snow between elevations of 3,000 and 9,000 feet had decreased by an average of 20 percent or more. "I was surprised by the result," said Mote, who works with a group of UW scientists called the Climate Impacts Group. "There's already a clearer regional signal of warming in the mountains than we expected" (Gordon 2003).

"If you think the water fights we have now are intense . . . you ain't seen nothing yet," University of Washington Professor Ed Miles said during the 2004 annual meeting of the American Association for the Advancement of Science in Seattle (Welch 2004). Miles presented evidence that moisture in snow that nourishes the West's network of rivers (and thus its farms and cities) has been steadily declining since at least World War II.

Further Reading

Gordon, Susan. "U.S. Pacific Northwest Gets Reduced Supply of Snow, Climate Study Says." *Tacoma News Tribune*. February 7, 2003, n.p. (LEXIS).
Welch, Craig. "Global Warming Hitting Northwest Hard, Researchers Warn." *Seattle Times*, February 14, 2004. http://seattletimes.nwsource.com/html/localnews/2001857961_warming14m.html (no longer available).

The pattern was not linear; some years (such as 2007–2008) were still notable for heavy snows. This study indicated, however, that since 1950 snowpack water content on April 1 has declined on average in eight of the nine mountainous areas—from 10 percent in the Colorado Rockies to 40 percent in the Oregon Cascades. The southern Sierra Nevada range was alone in not showing a decline.

Jonathan Overpeck, a climate scientist at the University of Arizona in Tucson, said the new study "closes the circle" in terms of understanding what is happening to the climate of the West. "Almost all of the models we've seen in recent years show the area becoming warmer and more arid due to climate change, but the question was always whether we could believe them," he said. "Now someone has done the statistical analysis to connect the dots so they can say with real confidence that this is happening because of greenhouse gases" (Kaufman 2008).

Great Plains: Warming and Drought in the Past

Paleoclimatic precedent exists for the kind of substantial drought that has become common in the United States West in recent years. Prevailing hot southwest winds during the growing season in the Great Plains of the United States 800 to 1,000 years ago spawned a drought that destroyed the Anasazi civilization in the Southwest. During this period, the prevailing winds shifted from the south to the southwest, bringing in moisture from the Gulf of Mexico, which is hot and dry like now.

A team of scientists (Sridhar Venkataramana, David B. Loope, James B. Swinehart, Joseph A. Mason, Robert J. Oglesby, and Clinton M. Rowe) from the University of Nebraska at Lincoln and the University of Wisconsin have discerned a major wind shift by studying the shapes of dunes in the Sand Hills, which once constituted a sea bottom, in north-central Nebraska. During the Medieval Warm Period, the drought killed the dunes' vegetative cover. The winds caused the dunes to migrate, leaving today's patterns, which are again anchored by grasses. As the team noted, "Longitudinal dunes built during the Medieval Warm Period (800 to 1,000 years before now) record the last major period of sand mobility. These dunes are oriented northwest to southeast and are composed of cross-strata with bipolar dip directions" (Sridhar et al. 2006, 345).

"Such a westward shift [in prevailing winds] would . . . greatly reduce the flow of moist air into the central Great Plains, thereby generating severe drought," the scientists wrote in July 2006 (Sridhar et al. 2006, 346). Although the "dry line" now brings spring and summer thunderstorms that traverse Nebraska from the Rocky Mountains, "during the Medieval warm Period the mean position of the dry line moved much further east, such that the Sand Hills were most often in the dry, hot air with greatly reduced precipitation" (Sridhar et al. 2006, 346). A similar wind shift coupled with depletion of the Ogallala Aquifer could make much of Nebraska too dry for agriculture in coming decades. Large parts of this aquifer, the largest body of underground water in North America, are being drawn down by feet per year while precipitation only recharges it in inches per year.

Such dune patterns can be used as proxies for atmospheric circulation that governs weather and climate over periods decades to centuries long but leaves no records. Approximately half the Sand Hills' annual precipitation falls during May, June, and July, nourishing a grassland ecosystem. During the Dust Bowl years of the 1930s (and to a lesser extent during droughts in the 1950s), part of this ecosystem broke down, producing what scientists call "isolated blowouts" (Sridhar et al. 2006, 345). Historical accounts indicate that some dune crests lost their grass to drought occasionally even during the generally cooler 19th century. These accounts paled, however, beside the sustained drought that accompanied the Medieval Warm Period.

Megadrought Projected for the U.S. West

In early 2015, NASA issued a warning: if current rates of increase in greenhouse gas emissions continue, most of North America, except for the Arctic and near-Arctic, would suffer a "megadrought" during the balance of the 21st century that will make what the U.S. Southwest has thus endured look like an appetizer. According to NASA, the drought between 2050 and 2100 will be the worst in at least 1,000 years since the Chaco drought that devastated the Anazasi and Mayan civilizations between 1100 and 1200 CE. Jason Smerdon, a climate scientist at Columbia University's Lamont–Doherty Earth Observatory in Palisades, New York, coauthored this study and said the impending megadrought will make the Chaco drought look "quaint" (Underwood 2015). In addition, in the midst of the United States' worst wildfire season in 2015, NOAA released a study projecting conditions six times as bad by 2050 because of global warming, with droughts and longer fire seasons aggravating fire conditions (Rice 2015, August 28).

"Future precipitation trends, based on climate model projections for the coming fifth assessment from the Intergovernmental Panel on Climate Change, indicate that droughts of this length and severity will be commonplace through the end of the century unless human-induced carbon emissions are significantly reduced," the team wrote. Drought tends to compound the warming of the atmosphere because plants stressed by lack of water photosynthesize less and take in only half of the carbon dioxide they would if they were adequately hydrated (Schwalm et al. 2012).

The effects of the great drought in the U.S. West exceed that of any other on record, including the Dust Bowl of the 1930s, which mainly gripped the center of the continent. "While that drought saw intervening years of normal rainfall, the years of the turn-of-the-century drought were consecutive. More seriously still, long-term climate records from tree-ring chronologies show that this drought was the most severe event of its kind in the western United States in the past 800 years," Schwalm and colleagues wrote (2012).

Most significantly, this drought is probably not cyclical. A warming atmosphere is locking in changes in upper-air circulation that may freeze it in place, with a few exceptions (see "Atmospheric Circulation Changes" and the discussion of Hadley cells). As Schwalm and colleagues wrote, "These climate-model projections suggest that what we consider today to be an episode of severe drought might even be classified as a period of abnormal wetness by the end of the century and that a coming megadrought—a prolonged, multidecade period of significantly below-average precipitation—is possible and likely in the American West" (Schwalm et al. 2012).

Scientists are measuring reductions in snowpack across western North America's higher elevations, which are important because they provide agricultural and urban sustenance. As noted by another team writing in *Science* in 2011,

In western North America, snowpack has declined in recent decades, and further losses are projected through the 21st century. Here, we evaluate the uniqueness of recent declines using snowpack reconstructions from 66 tree-ring chronologies in key runoff-generating areas of the Colorado, Columbia, and Missouri River drainages. Over the past millennium, late 20th-century snowpack reductions are almost unprecedented in magnitude across the northern Rocky Mountains and in their north-south synchrony across the cordillera. Both the snowpack declines and their synchrony result from unparalleled springtime warming that is due to positive reinforcement of the anthropogenic warming by decadal variability. The increasing role of warming on large-scale snowpack variability and trends foreshadows fundamental impacts on streamflow and water supplies across the western United States. (Pederson 2011, 332)

Other studies support these conclusions. In 2010, Jonathan Overpeck and Bradley Udall wrote in *Science* (2010),

Signs of climate change in western North America . . . include soaring temperatures, declining late-season snowpack, northward-shifted winter storm tracks, increasing precipitation intensity, the worst drought since measurements began, steep declines in Colorado River reservoir storage, widespread vegetation mortality, and sharp increases in the frequency of large wildfires. These shifts have taken place across a region that also saw the nation's highest population growth during the same period. . . . The climate of the western United States could become much drier over the course of this century. (Overpeck and Udall 2010, 1642)

CO_2 Emissions and Tree Mortality

Mortality rates have risen substantially even in previously healthy conifer forests of the western United States averaging more than 200 years of age, doubling during two to three decades because new trees often fail to replace those that die. Drought provoked by rising temperatures is a major reason for rising death rates of pine, fir, hemlock, and others, according to a study released early in 2009. The increasing mortality of trees also reduces the ability of the forests to absorb carbon dioxide (Van Mantgem et al. 2009; Pennisi 2009).

If greenhouse gas emissions continue to increase along current trajectories throughout the 21st century, there is an 80 percent likelihood of at least one decades-long megadrought in the Southwest and Central Plains between the years 2050 and 2099 ("NASA Study" 2015). The high-emissions scenario projects an atmospheric carbon dioxide level of 1,370 parts per million (ppm) by 2100, whereas the moderate emissions scenario projects 650 ppm by 2100, compared to around 400 ppm in 2015 ("NASA Study" 2015). "We can't really understand the full variability and the full dynamics of drought over western North America by focusing only on the last century or so," Cook said. "We have to go to the paleoclimate record, looking at these much longer time scales, when much more extreme and extensive drought events happened, to really come up with an appreciation for the full potential drought dynamics in the system" ("NASA Study" 2015).

"Summers are getting longer," said Nathan L. Stephenson, of the U.S. Geological Survey (USGS), a coauthor of the study with Phillip van Mantgem, also of USGS. "Trees are under more drought stress" (Navarro 2009). The recent warming in the West "has contributed to widespread hydrologic changes, such as a declining fraction of precipitation falling as snow, declining water snowpack content, earlier spring snowmelt and runoff, and a consequent lengthening of the summer drought," the scientists wrote (Eilperin 2009). "It's very likely that mortality rates will continue to rise," said Stephenson, adding that the death of older trees is rapidly exceeding the growth of new ones, which is analogous to a human community in which the deaths of old people surpass the number of babies being born. "If you saw that going on in your home town, you'd be concerned" (Eilperin 2009).

The Colorado River by 2004 was carrying only half as much water compared to the Dust Bowl drought of the 1930s. Utah's Lake Powell—the second biggest human-made lake in the United States—had lost nearly 60 percent of its water by 2004. Assuming current trends, the lake may lose its ability to generate electricity in three years' time (Lean 2004, 20). After nine years of drought that caused Lake Powell's water level to fall more than 100 feet, heavy snows during 2006–2007 recouped 50 feet of water (Clark 2008, 1D).

If global warming continues to intensify, the danger exists that decreasing winter snowpack, intensifying summer dry conditions, and increasing forest fires may reduce vegetation cover and regional soil-biosphere water-holding capacity. If these conditions reach sufficient intensity and geographical scale, they may become self-perpetuating; if so, we will have suddenly entered a long-term megadrought in the western United States. Weather would continue to fluctuate from year to year, but water supplies would be much more limited than in prior decades and dust storms may become frequent. We cannot say what level of global warming is needed to cause such a megadrought, but the likelihood increases with increase of greenhouse gases and global warming (Hansen 2006, 31–32).

Not If, but When

In Canada's province of Alberta, experts have warned that the area is headed for a massive drought worse than the Dust Bowl conditions of the 1930s, largely because of dwindling supplies from mountain snowpack coupled with a rising number of people and livestock using water. "I see a disaster shaping up in Alberta and it's not a question of if, it's a question of when," said University of Alberta ecologist David Schindler, a water researcher and longtime critic of Alberta's water policies. "There is going to be a major drought coinciding with global warming and record numbers of people and livestock on the landscape. It's something that gives me nightmares" (Semmens 2003). Glacier cover in the Canadian Rockies is nearing its lowest point in at least 10,000 years. Statistics Canada reported that 1,300 glaciers in the country have lost between 25 percent and 75 percent of their mass since 1850. Most losses, however, have been recorded in the last 50 years. The report said that most of losses can be attributed to global warming (Paraske-vas 2003).

Schindler cited research by several Alberta scientists that indicates the average temperature for western Canada is the highest in at least 10,000 years and is expected to continue climbing. Such conditions will lead to continued glacial recession in the Rocky Mountains, less rain and snowfall, and increasing water evaporation. With some of Alberta's rivers already running as much as 80 percent lower than 100 years ago, Schindler said Alberta is also overdue for a drought that tends to hit the province every 100 to 150 years (Semmens 2003).

Water Levels Fall on Lake Mead

Even at 140 feet below its capacity level of 1,220 feet, Lake Mead—formed by the Colorado River backing up behind Hoover Dam—is a formidable body of water. Even in 2014, it drew 7 million tourists. It still covers more than 242,000 acres and reaches a depth of 300 feet in some places (Rojas 2015). Las Vegas, Nevada, which draws 90 percent of its water from Lake Mead, has been reducing its usage even as its population has risen; usage went down 30 percent between 2002 and 2015. "We came into the 21st century with Lake Mead essentially full, and over the last 14 years, it has declined about 130 feet," said John Entsminger, general manager of the Southern Nevada Water Authority. "What we've learned here in Las Vegas is it sure looks like the 21st century in the Western U.S. is a time when everyone is going to have to use less water" (Rojas 2015).

With weather patterns in a warming world favoring a drier American West, a study by scientists at the Scripps Institution of Oceanography indicates that Lake Mead, which spans the border of Nevada and Arizona and is a major source of water for Phoenix as well as Las Vegas, could run so low that water pumps may become useless. The study has become a center of controversy between scientists at Scripps and others at the U.S. Bureau of Reclamation who assert that the institution's climate models are too crude to forecast the future water level of a single large lake.

The Scripps study found that Lake Mead's water supply has a 50-percent chance of becoming unusable by 2021 if the demand for water remains at current levels and if global warming trends conform to midrange models. Researchers Tim P. Barnett and David W. Pierce of Scripps said that demand for the lake's water exceeds the amount added each year by runoff, even with an occasional snowy winter (such as 2007–2008 and 2015–2016). "We were really sort of stunned," Barnett said. "We didn't expect such a big problem basically right on our front doorstep. We thought there'd be more time." He added, "You think of what the implications are, and it's pretty scary" (Barringer 2008). By 2015, Lake Mead was down to 38 percent of its capacity, the lowest in its history (it was filled during the 1930s). In California in 2015, hydropower contributed 7 percent of the state's electricity, down from 23 percent in 2011 because of persistent drought. A new generation of solar and wind power technologies were filling that gap ("Lake Mead's Level" 2015; "Energy Is Another" 2015).

Other research has found that the Colorado River watershed, of which Lake Mead is a part, has had a long-standing tendency toward drought that makes the last century look unusually wet. Climate models also indicate that a warmer climate favors persistent drought in this area. The projected drying trends "are really quite robust and in many cases scary when you compare them to even the severe mega-droughts of the 1100s ands 1200s," said Benjamin Cook, of the NASA Goddard Institute for Space Studies in New York City (Underwood 2015).

"Natural droughts like the 1930s Dust Bowl and the current drought in the Southwest have historically lasted maybe a decade or a little less," said Ben Cook, climate scientist at NASA's Goddard Institute for Space Studies and lead author of the study. "What these results are saying is we're going to get a drought similar to those events, but it is probably going to last at least 30 to 35 years," Cook said (Rice 2015, February 13).

"The study is strong scientifically," said Jonathan Overpeck, at the University of Arizona's Institute of the Environment. "It strengthens our understanding of what is ahead, and it isn't pretty. . . . The future megadroughts will be hotter and more severe" than those in the past (Underwood 2015). Under such conditions, even large rivers such as the Colorado may run dry, according to this study. "They do the best job yet in linking likely future change in drought severity to that of the last 1,000 years as recorded by tree rings," Overpeck said. "The results are striking, and highlight how future temperature increases will trump precipitation change in driving totally unprecedented levels of drought unless we make dramatic reductions in greenhouse-gas emissions" (Rice 2015, February 13).

Further Reading

AghaKouchak, Amir, et al. "Water and Climate: Recognizing Anthropogenic Drought." *Nature* 524 (August 27, 2015): 409–411. http://www.nature.com/news/water-and-climate -recognize-anthropogenic-drought-1.18220.

"Alberta Wildfires Prompt Oil firms to Suspend Production and Evacuate Staff." *The Guardian* (U.K.), May 26, 2015. http://www.theguardian.com/world/2015/may/26/alberta -wildfires-oil-production-suspended-evacuations.

Barringer, Felicity. "Lake Mead Could Be Within a Few Years of Going Dry, Study Finds." *The New York Times*, February 13, 2008. http://www.nytimes.com/2008/02/13/us/13mead .html.

Belmecheri, Soumaya, et al. "Multi-Century Evaluation of Sierra Nevada Snowpack; California Snowpack Lowest in Past 500 Years." *Nature Climate Change*, September 2015. http://www.nature.com/nclimate/journal/vaop/ncurrent/full/nclimate2809.html. doi: 10.1038/nclimate2809.

Carlton, Jim. "Some in Santa Fe Pine for Lost Symbol, but Others Move On." *Wall Street Journal*, July 31, 2006, A1, A8.

Clark, Jayne. "Lake Powell on the Rise." *USA Today,* May 9, 2008, 1D, 2D.

Diffenbaugh, Noah S., and Christopher B. Field. "A Wet Winter Won't Save California." *The New York Times*, September 18, 2015. http://www.nytimes.com/2015/09/19/opinion/a -wet-winter-wont-save-california.html.

Eilperin, Juliet. "Study Ties Tree Deaths to Change in Climate." *Washington Post,* January 23, 2009, A8. http://www.washingtonpost.com/wp-dyn/content/article/2009 /01/22/AR2009012202473_pf.html.

"Energy Is Another Victim of Drought." *Omaha World-Herald*, May 1, 2015, 3A.

Fears, Darryl. "Scientists Say California Hasn't Been This Dry in 500 Years." *Washington Post*, September 14, 2015. http://www.washingtonpost.com/news/energy-environment/wp /2015/09/14/scientists-say-its-been-500-years-since-california-was-this-dry/?wpmm =1&wpisrc=nl_evening.

Gillis, Justin. "Climate Change Intensifies California Drought, Scientists Say." *The New York Times*, August 20, 2015. https://www.nytimes.com/2015/08/21/science/climate-change -intensifies-california-drought-scientists-say.html.

Hansen, James E. "Declaration of James E. Hansen." *Green Mountain Chrysler-Plymouth-Dodge-Jeep, et al., Plaintiffs v. Thomas W. Torti, Secretary of the Vermont Agency of Natural Resources, et al., Defendants.* Case Nos. 2:05-CV-302 and 2:05-CV-304, Consolidated.

United States District Court for the District of Vermont. August 14, 2006. http://www
.giss.nasa.gov/~dcain/recent_papers_proofs/vermont_14aug20061_textwfigs.pdf (no
longer available).

"Intense Fires in Northern Canada." NASA Earth Observatory, June 3, 2015. http://
earthobservatory.nasa.gov/IOTD/view.php?id=85972&src=eoa-iotd.

Kaufman, Marc. "Decline in Snowpack Is Blamed on Warming." *Washington Post* February 1, 2008, A1. http://www.washingtonpost.com/wp-dyn/content/article/2008/01/31
/AR2008013101868_pf.html.

"Lake Mead's Level Hits Landmark Low." *Omaha World-Herald*, May 1, 2015, 3A.

Lean, Geoffrey. "Worst U.S. Drought in 500 Years Fuels Raging California Wildfires." *The
Independent* (London), July 25, 2004, 20.

"Metro Vancouver Air Quality Comparable to Beijing: Health Authorities Advise Caution
Due to Smoke from Hundreds of Wildfires Across B.C." CBC News, July 6, 2015. http://
www.cbc.ca/news/canada/british-columbia/metro-vancouver-air-quality-comparable-to
-beijing-1.3140735.

Mooney, Chris. "Alaska's Terrifying Wildfire Season and What It Says about Climate Change."
Washington Post, July 26, 2015. http://www.washingtonpost.com/news/energy
-environment/wp/2015/07/26/alaskas-terrifying-wildfire-season-and-what-it-says
-about-climate-change/?wpisrc=nl_headlines&wpmm=1.

Moyer, Justin W. "Drought-Stricken California's San Joaquin Valley Is Sinking, NASA Says."
Washington Post, August 20, 2015. http://www.washingtonpost.com/news/morning-mix
/wp/2015/08/20/drought-stricken-californias-san-joaquin-valley-is-sinking-nasa-says/
?wpmm=1&wpisrc=nl_headlines.

Nagourney, Adam. "As California Drought Enters 4th Year, Conservation Efforts and Worries Increase." *The New York Times*, March 17, 2015. http://www.nytimes.com/2015/03
/18/us/as-california-drought-enters-4th-year-conservation-efforts-and-worries
-increase.html.

"NASA Study Finds Carbon Emissions Could Dramatically Increase Risk of U.S. Mega-
droughts." NASA press release, February 12, 2015. http://www.nasa.gov/press/2015
/february/nasa-study-finds-carbon-emissions-could-dramatically-increase-risk-of-us/#
.VN3kU8Ysou0.

Navarro, Mireya. "Environment Blamed in Western Tree Deaths." *The New York Times*, January 23, 2009. http://www.nytimes.com/2009/01/23/us/23trees.html.

Overpeck, Jonathan, and Bradley Udall. "Dry Times Ahead." *Science* (June 25, 2010):
1642–1643.

Paraskevas, Joe. "Glaciers in the Canadian Rockies Shrinking to Their Lowest Level in
10,000 Years." *National Post* (Canada), December 4, 2003, A8.

Pederson, Gregory T., et al. "The Unusual Nature of Recent Snowpack Declines in the North
American Cordillera." *Science* 15 (July 2011): 332–335.

Pennisi, Elizabeth. "Western U.S. Forests Suffer Death by Degrees." *Science* 323 (January 23,
2009): 447.

Rice, Doyle. "Megadrought May Plague Parts of USA." *USA Today*, February 13, 2015. http://
www.usatoday.com/story/weather/2015/02/12/western-plains-drought-climate-change
/23298093/

Rice, Doyle. "'Study:' Risk of Wildfires Rises." *USA Today*, August 28, 2015, 4A.

Rojas, Rick. "Drought's Extremes Tallied at Record-Low Lake Mead." *The New York Times*,
May 5, 2015. http://www.nytimes.com/2015/05/05/us/droughts-extremes-can-be
-measured-at-record-low-lake-mead.html.

"Saskatchewan First Nations Evacuate 13,000, Declare Wildfire State of Emergency." Indian Country Today Media Network, July 6, 2015. http://indiancountrytodaymedianetwork .com/2015/07/06/saskatchewan-first-nations-evacuate-13000-declare-wildfire-state -emergency-160973.

Schwalm, Christopher, Christopher A. Williams, and Kevin Schaeffer. "Hundred-Year Forecast: Drought." *The New York Times*, August 12, 2012. http://www.nytimes.com/2012 /08/12/opinion/sunday/extreme-weather-and-drought-are-here-to-stay.html.

Semmens, Grady. "Ecologists See Disaster in Dwindling Water Supply." *Calgary Herald*, November 27, 2003, A14.

"Smoke Blankets British Columbia." NASA Earth Observatory, July 8, 2015. http:// earthobservatory.nasa.gov/IOTD/view.php?id=86190&src=eoa-iotd.

Sridhar, Venkataramana, et al. "Large Wind Shift on the Great Plains during the Medieval Warm Period." *Science* 313 (July 21, 2006): 345–347.

Underwood, Emily. "Models Predict Longer, Deeper U.S. Droughts." *Science* 347 (February 13, 2015): 707.

Van Mantgem, Phillip J., et al. "Widespread Increase of Tree Mortality Rates in the Western United States." *Science* 323 (January 23, 2009): 521–524.

Williams, A. P., et al. "Contribution of Anthropogenic Warming to California Drought during 2012–2014." *Geophysical Research Letters*, August 28, 2015. http://onlinelibrary.wiley .com/doi/10.1002/2015GL064924/abstract. doi: 10.1002/2015GL064924.

Zacharias, Yvonne. "Global Warming Exacerbates B.C. Wildfire Severity, Scientist Says." *Vancouver Sun*, July 6, 2015. http://www.vancouversun.com/technology/global+warming+ exacerbates+wildfire+severity+scientist+says/11192869/story.html.

See also: Atmospheric Circulation; Australia, Heat and Drought in; Desertification; Drought and Deluge, Scientific Analysis of; Drought, Worldwide; El Niño and La Niña; Heat Waves, Evidence of; Heat Waves, Forecasts of; Wildfires

DROUGHT, WORLDWIDE

Droughts usually afflict more people than any other kind of weather disaster because they cover large areas and may persist for several years. According to the United Nations report *The Global Climate 2001–2010, A Decade of Extremes*, the decade from 2001 to 2010 witnessed droughts across Earth. "Some of the highest-impact and long-term droughts struck Australia, East Africa in 2004 and 2005, resulting in widespread loss of life; and the Amazon Basin in 2010 with negative environmental impacts" ("Unprecedented" 2013). Droughts persisted into the next decade (see "Atmospheric Circulation Changes" and its discussion of Hadley cells).

By 2010 and 2011, China's major wheat-growing regions were being devastated by their worst drought in 60 years. Some areas such as Shandong Province, a major grain producer, were entering their worst droughts in 200 years, having received less than half an inch of rain in six months. In some areas, drinking-water supplies also were imperiled. The drought affects Beijing southward through Hebei, Henan and Shandong, and Jiangsu provinces north of Shanghai. By March 2011, timely rains and stepped-up irrigation saved much of the wheat crop.

Drought and Syria's Civil War

Scientists have associated Syria's civil war and the rise of the Islamic State in Iran and Syria (ISIS) with a prolonged and intensifying drought in the region that began in 2007. The drought is seen as one manifestation of global warming. According to Elizabeth Kolbert, writing in *The New Yorker* (2015, 23), "The country [in 2008] experienced its driest winter on record. Wheat production failed, many small farmers lost their herds, and prices of basic commodities more than doubled." Within months, as the drought continued, hundreds of thousands of people abandoned their homes and farms in the countryside and moved to Damascus, Homs, Aleppo, and other cities, crowding them with more than a million refugees from war in Iraq. By 2015, more than 4 million Syrian refugees had moved to Turkey, Lebanon, and Jordan, as well as several European countries. By 2016, NASA had issued a report using tree rings to determine that the drought in this region from 1998 to 2012 was the worst in at least 900 years ("NASA" 2016).

Researchers from the University of California–Santa Barbara and Columbia University have linked civil unrest beginning in 2007 to poverty and the collapse of farming in Syria and the migration of 1.5 million farmers to cities. Because of the civil war, weather records after 2010 have been scarce. Droughts have become more frequent and intense in Syria, and three of Syria's longest droughts have occurred over the last 30 years as temperatures rose and winter precipitation declined.

"There are various things going on, but you're talking about 1.5 million people migrating from the rural north to the cities," said climate scientist Richard Seager of Columbia University and coauthor of the study in the *Proceedings of the National Academy of Sciences.* "It was a contributing factor to the social unraveling that occurred that eventually led to the civil war" (Borenstein 2015). The study's lead author, Colin Kelley, said that climatic change combined with oppression by the regime of Bashar al-Assad, immigration of at least 1 million refugees from Iraq, and political instability across the region led to the civil war. However, said Seager, this is the "single clearest case" ever presented by scientists of climate change playing a part in conflict because "you can really draw a blow-by-blow account with the numbers" (Borenstein 2015).

Kelley and colleagues summarized the situation in the *Proceedings of the National Academy of Sciences* (2015):

Before the Syrian uprising that began in 2011, the greater Fertile Crescent experienced the most severe drought in the instrumental record. For Syria, a country marked by poor governance and unsustainable agricultural and environmental policies, the drought had a catalytic effect, contributing to political unrest. We show that the recent decrease in Syrian precipitation is a combination of natural variability and a long-term drying trend, and the unusual severity of the observed drought is here shown to be highly unlikely without this trend. Precipitation changes in Syria are linked to rising mean sea level pressure in the Eastern Mediterranean, which also shows a long-term trend. There also has been a long-term warming trend in the Eastern Mediterranean, adding to the drawdown of soil moisture. No natural cause is apparent for these trends, whereas the observed drying and warming are consistent with model studies of the response to increases in greenhouse gases. Furthermore, model studies show an increasingly drier and hotter future mean climate for the Eastern Mediterranean. Analyses of observations and model simulations indicate that a drought of the severity and duration of the recent Syrian drought, which is implicated in the current conflict, has become more than twice as likely as a consequence of human interference in the climate system.

Syria is a pointed example of a country where climate-related crises have been fundamental to violent conflict. A paper in the *Proceedings of the National Association of Sciences* traces conflict and climate generally to societies with existing social and political conflicts that fracture along ethnic lines. The authors of this paper assigned climate change a role as a triggering mechanism:

This overall state of affairs is likely to be exacerbated by anthropogenic climate change and in particular climate-related natural disasters. Ethnic divides might serve as predetermined conflict lines in case of rapidly emerging societal tensions arising from disruptive events like natural disasters. Here, we hypothesize that climate-related disaster occurrence enhances armed-conflict outbreak risk in ethnically fractionalized countries. (Schleussner et al. 2016)

"This debate comes up time and again—is climate change really something like a trigger for violent conflict?" said Hans Joachim Schellnhuber, director of the Potsdam Institute for Climate Impact Research and coauthor of the new paper (Harvey 2016). This study applied climatic criteria to a list of armed conflicts bertween1980 and 2010, analyzing each disaster, which "found a significant link between climate disasters and the outbreak of violent conflict specifically in countries with high degrees of ethnic fractionalization. . . . about 23 percent of armed conflicts in highly ethnically divided nations coincided with climate-related disasters." "We cannot explain the full complexity of the emergence of violent conflict, but here we have found something really robust, a factor that really matters," Schellnhuber said (Harvey 2016).

Further Reading

Borenstein, Seth. "Syria's Civil War Linked Partly to Drought, Global Warming." Associated Press March 3, 2015. http://hosted.ap.org/dynamic/stories/U/US_SCI _WARMING_DROUGHTS?SITE=AP&S.

Harvey, Chelsea. "How Climate Disasters Can Drive Violent Conflict around the World." *Washington Post*, July 25, 2016. https://www.washingtonpost.com/news /energy-environment/wp/2016/07/25/how-climate-disasters-can-drive-violent -conflict-around-the-world/?wpisrc=nl_headlines&wpmm=1.

Kelley, Colin, et al. "Climate Change in the Fertile Crescent and Implications of the Recent Syrian Drought." *Proceedings of the National Academy of Sciences* 112(8) (March 2, 2015). doi: 10.1073/pnas.1421533112.

Kolbert, Elizabeth. "Unsafe Climates." *The New Yorker*, December 7, 2015, 23–24.

"NASA: Recent Mideast Drought Worst in 900 Years." *Omaha World-Herald*, March 4, 2016, 7A.

Schleussner, Carl-Friedrich, et al. "Armed-Conflict Risks Enhanced by Climate-Related Disasters in Ethnically Fractionalized Countries." *Proceedings of the National Association of Sciences* 113(30), July 25, 2016. http://www.pnas.org/content/early/2016 /07/20/1601611113. doi: 10.1073/pnas.1601611113.

Heat waves in Russia during the summer of 2010 and floods following severe drought in Australia also have affected grain markets. China produces a sixth of the world's wheat, and shortages were driving up prices worldwide and influencing geopolitical events such as the revolution in Egypt. "China's grain situation is critical to the rest of the world—if they are forced to go out on the market to procure adequate supplies for their population, it could send huge shock waves through the world's grain markets," said Robert S. Zeigler, International Rice Research Institute director general in Los Banos in the Philippines, near Manila (Bradsher 2011).

Global warming is contributing to a rise in both the magnitude and the frequency of wintertime droughts in the Mediterranean region. Of the 12 driest winters the area has experienced since 1902, 10 occurred in the 20 years before 2010, according to a records reaching back to 1902 as compiled by Martin Hoerling and colleagues at the U.S. National Oceanic and Atmospheric Administration (NOAA) in Boulder, Colorado.

Given deforestation and other land-use changes in the Amazon River valley, the area no longer functions generally as "the lungs of the world." The area acts as a carbon sink (absorber) only when temperature and precipitation patterns are healthy for the forest. In years of unusual drought (such as during the El Niño episodes of the middle and late 1990s and later), unusually hot, dry weather turns the Amazon into a net producer of carbon dioxide. In a usual year, the Amazon absorbs about 700 million tons of carbon dioxide, but in the recent El Niño year of 1995, the Amazon forest *added* 200 million tons of carbon dioxide to the atmosphere. According to A. Lindroth, A. Grelle, and A. S. Moren writing in *Nature*, a decrease in soil moisture can turn a forest that consumes carbon dioxide into one that adds the gas to the atmosphere; this variation occurs in temperate-zone forests as well as those of the tropics, the authors assert (Lindroth et al. 1998).

Too Warm and Dry for Truffles

Truffles can be worked into all sorts of foods—from cheese and sausage to spaghetti and popcorn. Fresh truffles, however, are rare and expensive mushrooms that are used in high-end French cooking as well as in stuffing for holiday chickens and turkeys. Truffles have become even more expensive than usual in recent years (prices rose 10-fold in 12 years ending in 2014), and global warming may be to blame. The prized fungi are becoming scarcer in their Mediterranean habitat. Prices for Black Périgords, the crème de la crème among truffles, had risen to more than $1,200 a pound by 2014, according to writer David Jolly of *The New York Times* (Jolly 2012). Few people can afford to buy a pound, and enough to add class to a small turkey can easily cost about $100, usually more than the bird itself. Of course, few people buy black truffles by the pound. Still, even a single black truffle big enough for bits to be slipped under the skin of a turkey and the rest added to the stuffing can easily cost €100 (approximately $110 U.S.).

Jolly described the appeal of truffles as a status symbol: "Like fine wine, truffles are a global luxury, with an appeal to the wealthy that keeps prices high even with Europe in recession. Macau billionaire Stanley Ho paid $417,200 at auction in 2010 for a nearly 1.3-kilogram Italian white truffle, a variety more treasured than even the Black Périgord" (Jolly 2012). The yearly black truffle harvest in the south of France, Italy, and Spain brought in 1,000 tons during the 1930s (when they were much less expensive) but only 50 tons after 2010. Land-use changes, including deforestation, bear some of the blame, but recently scientists writing in *Nature Climate Change* linked climate change to the shrinking harvest. Jolly wrote: "They found that the harvest of the French and Spanish black truffle correlates closely with summer rains, while the truffle habitat has suffered over the past few decades from hotter summers and less precipitation. That trend is expected to continue, according to most climate models" (2012).

"The scientists say that the exact reasons hotter, drier summers should reduce yields is unknown, but that it may be because the subterranean fungus and its host trees, mostly oaks and hazelnuts, end up competing for water when rainfall is scarce. 'If we know the reason, maybe we can adapt and compensate,' said Ulf Büntgen, the paleoclimatologist who led the study" (Jolly 2012). A major question in this research concerns the future habitat of truffles: will they move northward in a warmer world like some other species? Büntgen said that the Burgundy truffle had already been found north of the Alps (Büntgen 2012).

Further Reading

Büntgen, Ulf, et al. "Drought-Induced Decline in Mediterranean Truffle Harvest." *Nature Climate Change* 2 (December 2012): 827–829.

Jolly, David. "$1,200 a Pound, Truffles Suffer in Heat." *The New York Times*, December 20, 2012. http://www.nytimes.com/2012/12/21/business/global/is-climate-change-shrinking-the-luxury-truffle-crop.html.

Clues from Prior Droughts

Tree-ring studies of the last 1,200 years indicate that the Medieval Warm Period, a time of unusual warmth in parts of the world, was punctuated by droughts that were longer and more intense than the one currently afflicting the U.S. West. "Whether increased warmth in the future is due to natural variables or greenhouse [gases], it doesn't matter," said Edward R. Cook of Columbia University's Lamont–Doherty Earth Observatory (Cook et al. 2004).

"If the world continues to warm, one has to worry we could be going into a period of increased drought in the western U.S. I'm not predicting that, [but] the data suggests that we need to be concerned about this" (Boxall 2004). "Compared to the earlier 'mega-droughts' that are reconstructed to have occurred around AD 936, 1034, 1150, and 1253, the current drought does not stand out as an extreme event because it has not yet lasted nearly as long," the authors wrote. "This is a disquieting result because future droughts in the West of similar duration to those seen prior to AD 1300 would be disastrous," said Cook (Boxall 2004; Cook et al. 2004). "If we are just at the beginning of dramatic warming . . . we can simply expect larger, more severe fires," said Grant A. Meyer, a coauthor of the study published in *Nature* ("New Research Links" 2004).

Irrigation water may play out just as intensifying drought arrives on the Great Plains. The Ogallala Aquifer supplies farms and ranches from Nebraska to Texas with 20 billion gallons of water more each *day* than is being replenished (Ayres 1999, 100). People in the area are just now realizing that the water that nourishes their way of life is a finite and rapidly diminishing resource. This knowledge has been helped along in recent years by intense drought that has rivaled the worst years of the 1930s Dust Bowl.

By 2006, several years of intense drought and demands from irrigation had caused the Ogallala Aquifer to decline 30 feet in six years in parts of southwest Nebraska. "The rate of decline is unprecedented," said Mark Burbach of the University of Nebraska–Lincoln School of Natural Resources (Hendee 2006, A1). Elsewhere across Nebraska, groundwater had declined five to 25 feet in the same period. Groundwater in parts of southwestern Nebraska declined more than 550 feet in 50 years after the advent of large-scale irrigated agriculture.

Writing in the *Handbook of Weather, Climate, and Water: Atmospheric Chemistry, Hydrology, and Societal Impacts* (2003), hydrological scientist Donald A. Wilhite said:

Projected changes in climate because of increased concentrations of carbon dioxide and other atmospheric trace gases suggest a possible increase in the frequency and intensity of severe drought in the Great Plains region. In a region where the incidence of drought is already high, any increase in drought frequency will place even greater pressure on the region's already limited water supplies. (Wilhite 2003, 756)

Further Reading

Ayres, Ed. *God's Last Offer: Negotiating a Sustainable Future.* New York: Four Walls Eight Windows, 1999.

Boxall, Bettina. "Epic Droughts Possible, Study Says." *Los Angeles Times*, October 8, 2004, A17.

Bradsher, Keith. "U.N. Food Agency Issues Warning on China Drought." *The New York Times*, February 8, 2011. http://www.nytimes.com/2011/02/09/business/global/09food.html.

Cook, Edward R., et al. "Long-Term Aridity Changes in the Western United States." *Science* 306 (November 5, 2004): 1015–1018.

Hendee, David. "Peril Is Seen to State's Water Table." *Omaha World-Herald*, October 22, 2006, A1, A2.

Lindroth, A., A. Grelle, and A. S. Moren. "Long-Term Measurements of Boreal Forest Carbon Balance Reveal Large Temperature Sensitivity." *Global Change Biology* 4 (April 1998), 443–450.

"New Research Links Global Warming to Wildfires across the West." *Omaha World-Herald*, November 5, 2004, 11A.

"Unprecedented Climate Extremes Mark Decade 2001–2010." Environment News Service, July 17, 2013. http://ens-newswire.com/2013/07/17/unprecedented-climate-extremes -mark-decade-2001-2010/.

Wilhite, Donald A. "Drought in the U.S. Great Plains." Pp. 743–758 in Thomas D. Potter and Bradley R. Colman, eds., *Handbook of Weather, Climate, and Water: Atmospheric Chemistry, Hydrology, and Societal Impacts.* Hoboken, NJ: Wiley Interscience, 2003.

See also: Atmospheric Circulation; Australia, Heat and Drought in; Desertification; Drought, United States; Drought and Deluge, Scientific Analysis of; El Niño and La Niña; Extreme Weather; Heat Waves, Evidence of; Heat Waves, Forecasts of; Wildfires

DROUGHT AND DELUGE, SCIENTIFIC ANALYSIS OF

Although warmer air generally holds more moisture, added rainfall is rarely distributed evenly in time or space. Droughts become more intense as heat intensifies evaporation. Droughts and deluges often alternate as well. More often than in the past, multiyear droughts are being punctuated by brief, heavy deluges as the hydrological cycle becomes more erratic.

Roughly half the United States was under serious drought conditions during the summer of 2002. The drought was occasionally punctuated by torrential rains, however. On September 13, 2002, for example, drought-stricken Denver was inundated by floods from a fast-moving thunderstorm that caused widespread flooding. Similar events took place south of Salt Lake City. Ten days later, a flooding cloudburst inundated similarly drought-stricken Atlanta. A year that had ended in Omaha with the most intense drought since the Dust Bowl years also had included (on

August 7), a one-day storm that dumped 10.5 inches of rain on Omaha, the most intense deluge in the city's history. Beginning in early September, however, Omaha and its hinterland endured a rainless spell of between 44 days (in Omaha) and 49 days (in Lincoln), the fourth-longest on record (Range 2000, 11).

At times, the swift passage from drought to deluge can mimic Robert Frost's legendary duality of fire and ice. During November 2003, for example, the Los Angeles area was scorched by its worst wildfires on record; these were driven by hot, desiccating Santa Ana winds that pushed temperatures to near 100°F. Less than two weeks later, parts of the Los Angeles Basin were pounded by a foot of pearl-sized hail. The island of Hispaniola (the Dominican Republic and Haiti) was seared by drought during 2003 and then drowned in floods that killed at least 2,000 people in May 2004.

Examples abound of increasing extremes in precipitation. November (2002), December (2002), and January (2003) were Minneapolis–St. Paul's driest in recorded history. These followed the wettest June through October there in more than 100 years. In December 2002, Omaha recorded its first month on record with no measurable precipitation. In March 2003, having endured its driest year in recorded history during 2002, Denver recorded 30 inches of snow in *one* storm. Snowfall on the drought-parched Front Range totaled as much as eight feet in the same storm.

Deluges amid Droughts

Extreme weather is not a novelty in much of the world, but warming tends to aggravate extremes of both drought and deluge. Reports of an intensifying hydrological cycle have been plentiful outside the United States. In some areas, deluges may punctuate severe droughts. Near Ogallala, Nebraska, at the northeastern edge of what was once called the "Great American Desert," on July 6, 2002, received as much as 10 inches of rain during a drought, running off the hardened soil, washing out sections of Interstate 80, killing a truck driver, and provoking evacuation of residents. Both approaches of a bridge over the South Platte River were washed out. The rainfall was two to three times the amount that previously had fallen in the area during the entire year of 2002.

Nearly a year later, on the night of June 22, 2003, a stagnant supercell thunderstorm dumped 12 to 15 inches of rain (half the area's annual average) south and east of Grand Island, Nebraska, an area that also was suffering intense drought at the time. The same storm spawned several tornadoes, killing one person and injuring several others. This storm, which destroyed large parts of Aurora, Nebraska, produced hail that was among the largest ever reported in the United States, as well as a tornado that stood virtually in one place for half an hour, devastating the town of Deshler.

On a larger scale, even episodic droughts are being punctuated by periods of flooding rain as the hydrological cycle intensifies. California, for example, had suffered several years of parching drought by the winter of 2014–2015, when a few days of flooding rains in December 2014 pounded Northern California with as much as 10 inches in a few days. With the beginning of January, the drought returned. For the first time since records have been kept, San Francisco had not one drop of rain in January 2015. When it rained, however, water ran off parched, denuded hills scarred by fires, raising "the potential for landslides, debris flows, sharp rises in streams and rivers, and urban flooding" (Rice 2015).

Further Reading

Rice, Doyle. "When It Rains, It Pours This Winter." *USA Today*, February 5, 2015, 4A.

Fifteen months later, Denver's weather let loose again. On June 9, 2004, suburbs north and west of the city received as much as three *feet* of hail. Residents had to use shovels to free their cars. The summer of 2003 was unusually dry in the Pacific Northwest; during the third week in October, however, Seattle recorded its wettest day on record with 5.02 inches of rain. The Pacific Northwest is accustomed to rain, but it has not heretofore been like this: a storm December 1–3, 2007, dumped a foot of water on parts of western Washington and coastal Oregon, and brought winds gusting to 129 miles an hour.

On the night of July 27, 2004, Dallas, Texas, recorded a foot of rain and widespread flooding, even as the U.S. West continued to endure its worst multiyear drought in at least 500 years. At the same time, in December 2006 western suburbs of Denver recorded four feet of snow, even though the city's average temperature for the month was 1.4 degrees above average.

After one of its driest summers on record, Seattle recorded its wettest month on record (15.63 inches at the Seattle–Tacoma airport) during November 2006. After an El Niño set in at Christmas the same year, the weather in Seattle again became unusually dry. In El Paso, Texas, after one inch of rainfall since January 2006, three days of heavy rains produced flooding (July 30 to August 1, 2006) followed by a monsoon of unprecedented proportions. Similarly, in Greece during August 2007, 100-degree temperatures and hot and dry winds played a role (along with arson) in provoking wildfires that ravaged Greece. Two weeks later, torrential rains pelted the same areas, eroding hillsides stripped of their foliage by the fires.

Sometimes even deep snow can be tied to an intensifying hydrological cycle addled by warmth. Oswego County, New York, received almost no lake-effect snow during December 2006 and January 2007, but was buried under more than 110 inches in seven days during February as a relentless cold wind crossed an unfrozen Lake Ontario. Within eight days, some areas near Oswego were blitzed by 10 feet of snow. In nearby Redfield, the National Weather Service reported that 141 inches had fallen in 10 days, a state record for a single storm, which was spurred by bitterly cold air traversing relatively warm lake water before hitting cold earth.

By late 2009, El Niño conditions had alleviated much of the drought conditions in the continental United States. Indeed, some areas such as large parts of the Southeast had gone from drought to flooding conditions. North Platte, Nebraska, plagued by drought in 2008, experienced record October snowfall of 30 inches in 2009, more than the area's average for an entire winter (Dorell 2009).

The winter of 2009–2010, which was famously cold and snowy across much of the Midwest and eastern United States, was mild and dry in Glacier National Park, which lost two more of its two dozen remaining glaciers in an area that 160 years previously had 150 bodies of moving ice. A glacier is defined as at least 25 acres of moving ice. During late September and early October 2010, the East Coast of the United States flipped from drought conditions to floods nearly overnight. Parts of North Carolina received 22 inches of rain in two days; Jacksonville, North Carolina was inundated by 12 inches in six hours. Following a protracted drought, for example, the U.S. Southeast received torrential rains in late summer 2009. September 19 through 21, Atlanta was doused with 20 inches of rain in three days, with as much as 15 inches falling in some areas during one day. With two-thirds of the United States (and all of Texas) in drought, areas near Austin, Texas, received more than 10 inches of rain in a few hours on July 10, 2012.

After several months of intense drought, 13.13 inches of rain fell at the official recording station in Pensacola, Florida, on June 9, 2012. As much as 18 inches was measured by other observers in the area the same day. On January 9, 2012, drought-stricken Houston, Texas was doused with several inches of rain in severe thunderstorms and possible tornadoes. Whereas during the summer of 2012, low water levels had impeded barge traffic on the Mississippi River; the next summer barges were riding so high at St. Louis that they were slipping their moorings. Some local wags called it "weather whiplash" (Rice 2013).

Elements of both drought and deluge sometimes were part of the same weather event. In early September 2011, the eastern side of Tropical Storm Lee drenched much of the southeastern United States with intense rain while dry winds from the north western side of the storm drove fires in drought-stricken Texas, causing at least 1,500 homes to burn near Austin, part of Texas's worst fire season in recorded history, which had burned 3.6 million acres by early September. Later, Lee drenched the northeast United States, already drowning on record rainfall, partially from Hurricane Irene less than two weeks earlier.

The drought-suffering town of Sidney, Nebraska, was struck by a storm so intense on September 9, 2013, that some if its 7,000 residents called it a hurricane. "We couldn't buy a drop of rain for six weeks, and it all came in a single cloud," said City Manager Gary Person (Gaarder 2013). Two to five inches of rain fell, depending on location, with wind gusts up to 78 miles an hour. In 2013, a drought that had ranked as one of the history's worst in the U.S. Midwest drowned in flooding rains. Corn and soybeans that had been parched during the summer of 2012 were stunted anew by the deluge. "This is the worst spring I can remember in my 30 years farming," said Rob Korff, who had planted 3,500 acres of corn and soybeans in northwestern Missouri. "Just continuous rain, not having an opportunity to plant. It can still be a decent crop, but as far as a good crop or a great crop, that's not going to happen" (Eligon 2013). Many areas from eastern Minnesota southward through Nebraska, Illinois, Oklahoma, and Missouri received three times their average rainfall during 2013's planting season. Even so, many farmers, reflecting on their recent experiences, realized that nature could turn off the spigot and turn on the heat and that drought could replace deluge within a few weeks.

Texas and Oklahoma entered 2015 in a state of debilitating drought, but by May the states were inundated in record rainfall with 20 inches or more in a few locations within days. Some towns were swamped. The Blanco River crested at 40 feet on May 24, 27 feet above flood stage, sweeping several hundred homes off their foundations. The Blanco near San Marcos, Texas, rose 26 feet in one hour ("Flash Floods" 2015). One of the most spectacular flash floods in recorded history anywhere inundated the area the night of May 23 at the end of an acute, five-year drought in the Texas Hill Country about 30 miles southwest of Austin. Observers called it a tsunami, a 20-foot wall of water that leaped the river's banks and swept houses completely off their foundations. "Bridges that had never seen water flow over them had 20 feet of water over them," said Will Conley, a Hays County Commissioner (Gaskill and Wines 2015). Three days later, 10 inches of rain paralyzed Houston.

During mid-August 2016, a nearly stagnant low pressure system dumped as much as 30 inches of rain on parts of Louisiana in 72 hours (August 13–15) with more expected; the town of Watson had 31 inches. The deluge was compared to that of Hurricane Katrina but without the wind. It was the area's second "500 year rainfall" in six months. At the same time, fire and drought continued to plague parts of the Amazon and Southern California.

Within weeks of the Louisiana deluge, scientists at NOAA, working with Climate Central, compiled a study that linked the event to a warming climate, which made a stalled low pressure system's moisture more intense. Scientists long had maintained that warmer air holds more moisture but shied away from attributing specific events to its influence. By 2016, "rapid response" attribution studies at NOAA were making attribution possible. The Louisiana floods provided a case study that used historical rainfall records and computer models. "It's probably [close] to a doubling of the probability" of such an event, or a 100-percent increase, said Heidi Cullen, chief scientist for Climate Central. "Climate change played a very clear and quantifiable role" (Fountain 2016).

Drought and Deluge: Europe and Asia

Similar reports of an intensifying hydrological cycle also have been plentiful outside the United States. The summer of 2002 featured a number of climatic extremes, especially vis à vis precipitation. Excessive rain deluged Europe and Asia, swamping cities and villages and killing at least 2,000 people while drought and heat scorched the U.S. West and eastern cities. Climate skeptics argued that weather is always variable, but other observers noted that extremes seemed to be more frequent than before.

In Europe, on September 10, 2002, six months' worth of rain fell in a few hours in the Gard, Herault, and Vaucluse departments in the south of France, drowning at least 20 people. In the village of Sommieres, near Nimes, a usually tiny stream exploded to a width of 300 meters, cutting off road traffic.

During August 2002, Prague in the Czech Republic was inundated by flooding rains that forced 200,000 people to evacuate—the worst flooding in 200 to 500 years, depending on who was keeping the tally. The rains followed springtime drought in the same area. Debate ranged over whether this was global warming provoking more intense rainstorms or a chance natural event. During the same month, torrential rains killed 900 people in China and another 700 in Southeast Asia, India, and Nepal. Deforestation and the spread of pavement also were cited as reasons for increasing flooding in urban areas.

Also during the summer of 2002 near the Black Sea, a large tornado and heavy rains left at least 37 people dead and hundreds of vacationers stranded. During the same week in China's southern province of Hunan, 70 people died after rains caused landslides and floods. South Korea mobilized thousands of troops after a week that saw two-fifths of the average annual total rainfall (Townsend 2002).

During the summer of 2007, Europe experienced the worst of all weather worlds—massive flooding in England as well as searing heat, drought, and wildfires south of the storm track. In the Canary Islands, fires fanned by 40-mile-per-hour winds and temperatures of 110°F forced at least 13,000 of tourists to run for their lives. Two-thirds of the Palmitos wildlife park on Gran Canaria Island was destroyed by a wildfire as many toucans, orangutans, and other animals perished. That fire was set by Juan Antonio Navarro Armas, an employee of the park who had lost his job.

Wildfires provoked by record heat also scorched the Balkans and areas eastward. Temperatures in drought-plagued Romania and Greece reached 110°F (Anderson 2007). An unusually sharp boundary between very hot and very cold air caused temperatures to fall suddenly from 50°F to 5 degrees in the highland Alps within a few hours, killing six mountain climbers who were caught at high elevations in light clothing. In Hungary, 500 elderly people died of heat stress.

In April 2007, Australia considered banning irrigation in the fertile Murray–Darling River basins, which grow 40 percent of the country's agricultural produce. Australia was in the midst of its worst drought on record. If substantial rain did not arrive in eight weeks, the irrigation ban could ruin farmers already heavily in debt after six consecutive dry years. At the same time, April 15, 2007, a late-season nor'easter dumped 7.46 inches of rain on Central Park in New York City, the second-wettest day since record keeping began there in 1869. (The wettest day was 8.28 inches on September 23, 1882.)

The Rio Negro, a major Amazon tributary, went from a record high to record low levels between 2009 and 2010, a severe deluge–drought cycle. According to an Associated Press report,

> Floating homes along the Rio Negro now rest on muddy flats, and locals have had to modify boats to run in shallower waters in a region without roads. Some riverbanks have caved in, although no injuries have been reported. Enormous fields of trash and other debris have been revealed by the disappearing waters.

The drought is hurting fishing, cattle, agriculture and other businesses, prompting authorities to declare a state of emergency in nearly 40 municipalities. (Azzioni 2010)

As floods devastated Pakistan in the summer of 2010, an extremely rare deluge inundated the town of Leh, in Ladakh, India, which sits in what is usually one of the driest deserts on the planet. The village is located in what is usually a high-altitude desert protected from most precipitation by surrounding mountains. The average rainfall there in August is 15 millimeters, barely a fraction of an inch. In the early morning of August 6, 2010, however, a half-hour deluge swept much of the village away, killing 150 people and leaving several hundred missing. The storm was so intense and isolated that it missed a weather station in the valley and went unmeasured.

A quarter-million people fled massive flooding in southwest Japan from record rains that killed at least 28 people during July 2012. The rain "was like a waterfall," Yoko Yoshika said in Yamaguchi prefecture (state). "It was horrible" ("Waterfall-Like Rain" 2012). "Our region gets hit with heavy rain every year, but I have never experienced anything like this," city employee Kumi Takesue said. "Rice paddies and roads all became water so you couldn't tell what was what," she said, adding that she had to wade in knee-high water, even near her home, which was not as hard hit as other areas ("Waterfall-Like Rain" 2012). The intense rain occurred as far north as the ancient capital of Kyoto, where rainfall exceeded 90 millimeters (3.5 inches) per hour—a condition in which rain cascades in such torrents that seeing ahead becomes impossible.

Deluges during California's Drought

In mid-December 2014, California, which was suffering a four-year parching drought, briefly drowned in flooding. On December 11, 2014, the city of Venado 65 miles northwest of San Francisco received 8.5 inches of rain in one day. The U.S. Southwest was suffering a multiyear drought by 2014 when the remains of Hurricane Norbert pumped up the monsoon and gave Phoenix its wettest single day on record at almost 3 inches. The evening news suddenly was filled with images of cars drowning up to their rooftops there as well as in Las Vegas, Nevada, and Riverside, California, east of Los Angeles. A few days later, the drought returned. Omaha, Nebraska, entered 2014 with intense drought but in June experienced two 200-year deluges within two weeks, each about six inches of flooding rain, although amounts varied considerably across the area.

Following five years of withering drought, El Niño conditions on October 17, 2015, unleashed torrential rains on California, propelling mudslides down fire-scarred hillsides that buried cars and houses. Television reports showed cars swallowed up to their roofs in solidified mud as workers in bulldozers struggled to clear interstate highways. Some of them became stuck in a four-to-five foot wall of viscous mud. Traffic stopped for several hours on major highways. More than three inches of rain fell in parts of the Los Angeles Basin in one hour along with hail the size of golf balls. Funnel clouds also were sighted in the area. Firefighters quickly switched from fighting blazes to performing swift-water rescues.

Ian Lovett of *The New York Times* reported, "Near Tehachapi, north of here [Los Angeles], the hills beside State Route 58 suddenly caved in. Jessica Rose, who was driving to her grandfather's funeral in Los Angeles, pulled off the highway and onto an overpass just before it hit. 'I saw the mudslide—it looked like a brown waterfall coming down the mountain,' Ms. Rose, 22, said. 'It was like Niagara Falls but brown, like the chocolate waterfall from Willy Wonka. Cars were all crashing into each other because there was 15 feet of water. Semi trucks were stuck in it. It was the most serious thing I've ever seen'" (Lovett 2015).

Further Reading

Lovett, Ian. "A 'Wall of Mud' in California, and Warnings to Heed El Niño." *The New York Times*, October 17, 2015. http://www.nytimes.com/2015/10/17/us/a-wall-of-mud-in-california-and-warnings-to-heed-el-nino.html.

In January 2013, following Australia's hottest weather on record, parts of the country were inundated by the remains of a tropical storm that dropped as much as 53 inches of rain in three days between Brisbane and Sydney. Flooding killed five people when their cars were swept away in a deluge that dropped as much as 14 inches on parts of southeast Queensland, Australia, in three hours on May 1, 2015, in and near Caboolture, 27 miles north of Brisbane. That deluge followed a cyclone-strength storm the previous week along Australia's east coast for three days, killing eight, cutting road and rail links, closing shipping ports, and causing millions of dollars of damage to Sydney and other cities.

Based on research reaching a decade earlier, scientists anticipated in 1995 that the "available evidence suggests that a warmer world is likely to experience an increase in the frequency of heavy precipitation events, associated with a more intense hydrological cycle and the increased water-holding capacity of a warmer atmosphere" (Gillis 2014). In a paper published in 1995 A. M. Fowler and K. J. Hennessy of the University of Auckland (New Zealand) warned that heavier precipitation would test dams, roads, and culverts around the world, provoking more damaging flash floods (Fowler and Hennessy 1995). Two decades later, the U.S. National Climate Assessment reported that "large increases in heavy precipitation have occurred in the Northeast, Midwest, and Great Plains, where heavy downpours have frequently led to runoff that exceeded the capacity of storm drains and levees, and caused flooding events and accelerated erosion" (Gillis 2014).

Linear Temperature Rise but Exponential Precipitation Changes

"If you warm up the air, [it] can hold more moisture," said David Easterling of the U.S. National Climatic Data Center. "And the amount it increases is not linear; it goes up exponentially" (Barringer 2007). The effects of rising greenhouse gas levels on the hydrological cycle is at least as important as temperature—and perhaps more insidious because many models indicate that, as one survey article in *Science* phrased it, "a strengthening greenhouse will increase precipitation where it is already relatively high and decrease it where it is already low. [This] mechanism of water-cycle amplification has been operating for the past half-century. The result also suggests that the water cycle is intensifying quickly under global warming—twice as fast as climate models have been predicting" (Kerr 2012).

Warmer Air, Heavier Cloudbursts

Writing in *Nature Geoscience*, Conrad Wasko and Ashish Sharma tied warming temperatures to more intense, unpredictable, and destructive downpours. They analyzed data sets of precipitation measurements for 500 heavy rainfall events from each of 79 sites across Australia, investigat[ing] the relationship between temporal patterns of precipitation intensity within storm bursts and temperature variations. They found "intense peak precipitation and weaker precipitation during less intense times . . . at higher temperatures, regardless of the climatic region and season. We suggest invigorating storm dynamics could be associated with the warming temperatures expected over the course of the 21st century, which could lead to increases in the magnitude and frequency of short-duration floods" (Wasko and Sharma 2015). In addition, "Civil engineers from the University of New South Wales (UNSW) Water Research Centre have analyzed close to 40,000 storms across Australia spanning 30 years and have found warming temperatures are dramatically disrupting rainfall patterns, even within storm events" (Gough 2015).

"These more intense patterns are leading to more destructive storms, which can significantly influence the severity of flood flows," said lead author and doctoral candidate Conrad Wasko from the UNSW School of Civil and Environmental Engineering. "The climate zones we studied in Australia are representative of most global climates, so it's very likely these same trends will be observed around the world." "The fact that these increases represent changes only in the intensity of rainfall within storms, and not the total volume of rain—which will also increase as temperatures warm—is concerning," says Professor Sharma. "These results highlight the need for local councils to think about redesigning sewage and road infrastructure, and updating guidelines about where it's safe to build homes" (Gough 2015).

Further Reading

Gough, Myles. "Study Shows Flash Flooding Risks Increase as Peak Downpours Intensify." UNSW Newsroom, June 9, 2015. https://newsroom.unsw.edu.au/news

/science-tech/study-shows-flash-flooding-risks-increase-peak-downpours -intensify.

Wasko, Conrad, and Ashish Sharma. "Steeper Temporal Distribution of Rain Intensity at Higher Temperatures within Australian Storms." *Nature Geoscience* (June 2015), online letter. http://www.nature.com/ngeo/journal/vaop/ncurrent/full/ngeo2456.html.

Paul J. Durack and colleagues showed

that ocean salinity patterns express an identifiable fingerprint of an intensifying water cycle. Our 50-year observed global surface salinity changes, combined with changes from global climate models, present robust evidence of an intensified global water cycle. . . . This rate is double the response projected by current-generation climate models and suggests that a substantial (16 to 24 percent) intensification of the global water cycle will occur in a future 2 to 3 [°C] warmer world. (Durack et al. 2012)

In 2014, parts of the Florida panhandle received two feet of rain in 26 hours, the type of deluge usually associated with hurricanes. This time, however, it was part of a rather ordinary low-pressure system. Earlier in 2014, torrential rains had undermined a steep cliff in the Cascade foothills of western Washington, creating a landslide that buried an entire town. The National Climate Assessment the same year delineated large increases in very heavy precipitation events, mainly rainstorms within the previous 60 years: 71 percent (northeastern United States), 37 percent (Midwest), and 27 (Southeast). Since the 1970s, the amount of water vapor in the warming atmosphere had increased 3 percent to 4 percent. A warmer atmosphere not only holds more moisture but also accelerating evaporation adds more water vapor to the air. The result: precipitation potential increases more quickly than temperatures in a climate regime that experiences more intense droughts as well as an overall trend toward heavier precipitation. When it does rain, it pours.

In 2016, as several extreme downpours doused the Louisiana, Houston, West Virginia, and other U.S. locations, a study published in *Nature Climate Change* predicted that similar deluges would occur three times as often in the United States (and twice that in the Mississippi River Delta) by the year 2100 given probable warming. Models on which these projections were based were much more precise than earlier forecasts.

The study's lead author, Andreas Prein, a scientist at the National Center for Atmospheric Research, told Seth Borenstein of the Associated Press that earlier projections did not factor in small-scale weather events such as thunderstorms, which often deliver large amounts of rain. "It's much more likely that you'll get hit by very strong thunderstorms, very strong downpours in the future climate," Prein said. "What this means in the future is you might have a much higher potential for flash floods. This can have really big impacts." "The paper elegantly shows why these heavy downpours increase in frequency when the air is moist but decrease when the air is dry," said Stanford University climate scientist Chris Field. "With high warming through the century, this paper projects that most of the U.S. [gets] scary increases in the frequency of downpours" (Borenstein 2016).

According to Prein et al. (2016), "Extreme precipitation intensities have increased in all regions of the contiguous United States and are expected to further increase with warming at scaling rates of about 7 percent per [°C], suggesting a significant increase of flash flood hazards due to climate change. However, the scaling rates between extreme precipitation and temperature are strongly dependent on the region, temperature, and moisture availability, which inhibits simple extrapolation of the scaling rate from past climate data into the future." The study further shows that "[extreme] precipitation is increasing with temperature in moist, energy-limited, environments and decreases abruptly in dry, moisture-limited, environments" (Prein et al. 2016).

A Trillion Gallons of Extra Water

Kevin E. Trenberth of the National Center for Atmospheric Research and David R. Easterling with NOAA estimated that warming, on average, has "put more than a trillion gallons of extra water into the air over the contiguous 48 states, probably closer to two trillion" (Gillis 2014). This is true worldwide. The Indian

Three Feet of Rain in One Day

India's annual monsoon dry season usually alternates with heavy rains and has adapted to a drought–deluge cycle. Approximately 90 percent of India's precipitation falls between June and September during an average year, so heavy rain in Mumbai (Bombay) during late July is hardly unusual. On July 26 and 27, 2005, however, 37.1 inches of rain fell in Mumbai in 24 hours, the heaviest on record for an Indian city in one diurnal cycle. The deluge contributed to more than 1,000 deaths in and near Mumbai and surrounding Maharashtra state ("Record Rainfall" 2005). The deluge coincided with a high tide, and flooding was further aggravated by urban development's removal of mangrove swamps in and near Mumbai that had retained monsoon rains.

Metropolitan Mumbai, a city of 17 million, was shut down by the rain, and several people drowned in their cars. Mass transit and telephone services stopped. Other people were electrocuted by wires falling onto flooded streets. Tens of thousands of animals also died. Two years later, some of the worst monsoon rains in India in memory killed at least 2,800 people in India, Bangladesh, Nepal, and Pakistan in 2007. Several million people lost their homes.

Further Reading

"Record Rainfall Floods India." *The New York Times*, July 28, 2005, A12.

subcontinent has seen greater extremes in rainfall during the South Asian summer monsoon season over the past 30 years than before 1980 ("Monsoon Wet" 2014).

The Met Office, Great Britain's weather service, said it has detected "the tentacles of climate change" in a series of severe storms that produced floods in Britain during the winter of 2013–2014; torrential rains were driven by wind gusts as strong as 106 miles per hour. England experienced its wettest January on record dating to 1766. The Met Office said that "there is no definitive answer" on the role played by climate change in Great Britain's recent weather and floods but that there is "an increasing body of evidence that extreme daily rainfall rates are becoming more intense," probably due to a warming world. Met Office chief scientist Julia Slingo told the BBC that "all the evidence suggests there is a link to climate change" ("Storm" 2014). In 2013, the Met Office joined NOAA in saying that "events ranging from Superstorm Sandy flooding to U.S. heat waves to extreme rainfall in Australia and New Zealand had all been made more likely by climate change" ("Storm" 2014). According to Michael Mann, a Pennsylvania State University climate scientist, "The bottom line is this: we are indeed now seeing with our very eyes the impacts of climate change on severe weather, record heat, drought, more intense hurricane activity" ("Storm" 2014).

Extreme Precipitation and Scary Math

"Climate models have improved a lot since 10 years ago, when we basically couldn't say anything about rainfall," said Gabriele Hegerl, a climate researcher at the University of Edinburgh, Scotland. Hegerl and colleagues compiled data from weather stations in the Northern Hemisphere and compared them with precipitation models. The study covered the years 1951 through 1999. "We can now say with some confidence that the increased rainfall intensity in the latter half of the 20th century cannot be explained by our estimates of internal climate variability," she said (Schiermeier 2011).

The second study associated damaging floods in 2000 in England and Wales. Myles Allen of the University of Oxford and colleagues found that human-induced climate change "may have almost doubled the risk of the extremely wet weather that caused the floods" (Schiermeier 2011). "What has been considered a 1-in-100-years event in a stationary climate may actually occur twice as often in the future," said Allen (Schiermeier 2011).

Another study put this trend into scientific context:

Given that atmospheric water-holding capacity is expected to increase roughly exponentially with temperature—and that atmospheric water content is increasing in accord with this theoretical expectation, it has been suggested that human-influenced global warming may be partly responsible for increases in heavy precipitation. . . . Here we show that human-induced increases in greenhouse gases have contributed to the observed intensification of heavy precipitation events found over approximately two-thirds of data-covered parts of Northern Hemisphere land areas. These results are based on a comparison of observed and multimodel simulated changes in extreme precipitation over the latter half of the 20th century analyzed with an optimal fingerprinting technique. Changes in extreme precipitation projected by models, and thus the impacts of future changes in extreme precipitation, may be underestimated because models seem to underestimate the observed increase in heavy precipitation with warming. (Min et al. 2011, 378)

Intensity of Rainfall Increases

The frequency of torrential rainstorms in the U.S. Midwest has jumped 20 percent since the late 1960s, according to Amanda Staudt, a climate scientist with the National Wildlife Federation. "Global warming is making tragedies like these more frequent and more intense," Staudt said. "As climate continues to warm and we have even more moisture in the air, the trend toward increasingly intense weather events will continue. These are not random events," he said. "We are getting a systematic pattern of floods larger and more frequent than estimated by those calculations" (Pegg 2008).

Compared to historical averages, the Midwest has experienced two 500-year floods in 15 years—1993 and 2008. On August 4, 2009, Louisville, Kentucky, was deluged by 6.5 inches of rain in *one* hour, the most intense deluge in the city's history. A second storm the same day dropped another inch. A million dollars worth of books were destroyed in the city's downtown public library.

Around the world, growing seasons have been steadily lengthening, and many areas that suffered occasional droughts and heat waves now experience them more often. The broad deserts of Australia are spreading toward the thickly populated southeastern coast as heat waves and wildfires singe the urban area of Sydney, riding hot, dry winds from the interior that are much like Southern California's Santa Ana winds.

"It's a situation of the poor getting poorer and the rich getting richer when it comes to rainfall," said Yochanan Kushnir of the Lamont–Doherty Earth Observatory of Columbia University. "From a climate perspective, these changes are quite dramatic" (Kaufman 2007).

Increasing rainfall intensity has sometimes been made worse by construction that ignores flood plains. In November and December 2015, for example, monsoon rains in and near Chennai on India's southeastern coast brought the heaviest precipitation in a century of record keeping, including one 14-inch diurnal deluge on December 2. At least 270 people died in the resulting floods. The damage and human loss was compounded by extensive construction over flood plains. Chennai's airport, which was closed by the early December floods, had been built on an area near the Adyar River that is known to flood during monsoon rains. Some urban developments near Chennai have sprawled over areas containing lakes that had been important conduits of heavy rain. "We have forgotten the art of drainage," said Sunita Narain, director general of the Center for Science and Environment based in Delhi. "We only see land for buildings, not for water" (Najar 2015). Chennai received 48 inches of rain in November 2015, 90 percent more than average, a figure that may have been enhanced by warmer than usual water temperatures in the Bay of Bengal. Although the monsoon season abates farther north in India by November, the southeast coast of India experiences some of its heaviest rains at that time ("Records Fall" 2015).

Extremes, Not the Averages, Cause Damage

"It's the extremes, not the averages, that cause the most damage to society and to many ecosystems," said lead author Claudia Tebaldi, a scientist with the National Center for Atmospheric Research. "We now have the first model-based consensus on how the risk of dangerous heat waves, intense rains, and other kinds of extreme weather will change in the next century" ("Scientists Predict" 2006).

James E. Hansen, director of NASA's Goddard Institute for Space Studies, said that global warming is expected to cause an increase in the extremes of the hydrological cycle and thus the intensity of droughts and forest fires on the one hand and the intensity of heavy rainfall events and floods on the other. "Specifically," said Hansen "a tendency for increased drought in subtropical regions including the American Southwest is expected because of a slowdown in the tropospheric overturning circulation. An increase in the strength of storms driven by latent heat of vaporization, which includes tropical storms, is expected. Melt-back of mountain glaciers is expected and, overall, a decrease of the snowpack in most mountain ranges" (Hansen 2006, 24).

The world—especially the western United States, the Mediterranean region and Brazil—is likely to suffer more extended droughts, heavy rainfalls, and longer heat waves over the next century. These forecasts were part of a major international report on climate change scheduled for release in 2007 from the National Center for Atmospheric Research that describes nine computer models. It was published slightly early in the December 2006 issue of *Climatic Change*. "It's going to be a wild ride, especially for specific regions," said the study's lead author, Claudia Tebaldi, a scientist at the federally funded academic research center. Tebaldi described the western United States, Mediterranean nations, and Brazil as "hot spots" that will get the most extreme weather (Borenstein 2006).

Further Reading

Anderson, John Ward. "Europe's Summer of Wild, Wild Weather; Fires, Droughts and Floods Leave Wake of Destruction." *Washington Post,* August 2, 2007, A11. http://www.washingtonpost.com/wp-dyn/content/article/2007/08/01/AR2007080102347_pf.html.

Azzioni, Tales. "Drought Has Amazon Tributary at Record Low Levels." Associated Press in *Washington Post,* October 25, 2010. http://www.washingtonpost.com/wp-dyn/content/article/2010/10/25/AR2010102502661_pf.html (no longer available).

Barringer, Felicity. "Precipitation Across U.S. Intensifies over 50 Years." *The New York Times,* December 5, 2007. http://www.nytimes.com/2007/12/05/us/05storms.html.

Borenstein, Seth. "Future Forecast: Extreme Weather; Study Outlines a Climate Shift Caused by Global Warming." Associated Press, October 21, 2006, A-2. http://www.washingtonpost.com/wp-dyn/content/article/2006/10/20/AR2006102001454_pf.html.

Borenstein, Seth. "Study: Warming to Trigger 3 Times as Many Downpours in U.S." Associated Press, December 5, 2016. http://bigstory.ap.org/article/9aacfd5603cb475eb461960f4e05b948/study-warming-trigger-3-times-many-downpours-us.

Dorell, Oren. "Like Cloud Lifted, Drought Eases." *USA Today*, November 12, 2009, A-3.

Durack, Paul J., Susan E. Wiffels, and Richard J. Matear. "Ocean Salinities Reveal Strong Global Water Cycle Intensification during 1950 to 2000." *Science* 336 (April 27, 2012): 455–458.

Eligon, John. "After Drought, Rains Plaguing Midwest Farms." *The New York Times*, June 9, 2013. http://www.nytimes.com/2013/06/10/us/after-drought-rains-plaguing-midwest-farms.html.

"Flash Floods Wreak Havoc." *Omaha World-Herald*, May 25, 2015, 7A.

Fountain, Henry. "Scientists See Push from Climate Change in Louisiana Flooding." *The New York Times*, September 7, 2016. http://www.nytimes.com/2016/09/08/science/global-warming-louisiana-flooding.html.

Fowler, A. M., and K. J. Hennessy. "Potential Impacts of Global Warming on the Frequency and Magnitude of Heavy Precipitation." *Natural Hazards* 11 (1995): 283. https://link.springer.com/article/10.1007/BF00613411. doi: 10.1007/BF00613411.

Gaarder, Nancy. "Heavy Storm Hurdles through Sidney, Neb." *Omaha World-Herald*, September 11, 2013, 6B.

Gaskill, Melissa, and Michael Wines. "At Least 3 Are Killed and 12 Are Missing as Storms Ravage Texas and Oklahoma." *The New York Times*, May 25, 2015, A9.

Gillis, Justin. "Looks Like Rain Again. And Again." *The New York Times*, May 13, 2014. http://www.nytimes.com/2014/05/13/science/looks-like-rain-again-and-again.html?ref=science.

Hansen, James E. "Declaration of James E. Hansen." *Green Mountain Chrysler-Plymouth-Dodge-Jeep, et al., Plaintiffs v. Thomas W. Torti, Secretary of the Vermont Agency of Natural Resources, et al., Defendants*. Case Nos. 2:05-CV-302 and 2:05-CV-304, Consolidated. United States District Court for the District of Vermont. August 14, 2006. http://www.giss.nasa.gov/~dcain/recent_papers_proofs/vermont_14aug20061_textwfigs.pdf (no longer available).

Kaufman, Marc. "Southwest May Get Even Hotter, Drier; Report on Warming Warns of Droughts." *Washington Post*, (April 6, 2007, A-3, i). http://www.washingtonpost.com/wp-dyn/content/article/2007/04/05/AR2007040501180_pf.html.

Kerr, Richard A. "The Greenhouse Is Making the Water-Poor Even Poorer." *Science* 336 (April 27, 2012): 405.

Min, Seung-Ki, et al. "Human Contribution to More-Intense Precipitation Extremes." *Nature* 470 (February 17, 2011): 378–381.

"Monsoon Wet Spells Get Wetter." *Nature* 509 (May 1, 2014): 11.

Najar, Nida. "More Rescues as Rain Hits South India." *The New York Times*, December 4, 2015, A14.

Pegg, J. R. "Warming Climate Adds to U.S. Flood Fears." Environment News Service, July 2, 2008. http://www.ens-newswire.com/ens/jul2008/2008-07-02-10.asp (no longer available).

Prein, Andreas F., et al. "The Future Intensification of Hourly Precipitation Extremes." *Nature Climate Change*, December 2016. http://www.nature.com/nclimate/journal/vaop/ncurrent/full/nclimate3168.html.

Range, Stacey. "Climatologists Say Midlands in Dust Bowl-Like Drought." *Omaha World-Herald*, January 12, 2000, 18.

"Records Fall in 2015 Cyclone Season." NASA Earth Observatory, December 5, 2015. http://earthobservatory.nasa.gov/IOTD/view.php?id=87092&src=eoa-iotd.

Rice, Doyle. "'Weather Whiplash' Swamps Midwest." *USA Today*, April 22, 2013, 3A.

Schiermeier, Quirin. "Increased Flood Risk Linked to Global Warming; Likelihood of Extreme Rainfall May Have Been Doubled by Rising Greenhouse Gas Levels." *Nature* 470 February 17, 2011, 316. http://www.nature.com/news/2011/110216/full/470316a.html.

"Scientists Predict Future of Weather Extremes." Environment News Service, October 20, 2006. http://www.ens-newswire.com/ens/oct2006/2006-10-20-03.html.

"Storm with 106-mph Gusts Slams Flooded Britain." Associated Press, February 12, 2014. http://www.foxnews.com/world/2014/02/12/storm-with-106-mph-gusts-slams-flooded-britain/.

Townsend, Mark. "Monsoon Britain: As Storms Bombard Europe, Experts Say That What We Still Call 'Freak' Weather Could Soon Be the Norm." *The Observer* (London), August 11, 2002, 15.

"Waterfall-Like Rain Eases in Southwest Japan, but 28 Dead, Thousands of Homes Damaged." *Washington Post*, July 15, 2012. http://www.washingtonpost.com/world/asia_pacific/floods-in-southern-japan-kill-at-least-26-thousands-remain-cut-off/2012/07/16/gJQAxYWcnW_print.htm (no longer available).

See also: Atmospheric Circulation; Australia, Heat and Drought in; Desertification; Drought, United States; Drought, Worldwide; El Niño and La Niña; Extreme Weather; Heat Waves, Evidence of; Heat Waves, Forecasts of; Wildfires

EXTREME WEATHER

Extreme weather is not a novelty in much of the world. In recent years, however, daily weather reports have indicated an accelerating trend of wild weather. To cite one example of many, on October 5, 2013, in Nebraska and South Dakota, severe tornadoes and a blizzard occurred as part of the same storm system. An F4 tornado, usually a rare event in October, struck Wayne in northeastern Nebraska. The same day, as much as 45 inches of snow fell in the Black Hills of South Dakota, part of a blizzard that also reached western Nebraska. Although rare, this sort of weather does have precedent in the U.S. wild-weather annals. In 1913, a late March tornado killed more than 100 people in Omaha and was followed the next day and in the same place by snow. Today, however, extreme weather has become nearly routine.

Decades of Extremes

The first decade of the 21st century was the warmest on the instrumental record on Earth's land surface and oceans, according to *The Global Climate 2001–2010: A Decade of Extremes* by the World Meteorological Organization (2013). More national temperature records were reported broken than in any previous decade, according to the report. "Rising concentrations of heat-trapping greenhouse gases are changing our climate, with far reaching implications for our environment and our oceans, which are absorbing both carbon dioxide and heat," said World Meteorological Organization (WMO) Secretary-General Michel Jarraud ("Unprecedented" 2013).

The report analyzed global precipitation and temperature, as well as heat waves in Europe and Russia, Hurricane Katrina, Tropical Cyclone Nargis (in Myanmar, also known as Burma), Pakistan's floods, and droughts in Australia, the Amazon River basin, and others. "Many of these events and trends can be explained by the natural variability of the climate system. Rising atmospheric concentrations of greenhouse gases, however, are also affecting the climate," Jarraud wrote in his introduction to the report. "Detecting the respective roles being played by climate variability and human-induced climate change is one of the key challenges facing researchers today" ("Unprecedented" 2013).

Record high temperatures occurred in this decade despite the general absence of major El Niño events that caused global temperatures to rise as in the previous record warm year of 1998. This pattern continued to 2014, which posted a record high average worldwide—also in the absence of an El Niño. With the exception of a moderate to strong El Niño in 2009–2010, the decade was dominated by cooling La Niñas or the absence of both oceanic temperature patterns (a neutral state for world climate). By 2015, a strong El Niño had formed, and temperatures rose to levels previously unknown in the instrumental record.

Baseball Bats May Be Victims of Warming

The traditional wood used to make major-league baseball bats comes from the ash tree. Ashes are now being killed by a beetle species, the emerald ash borer, that may be encouraged by warming temperatures. Owners of bat factories in the ash country of northwestern Pennsylvania have made emergency plans if the white ash tree, the source of the best wood for bats, is "compromised." White ash seeds are being collected in Michigan in case natural species are endangered.

In 2007, Asian wasps were imported and set loose in ash forests to attack the shiny green borer (*Agrilus planipennis Fairmaire*), itself an Asian immigrant, which has killed upward of 25 million ash trees in Michigan, Illinois, Indiana, Ohio, and Maryland after it was first found near Detroit in 2002 (Davey 2007). By late June 2007, the ash borer was invading the choice baseball bat ash groves in Pennsylvania near the border with New York state.

Warming temperatures may be partially to blame for the ash borer's expansion because a longer growing season is causing the white ash's wood to become softer, making it easier for the beetles to eat. Such changes in wood density also make it less suitable for baseball bats. Warmer weather also may speed up the reproductive cycle of the beetle.

"We're watching all this very closely," said Brian Boltz, general manager of Larimer & Norton, owner of the mill in Russell, Pennsylvania, that each day saws, grades, and dries scores of billets destined to become Louisville Slugger bats. "Maybe it means more maple bats. Or it may be a matter of using a different species for our bats altogether" (Davey 2007). Although unlikely, the major leagues could turn to aluminum bats, which are already used in college and high school baseball programs.

As with most aspects of global warming, this one is open to debate. Dan Herms, an associate professor of entomology at Ohio State University, denies a link between the ash borer and climate change because the beetles survive in a wide range of temperatures in Asia (Davey 2007).

Further Reading

Davey, Monica. "Balmy Weather May Bench a Baseball Staple." *The New York Times*, July 11, 2007. http://www.nytimes.com/2007/07/11/us/11ashbat.html.

In August 2002, vineyards in northeastern Italy were badly pummeled by hail during the grape harvest, leading to widespread losses of wine grapes (Townsend 2002). During July 2001, a 15-year-old girl was killed by a rare lightning strike about 100 miles from Oslo, Norway, in an area where such storms have been extremely rare. Also in mid-August 2003, the Norwegian village of Atnadalen was flooded by a thunderstorm that dumped 116 millimeters (more than 4.5 inches) of rain in 24 hours, twice the village's monthly average in August.

During the summer of 2004, as if on cue to support the predictions of climate models, the usual monsoon in South Asia brought deluges that killed more than 2,000 people and left millions homeless in Bangladesh, Vietnam, China, India, and Nepal. After five years of intense drought, Southern California was battered in late December 2004 and throughout January and February 2005 by incredible amounts of rain and snow. Los Angeles received 17 inches of rain between December 27, 2004, and January 10, 2005, a record amount for two weeks. Some areas in the Sierra Nevada reported more than 10 feet of snow during the same period. At the same time, the usually wet Pacific Northwest experienced its worst drought in almost 30 years. On June 9, 2008, as large parts of California suffered the state's worst drought on record (since 1895), flooding rains repeatedly doused large parts of the Midwest, including Indiana, which was one-third underwater.

Episodic versus Systematic

Reports of wild weather, including droughts and deluges, pose a problem for scientists because they are episodic and not systematic, even though they may be frequent and intense. Scientists tend to distrust anecdotal stories and instead seek consistent evidence that strongly supports a given idea or testable hypothesis. Thus, even though the hydrological cycle may seem to be changing, precipitation measurements that support this idea have been difficult to assemble on a worldwide basis. "In many parts of the world," according to one scientific source, "We still cannot reliably measure true precipitation. . . . Due to rain gauge undercatch . . . precipitation is believed to be underestimated by 10 to 15 percent" (Potter and Colman 2003, 144). Ground-level precipitation is only rarely measured over the oceans that cover two-thirds of Earth, and satellite estimates do not measure local variability. Thus, "proof" of global hydrological cycle changes is elusive.

However, preponderance of evidence plays an important role in interpreting climate change. "My thinking has evolved," said Gavin Schmidt, a climate modeler at NASA's Goddard Institute for Space Studies in New York. Thanks to advances in statistical tools, climate models. and computer power, "attribution of extremes is hard—but it is not impossible," he said (Schiermeier 2011, 148). Atmospheric scientists have advocated "creation of a database of frequency and intensity using hourly precipitation amounts" (Trenberth et al. 2003, 1213). "Atmospheric moisture amounts are generally observed to be increasing . . . after about 1973, prior to which reliable moisture soundings are mostly not available" (Trenberth et al. 2003, 1211). Annual mean precipitation amounts over the United States have been increasing at 2 percent to 5 percent per decade, with "most of the increase related to temperature and hence in atmospheric water-holding capacity. . . . There is clear evidence that rainfall rates have changed in the United States. . . . The prospect may be for fewer but more intense rainfall—or snowfall—events" (Trenberth et al. 2003, 1211, 1212). Individual storms may be further enhanced by latent heat release that supplies more moisture.

According to Thomas Karl, director of the National Climatic Data Center (NCDC) in the U.S. National Oceanic and Atmospheric Administration (NOAA), "It is likely that the frequency of heavy and extreme precipitation events has increased as global temperatures have risen." This, he said, "is particularly evident in areas where precipitation has increased, primarily in the mid- and high latitudes of the Northern Hemisphere" (Hume 2003, A13). Studies at the Goddard Institute for Space Studies and Columbia University indicate that the frequency of heavy downpours has indeed increased and suggest that the trend will intensify. In the U.S. corn belt, for example, the average number of extreme precipitation events is predicted to jump by 30 percent over the next 30 years and 65 percent over the next century.

Statistical Support Deluges

The first statistical support linking tropical deluges to a warming climate was published in August 2008 by Richard P. Allan of the University of Reading in England and Brian J. Soden at the University of Miami in the online journal *Science Express*. The new paper was important "because it uses observations to demonstrate the sensitivity of extreme rainfall to temperature," said Anthony J. Broccoli, director of the Center for Environmental Prediction at Rutgers University. "Such changes in extreme rainfall are quite important in my view, as flash flooding is produced by the extreme rain events," Broccoli added. "In the United States, flooding is a greater cause of death than lightning or tornadoes, and presumably poses similar risks elsewhere" (Revkin 2008). The study analyzed two decades of NASA satellite data through several El Niño cycles. Moreover, according Allan and Soden, the "observed amplification of rainfall extremes is found to be larger than that predicted by models, implying that projections of future changes in rainfall extremes in response to anthropogenic global warming may be underestimated" (Allan and Soden 2008, 1484).

Generally, higher temperatures enhance evaporation and provide some compensatory cooling when water is available. Increased evaporation also intensifies drought, which compounds itself to some degree as moisture is depleted, leading "to increased risk of heat waves and wildfires in association with such droughts; because once the soil moisture is depleted then all the heating goes into raising temperatures and wilting plants" (Trenberth et al. 2003, 1212).

In midlatitude mountain areas, wrote Trenberth et al., the

> winter snowpack forms a vital resource, not only for skiers, but also as a freshwater resource in the spring and summer as the snow melts. Yet warming makes for a shorter snow season with more precipitation falling as rain rather than snow, earlier snowmelt of the snow that does exist, and greater evaporation and ablation. These factors all contribute to diminished snowpack. In the summer of 2002, in the western parts of the United States, exceptionally low snowpack and subsequent low soil moisture likely contributed substantially to the widespread intense drought because of the importance of recycling [in the hydrological cycle]. Could this be a sign of the future? (Trenberth et al. 2003, 1212)

Examining several extreme weather events in 2012, scientists found anthropogenic climate change's footprints in approximately half. Other events, including a severe drought in the Midwest, appeared to be mainly part of natural patterns. The extreme weather events "would have likely occurred regardless of climate change," said Thomas Karl of the NCDC. "The importance of attribution research comes with understanding, however, the impact that climate change adds, or doesn't add, to any extreme event," he told Kenneth Chang of *The New York Times* (2013). The research included 19 studies by 18 teams of scientists under the aegis of *The Bulletin of the American Meteorological Society*.

Chang described a warming climate as one of several provocations driving severe weather events: Karl "likened climate change to someone habitually driving a bit over the speed limit. Even if the speeding itself is unlikely to directly cause an accident, it increases the likelihood that something else—a wet road or a distracting text message—will do so and that the accident, when it occurs, will be more calamitous." Even when global warming contributes to extreme weather, "natural variability can still be the primary factor in any individual extreme event," Chang wrote (2013).

Frequency of Floods May Rise

Writing *in Nature* about an increasing risk of floods in a changing climate, P. C. D. Milly and colleagues noted, "We find that the frequency of great floods increased substantially during the 20th century. The recent emergence of a statistically significant positive trend in risk of great floods is consistent with results from the climate model, and the model suggests that the trend will continue" (Milly et al. 2002, 514–515). An increasing risk of flooding in Britain and northern Europe during the 21st century was quantified in the January 31, 2002, edition of *Nature* by Tim Palmer of the European Centre for Medium-Range Weather Forecasts (Reading, England) and Jouni Raisanen of the Swedish Meteorological and Hydrological Institute (Norrkoping, Sweden).

The team analyzed the forecasts of 19 climate models to produce an "ensemble forecast." This study also revealed that an increased probability of extremely wet summers in the Asian monsoon region (Highfield 2002). This was the first time that such a probability forecast of weather extremes caused by climate change had been assembled, although its methodology has been used for weather and seasonal forecasts. "Our results suggest that the probability of such extreme precipitation events is already on the increase," said Palmer, but it was "extremely difficult to verify a small increase in a statistic about extreme seasonal weather, especially over a small area like the [United Kingdom]" (Highfield 2002). T. N. Palmer and J. Ralsanen wrote, "We estimate that the probability of total boreal winter precipitation exceeding two standard deviations above normal will increase by a factor of five over parts of the United Kingdom over the next 100 years. We find similar increases in probability for the Asian monsoon region in boreal summer, with implications for flooding in Bangladesh" (Palmer and Ralsanen 2002, 512).

According to a 2002 report by the France-based World Water Council, "The economic toll of floods, droughts, and other weather-related disasters has increased almost tenfold in the last four decades, a devastating pattern that must be halted with more aggressive efforts to mitigate the damage." The same report said that increasingly rapid and extreme climate changes point to a future of intensified natural disasters that will result in more human and economic misery in many parts of the world unless action is taken (Greenaway 2003). "Most countries aren't ready to deal adequately with the severe natural disasters that we get now, a situation that will become much worse," said William Cosgrove, vice president of the council (Gardiner 2003).

Floods Affect 100 Million People a Year

The World Water Council's statistics indicate that floods affected more than 1.5 billion people worldwide, or 100 million people a year, between 1971 and 1995. An estimated 318,000 were killed and more than 18 million left homeless. The economic costs of these disasters increased from some $35 billion U.S. in the 1960s to an estimated $300 billion U.S. in the 1990s (Greenaway 2003).

Global warming is causing changes in weather patterns as growing numbers of people migrate to vulnerable areas and increase the costs of individual weather events, said Cosgrove. "The forecast is that it's going to continue to get worse unless we start to take actions to mitigate global warming" (Gardiner 2003). Scientists cited by the World Water Council expect that climate changes during the 21st century will lead to shorter and more intense rainy seasons in some areas, as well as longer and more intense droughts in others, endangering some crops and species and causing a drop in global food production (Gardiner 2003).

Severe summer floods in Europe in 2002 may be an indicator of an emerging pattern, according to Jens H. and Ole B. Christensen, who modeled precipitation patterns in Europe under warming conditions of a type that may be prominent in the area by 2070 to 2100. "Our results," they wrote in *Nature*, "indicate that episodes of severe flooding may become more frequent, despite a general trend toward drier summer conditions" (Christensen and Christensen 2003).

Deluges are not consistent, however; in 2003, Europe experienced heat and drought of a generational order. Farmers whose crops drowned in 2002 watched them wither and die in 2003. As much as 80 percent of the grain crop died in eastern Germany, the site of some of the worst floods in 2002 ("Drought, Excessive Heat" 2003). In other words, the trend toward drought or deluge will intensify as warming intensifies the hydrological cycle. A warming atmosphere will contain more water vapor, providing "further potential for latent-heat release during the buildup of low-pressure systems, thereby possibly both intensifying the systems and making more water available for precipitation" (Christensen and Christensen 2003).

In a similar vein, Wilhelm May and colleagues expect that

> [in] Southern Europe, the reduction of precipitation throughout the year will cause serious problems for the water supply in general and for agriculture in particular, which is very vulnerable due to its dependency on irrigation. The elongation of droughts in this region will worsen the living conditions, in particular in summer in combination with the warmer temperatures. . . . The dryness in Southern Europe is accompanied by an intensification of heavy precipitation events. (May et al. 2002, 27)

May and colleagues also anticipate more heavy precipitation events farther north in Europe, such as in Scandinavia, where precipitation amounts are expected to increase as climate warms.

Further Reading

Allan, Richard P., and Brian J. Soden. "Atmospheric Warming and the Amplification of Precipitation Extremes." *Science* 321 (September 12, 2008): 1481–1484.

Chang, Kenneth. "Research Cites Role of Warming in Extremes." *The New York Times*, September 5, 2013. http://www.nytimes.com/2013/09/06/science/earth/research-cites-role-of-warming-in-extremes.html.

Christensen, Jens H., and Ole B. Christensen. "Severe Summertime Flooding in Europe." *Nature* 421 (February 20, 2003): 805.

"Drought, Excessive Heat Ruining Harvests in Western Europe." *Daytona Beach News-Journal* (Florida), August 5, 2003, 3A.

Gardiner, Beth. "Report: Extreme Weather on the Rise, Likely to Get Worse." Associated Press Worldstream, International News, London. February 27, 2003 (LEXIS).

Greenaway, Norma. "Disaster Toll from Weather Up Tenfold: Droughts, Floods Need More Damage Control, Report Says." *Edmonton Journal*, February 28, 2003, A5.

Highfield, Roger. "Winter Floods 'Five Times More Likely.'" *Daily Telegraph* (London), January 31, 2002, 8.

Hume, Stephen. "A Risk We Can't Afford: The Summer of Fire and the Winter of the Deluge Should Prove to the Naysayers That If We Wait Too Long to React to Climate Change We'll Be in Grave Peril." *Vancouver Sun*, October 23, 2003, A13.

May, Wilhelm, Reinhard Voss, and Erich Roeckner. "Changes in the Mean and Extremes of the Hydrological Cycle in Europe under Enhanced Greenhouse Gas Conditions in a Global Time-slice Experiment." Pp. 1–30 in Martin Beniston, ed., *Climatic Change: Implications for the Hydrological Cycle and for Water Management*. Dordrecht, Germany: Kluwer Academic Publishers, 2002.

Milly, P. C., et al. "Increasing Risk of Great Floods in a Changing Climate." *Nature* 415 (January 30, 2002): 514–517.

Palmer, T. N., and J. Ralsanen. "Quantifying the Risk of Extreme Seasonal Precipitation Events in a Changing Climate." *Nature* 415 (January 30, 2002): 512–514.

Potter, Thomas D., and Bradley R. Colman, eds. *Handbook of Weather, Climate, and Water: Dynamics, Climate, Physical Meteorology, Weather Systems, and Measurements*. Hoboken, NJ: Wiley-Interscience, 2003.

Revkin, Andrew C. "Tropical Warming Tied to Flooding Rains." *The New York Times*, August 8, 2008. http://www.nytimes.com/2008/08/08/science/earth/08rain.html.

Schiermeier, Quirin. "Climate and Weather: Extreme Measures." *Nature* 477 (September 7, 2011): 148–149.

Townsend, Mark. "Monsoon Britain: As Storms Bombard Europe, Experts Say That What We Still Call 'Freak' Weather Could Soon Be the Norm." *The Observer* (London), August 11, 2002, 15.

Trenberth, Kevin E., et al. "The Changing Character of Precipitation." *Bulletin of the American Meteorological Society* 84(9) (September 2003): 1205–1217.

"Unprecedented Climate Extremes Mark Decade 2001–2010." Environment News Service, July 17, 2013. http://ens-newswire.com/2013/07/17/unprecedented-climate-extremes-mark-decade-2001-2010/.

World Meteorological Organization. *The Global Climate 2001–2010: A Decade of Extremes*. 2013. http://library.wmo.int/pmb_ged/wmo_1119_en.pdf.

See also: Atmospheric Circulation; Australia, Heat and Drought in; Desertification; Drought, United States; Drought, Worldwide; El Niño and La Niña; Floods; Heat Waves, Evidence of; Heat Waves, Forecasts of; Temperatures, Global; Temperatures, Greenhouse Gas Levels and; Wildfires

FLOODS

The atmosphere's capacity to hold water—and thus to release it—increases with rising temperatures. Scientific theory as well as a barrage of daily weather reports have been supplying ample evidence that Earth's hydrological cycle is changing more quickly than its temperatures. What follows is a sample of local reports from the past two decades.

In 2010, Clean Air-Cool Planet surveyed 60 years of National Weather Service rainfall reports and found that "extreme precipitation events" (defined as more than an inch of rain or equivalent in one day) have become more common in nine U.S. northeastern states as temperatures have warmed. This report was released as waters receded in Rhode Island, Massachusetts, and Connecticut from the worst floods in a century. Skeptic Patrick Michaels of the Cato Institute dismissed the study as not indicative of trends outside that area. A day later, record rains (11 inches in 36 hours) killed more than 200 people in Rio de Janeiro.

The intensity of downpours has increased substantially across the United States as temperatures have risen, according to data developed by Environment America and released in a 2007 report titled *When It Rains, It Pours: Global Warming and the Rising Frequency of Extreme Precipitation in the United States*. Heavy rainfall episodes in 2006 were more than 50 percent higher in Louisiana, New York, Massachusetts, Rhode Island, and Vermont than 60 years previously, according to the report. In the Dallas–Fort Worth area, such storms increased 42 percent during the same period. On average, extreme rainstorms increased 22 percent to 26 percent across the country (Barringer 2007; "Extreme Rainstorms" 2007).

The Environment America report used data from 3,000 weather stations and a methodology originally developed by scientists at the National Climatic Data Center and the Illinois State Water Survey. Although the report concentrates on Texas, it extrapolates its findings to the United States as a whole. Nationwide, it says, extreme precipitation events have increased by 24 percent across the continental United States since 1948 ("Extreme Rainstorms" 2007).

Deluges across the United States

Examples abound of flooding rain events in the eastern, southeastern, and midwestern sections of the United States. Torrential rains fell over Chicago during the early morning of Saturday, July 23, 2011, as O'Hare International airport measured 6.91 inches of rain between 1 a.m. and 7 a.m. For the entire day, 8.20 inches provided Chicago's wettest day on record. Torrential rains inundated the Washington, D.C., area September 7, 2011, killing two people in Fairfax County (Virginia) and one in Anne Arundel County (Maryland), and trapping large numbers of people in their cars as Interstate 66 and the Capital Beltway closed. A 12-year-old boy was killed as he was "swept away by the flood-swollen waters of Piney Branch Creek in Vienna; a 60-year-old man in Great Falls who was killed near his stranded vehicle; and a 49-year-old man drowned in Pasadena, Md." (Klein et al. 2011). On August 6, 2011, Charlotte, North Carolina, had seven inches of rain in three hours. Roughly 13 to 15 inches of rain fell overnight on July 28, 2011, over some sections of eastern Iowa and northern Illinois. During one 24-hour period in late May, 2016, Washington County in southeastern Texas received between 17 and 30 inches of rain.

On May 1–2, 2010, Nashville, Tennessee, was flooded by 14 inches of rain in two days, equal to twice the former daily record—like having two record daily deluges back to back. Milwaukee, Wisconsin, received eight inches of rain in two hours on July 22, 2010, provoking widespread flash flooding, including the runways of the airport, which canceled all flights for several hours and stranded many travelers overnight. In March 2010, Providence, Rhode Island, recorded 16.32

inches of rain, a new record for any month. Boston's total rainfall for March was 14.83 inches, the second wettest month since weather records were first recorded in 1872. In Central Park, New York City, 10.68 inches set a record for March ("Northeast Storms" 2010).

On June 11, 2010, swollen rivers fed by a sudden deluge of six to eight inches within an hour or two swamped a campground in the Ouachita National Forest in southwestern Arkansas, killing 19 people. "This was such a huge, fast-moving event. I talked to somebody who [had] lived here all their [sic] life. They [had] never heard of anything like this," said U.S. Forest Service geologist John Nichols (Copeland and Welch 2010, 1A). The force of the flood ripped apart cement foundations and splintered large trees into kindling. Raymond Slade, a retired hydrologist with the U.S. Geological Survey, said the intensity of the rain was unusual even for an area called "flash flood alley"—"much greater than a 100-year rainfall" (Copeland and Welch, 2010, 6A). Three days after the deadly flash floods in Arkansas, one to three inches of rain an hour fell in the Oklahoma City area June 14; the National Weather Service said a few areas measured almost 10 inches in one day. Many area residents had to find safety in trees and on rooftops.

Record rainfall in the U.S. Northeast during August 2011 included 19 inches in New York City and Philadelphia, primarily from Hurricane Irene. On August 14–15, 2011, New York City was drenched with 7.8 inches of rain, a record for two days (6.0 inches the first day, 1.8 inches the second). Residents evacuated their homes and animals escaped from their pens at a zoo as floods fed by a steady torrential downpour dumped some 8 inches of rain on Duluth on June 20, 2012. Police officers helped track down a polar bear that got out of its enclosure overnight at the Lake Superior Zoo in Duluth. Many other zoo animals drowned (Krueger and Myers 2012).

A front stalled over Colorado's Front Range during the second week of September 2013, dropping seven inches of rain on Boulder in 24 hours, a record. The National Weather Service called the rainfall "biblical." One-quarter of the buildings on the University of Colorado–Boulder's campus flooded.

On June 12, 2007, heavy thunderstorms dumped 11 to 15 inches of rain in one night on parts of parched southwestern Nebraska, surpassing the entire previous year's worth of precipitation. The next week, similar amounts of rain drenched northern Texas, killing four people in flash floods. Overnight on May 6–7, 2015, a narrow swath of southeastern Nebraska 50 miles southwest of Omaha was doused by as much as 11 inches of rain, a deluge the local National Weather Service office called a once-in-a-thousand-years event (Gaarder et al. 2015). On October 30, 2015, Texas experienced epic rains and floods, the second such episode in a month. Austin's airport shut down as 14.75 inches drowned the city in one day, an all-time record.

On October 3 and 4, 2015, what local weather-forecasting officials called a "mind-boggling" downpour that could have been expected only once in 1,000 years poured 24 inches into Mount Pleasant, a suburb of Charleston, South Carolina, with only slightly less falling in many surrounding areas. The same day, 6.7 inches of rain falling in two hours contributed to the deaths of at least 16 people on the French Riviera (Bacon 2015; Onyanga-Omara 2015).

Flooding Rains Worldwide

Similar events have been experienced around the world. On July 21, 2007, locations in southern and central England witnessed their worst flooding on record after a month's worth of rain, with as much as 5 inches falling in one hour after the wettest June in England's history. One such location was Shakespeare's Stratford-on-Avon, which led some residents to remark that had become Stratford-under-Avon. The floods displaced thousands of people and fouled water for half a million. Once-in-a-century rains also flooded large parts of China, destroying 3.6 million homes and killing at least 500 people. At the same time, the western United States was scorched by a record number of wildfires provoked by heat and drought.

On November 19–20, 2009, the English village of Cockermouth was doused with 12.3 inches of rain, the largest one-day deluge in British history, leaving floodwaters up to eight feet deep. During the fall of 2010, the South American nation of Colombia experienced its worst rainy season in at least 30 years, leaving more than 130 people dead and a million homeless. By the end of July, 2010, 1,072 people had been killed by floods in China, according to Shu Qingpeng, deputy director of the Office of Flood Control and Drought Relief with the central government. Some 140 million people had been affected by floods in China during those seven months; a million homes had been destroyed, and economic damage was estimated at $31 billion (MacLeod 2010).

On August 21, 2010, a foot of rain fell in nine hours on the North Korean border with China, part of a longer deluge that led to the evacuation of almost 200,000 people. More than 10 inches of rain fell during the first week of August 2011 in and around Seoul, South Korea. In January 2011, near-record floods swept a large area of northeastern Australia, including Brisbane, a city of 2 million people, and forcing evacuation of 30,000 homes and businesses, killing at least 15 people. At the same time, floods rampaged through towns near Rio de Janeiro, Brazil, killing more than 600 people, most of them in landslides. Many were killed in their beds during the wee hours of the morning, including at least 250 in Teresoplis.

During July and August 2010, Pakistan mourned the deaths of more than 2,000 people in its worst monsoon deluge on record. One-fifth of the country was flooded by the raging Indus River, affecting 20 million people. Eight million lost or were driven from their homes. The floods followed record high temperatures; in July, 129°F in India was the highest ever recorded anywhere in Asia. "These should be added to the extensive list of extreme weather-related events, such as droughts and fires in Australia and a record number of high-temperature days in the eastern United States of America, as well as other events that occurred earlier in the year," said the World Meteorological Organization (WMO), a specialized agency of the United Nations. "While a longer time range is required to establish whether an individual event is attributable to climate change, the sequence of current events matches IPCC projections," the WMO said ("Extreme Weather" 2010).

In late July 2012, an epic downpour in Beijing, China's capital, killed 77 people as 18 inches of rain fell in 24 hours. Most of the people died because of flooding in streets that were inadequately drained. In some cases, people were killed by torrents gushing from the ceilings of ground-floor residences and buildings full of modern skyscrapers. Beijing is on the edge of the Gobi desert and had been suffering a lengthy drought, long enough that engineers of expensive new neighborhoods in a rapidly expanding city of 22 million people had skimped on sewage systems. The Beijing deluge was the city's heaviest since 1951. Some 80,000 people were stranded at Beijing International Airport. Elsewhere in China, 60 more people died in these storms.

In earlier times, when the city was much smaller, officials had paid attention to drainage because Chinese mythology ascribes natural disasters, including floods, as well as earthquakes, snowstorms, drought, and disease, to "a revocation of the mandate of heaven that is bestowed by divine powers on earthly rulers" (Fromer and Wong 2012). Observers noted that the ancient Forbidden City has an excellent drainage system and that it had remained relatively dry during the storms. From September 8–10, 2015, Tokyo received 20 inches of rain in three days from a dying typhoon. Television reports showed entire houses flowing swiftly down swollen rivers.

In the first week of August 2012, waves of tropical moisture flooded half of Manila, and its suburbs, which are home to 12 million people. One part of the city was destroyed by a landslide that killed eight people who were trapped in shanties in Quezon City, the Philippines. "Army troops and police dug frantically to save those buried, including four children, as surviving relatives and neighbors wept. All the victims were later dug up, including a man whose body was found near an entombed shanty's door," the Associated Press reported ("Torrential Rains Submerge" 2012). The deluge was part of a seasonal monsoon that had been intensified by a series of tropical cyclones. Given the relentless waves of rain and wind, the flooding was exceptional as rains pounded saturated ground a few days after Typhoon Saola, which had killed 53 people. "It's like a water world," said Benito Ramos, head of the government's disaster response agency. According to the Associated Press, Ramos said, "The rains flooded 50 percent of metropolitan Manila . . . and about 30 percent remained under waist- or neck-deep waters" a day later ("Torrential Rains Submerge" 2012).

During the first week of July 2013, a severe thunderstorm flooded expressways around Toronto, shutting down its transit system and airports and cutting power to several hundreds of thousands of Canadians. More than three inches—a month's worth of rain in a usual July—fell in about one hour as "passengers were trapped on the upper level of a double-decker commuter train for about four hours before the Toronto police marine unit arrived in boats. Motorists fled from cars that had become inundated on expressways. More rain was expected through the night, and there were fears that the banks of at least one river in the city might collapse" (Austen 2013).

The Associated Press reported that at least 52 people had drowned or "were electrocuted or died in other accidents as flooding from days of torrential rains during the first week of April 2013 swamped Argentina's low-lying capital and province of Buenos Aires" ("Torrential Rains Flood" 2013). Most of them died in a deluge of biblical proportions in La Plata as reported by its governor, Daniel Scioli. "It started to rain really hard in the evening, and began to flood," said Augustina Garcia Orsi, a student there. "I panicked. In two seconds, I was up to my knees in water. It came up through the drains—I couldn't do anything" ("Torrential Rains Flood" 2013).

Flooding contributed to a fire in Argentina's largest oil refinery, which was blamed on "an extraordinary accumulation of rainwater and power outages in the entire refinery complex" ("Torrential Rains Flood" 2013). "Such intense rain in so little time has left many people trapped in their cars, in the streets, in some cases electrocuted. We are giving priority to rescuing people who have been stuck in trees or on the roofs of their homes," Scioli said ("Torrential Rains Flood" 2013). Within a few hours, 16 inches of rain fell in La Plata, exceeding the previous record for all of April. Another squall brought another four inches two days later. La Plata's Spanish Hospital was closed and patients evacuated for several days after water flooded its basement, extinguishing power and telephones, and ruining x-ray machines and other diagnostic equipment.

Further Reading

Austen, Ian. "Canada: Storm Floods Toronto." *The New York Times*, July 8, 2013. http://www.nytimes.com/2013/07/09/world/americas/canada-storm-floods-toronto.html.

Bacon, John. "1,000-Year Storm Slams S.C." *USA Today*, October 5, 2015, 1A.

Barringer, Felicity. "Precipitation Across U.S. Intensifies Over 50 Years." *The New York Times*, December 5, 2007, A21. http://www.nytimes.com/2007/12/05/us/05storms.html.

Copeland, Larry, and William M. Welch. "'Walls of Water' Hit with Little Warning." *USA Today*, June 14, 2010, 1A, 5A, 6A.

Environment America. *When It Rains, It Pours: Global Warming and the Rising Frequency of Extreme Precipitation in the United States*. Boston: Environment America, 2007. http://www.environmentamerica.org/reports/ame/when-it-rains-it-pours-global-warming-and-rising-frequency-extreme-precipitation-united.

"Extreme Rainstorms a New Texas Trend." Environment News Service, December 3, 2007. http://www.ens-newswire.com/ens/dec2007/2007-12-04-095.asp (no longer available).

"Extreme Weather Events Signal Global Warming to World's Meteorologists." Environment News Service, August 17, 2010. http://www.ens-newswire.com/ens/aug2010/2010-08-17-01.html.

Fromer, Jacob, and Edward Wong. "An Epic Downpour Wipes Away a Capital's Sheen." *The New York Times*, July 27, 2012, A10.

Gaarder, Nancy, David Hendee, and Andrew J. Nelson. "Once-in-a-thousand-years Rain." Omaha *World-Herald*, May 8, 2015, A-1, A-4.

Klein, Allison, Fredrick Kunkle, and Tim Craig. "Torrential Rains Inundate D.C. Region: 3 Killed, Roads and Schools Closed." Washington Post, September 8, 2011. http://www.washingtonpost.com/local/tropical-storm-lee-inundates-dc-region/2011/09/08/gIQA7OHIDK_print.html.

Krueger, Andrew, and John Myers. "Duluth Mayor Declares State of Emergency Due to Flooding." Duluth *News Tribune*, June 20, 2012. http://www.duluthnewstribune.com/event/article/id/234914/.

MacLeod, Calum. "China, on Cusp as Superpower, Wrestles Plant." *USA Today*, August 10, 2010, 6A.

"Northeast Storms Growing More Fierce." *USA Today*, April 6, 2010, 3A.

Onyanga-Omara, Jane. "At Least 16 Killed in French Riviera Flash Floods." *USA Today*, October 5, 2015, 6A.

"Torrential Rains Flood Argentina, Overall Death Toll Rises to 52." Associated Press in Washington *Post,* April 3, 2013. http://www.washingtonpost.com/business/torrential-rains -flood-argentina-overall-death-toll-rises-to-31/2013/04/03/f2e584c8-9c6d-11e2-9219 -51eb8387e8f1_print.html (no longer available).

"Torrential Rains Submerge Half of Philippine Capital, Trigger Landslide That Kills 8 People." Associated Press, August 7, 2012. https://www.washingtonpost.com/world/torrential -rains-submerge-half-of-philippine-capital/2012/08/07/29c94b48-e0a6-11e1-a19c -fcfa365396c8_gallery.html?utm_term=.0cb49d9758b3.

See also: Atmospheric Circulation; Australia, Heat and Drought in; Desertification; Drought, United States; Drought, Worldwide; El Niño and La Niña; Heat Waves, Evidence of; Heat Waves, Forecasts of; Temperatures, Global; Temperatures, Greenhouse Gas Levels and; Wildfires

HEAT WAVES, EVIDENCE OF

The spread of heat around the world from rising levels of atmospheric carbon dioxide, methane, and other greenhouse gases is not linear. Heat is aggravated by many climatic conditions in addition to greenhouse gas levels, including weather patterns, especially persistent high pressure during summer months. In 2014, for example, world temperatures set a record high for the instrumental record (since the 1880s, at least), but many of the temperatures of the eastern and midwestern portions of the United States were below average because of persistent weather patterns (the polar vortex or Arctic Oscillation) that displaced Arctic air into that area.

In November 2016, the North Pole experienced melting in areas that had been frozen in late fall. According to an account by Chris Mooney and Jason Samenow in the *Washington Post* (2016), "This is the second year in a row that temperatures near the North Pole have risen to freakishly warm levels." During 2015's final days, the writers noted, the temperature near the pole spiked to the melting point thanks to a massive storm that pumped warm air into the region. "It's about [20°C or 36°F] warmer than normal over most of the Arctic Ocean, along with cold anomalies of about the same magnitude over north-central Asia," said Jennifer Francis, an Arctic specialist at Rutgers University. "The Arctic warmth is the result of a combination of record-low sea ice extent for this time of year, probably very thin ice, and plenty of warm and moist air from lower latitudes being driven northward by a very wavy jet stream." Francis's research indicates that the jet stream is becoming wavier and more elongated as the Arctic warms faster than the equator.

By 2015, persistent distortion of the jet stream was provoking unusual temperature readings in Alaska, including 86°F in Fairbanks and 91°F in the hamlet of Eagle during the third week of May. These were warmer afternoon temperatures than was being felt in much of the eastern and southern parts of the United States— and warmer than Dallas or Houston had been all year to that date ("Baked Alaska" 2015). July 2016 was the hottest month in Earth's history for any month dating to the late 1800s when detailed records were first kept. Persistent high pressure has played a major role in a hellish summer across Northern Africa into the Middle

East during the summer of 2016. Temperatures averaged 10 to 20 degrees above average in Baghdad, with high temperatures of 109°F or higher from June 19 to the end of August. Jeddah, Saudi Arabia, hit 126°F, an all-time high. The number of days 118°F or higher has doubled in Baghdad within a decade. On July 21, 2016, one location in Kuwait hit 129.2°F (55°C), and nearby Basra, Iraq, reached 129.0°F (54.9°C), the highest ever recorded outside of Death Valley, California, which holds the world record of 134.1°F (56.7°C), recorded in 1913 but since disputed. A reading of 129.2°F. recorded in Death Valley June 30, 2013 equaled the same temperature in Tirat Tsvi, Israel, June 22, 1942.

Record Highs Worldwide During Summer 2003

As warm seasons expand, some events meant for cooler weather collide with a new climatic reality. Heat is also relative to humidity, which impedes cooling through evaporation. In Chicago, for example, the annual marathon held on October 7, 2007, was an extremely humid day with temperatures in the upper 80s Fahrenheit. Shortly before noon, the temperatures were almost 25°F above average for the season. The heat combined with high humidity killed one runner, 35-year-old Chad Schieber of Midland, Michigan, and sickened several hundred others, forcing cancellation of the race for the first time in its 30-year history.

More than 300 runners left the race course in ambulances with nausea, heart palpitations, and dizziness, all symptoms of heat illness. Fifteen city buses, their air conditioners pumping at maximum, were commandeered as aid stations. Some of the runners had core body temperatures as high as 107°F (Davey 2007, A1). The same heat wave set records across the U.S. Midwest and East and was accompanied by stifling humidity. Detroit (90°F) and Indianapolis (at 91°F) had their latest 90-degree highs on the instrumental record (since the 1870s). Louisville, Kentucky, at 93°F, set an all-time record for November.

In Phoenix, Arizona, on July 14, 2003, the high temperature reached 116°F; the diurnal low the same day was 96°F. Windshields popped in the heat and flip-flop sandals stuck to pavement. In Milan, Italy, at about the same time, an African sirocco blew northward from the Sahara, making it "a little dangerous to wear stilettos [spike high-heeled shoes]; they sank half an inch into the melting tar, impaling one's legs to the spot where one's torso pitched forward, head first, like a stone being launched from a sling-shot" (Thurman 2003, 78). One day in June 2003, the temperature hit 40°C (104°F) in Milan. As stilettos were sinking into asphalt in Milan and windshields were popping in Phoenix, much of southern China was experiencing weeks of searing heat, causing drinking-water shortages and massive drop damage. Hardest-hit Hunan, Jiangxi, Fujian, and Zhejiang provinces experienced temperatures above 40°C day after day.

Anecdotal evidence in the United States indicated heat waves were becoming more common. From late June to July 8, 2012, the city of St. Louis, Missouri, recorded 10 straight daily highs over 100°F. On July 7, 2012, a day when the temperature in Washington, D.C., hit 105°F, the tires of a jet melted into the tarmac at Reagan National Airport. Also in 2012, Louisville, Kentucky, hit 108°F on

July 4, the same day that the high was 107°F in Madison, Wisconsin, and 104°F in Detroit. Summer 2012 arrived with record heat across the eastern two-thirds of the United States, record wildfires in Colorado (and other western states) that scorched several hundred homes, and, in the midst of it all, a hurricane-intensity line of thunderstorms cut power to 3 million homes and businesses from Ohio to Virginia. At the same time, Atlanta set an all-time high of 106°F on June 30— then matched it the next day. Nashville hit 107°F, and Columbia, South Carolina, hit 108°F—both all-time highs—on June 30, and St. Louis hit 106°F the same day.

The heat wave of 2012 rewrote the record books in the Rockies and Great Plains. On June 26, an all-time record high was set in Denver (105); McCook, Nebraska (115); and Sidney, Nebraska (111). Several other all-time records more and more than 7,000 daily records were set. The mountain West was singed by some of the worst wildfires in its history as entire suburbs burned in Colorado Springs. At the same time, record rains inundated parts of formerly drought-stricken Florida under slow-moving Tropical Storm Debby, including 30 inches over 24 hours in Live Oak, Florida, west of Jacksonville.

NOAA reported that January through June 2012 was the warmest on record in the United States to that time. The average temperature, 52.9°F was 4.5 degrees above average. The previous 12 months was 56.0°F, 3.2°F above average. Twenty-eight states and more than 100 cities posted their highest averages on record (January to June 2012); Chicago was 7°F above average, and New York City was 4°F above. Between January and June, 22,356 record highs were set during those six months, compared to 2,448 new lows—almost a 10:1 ratio.

Europe's Searing Summer of 2003

During June 2003, Europe was dominated by high pressure that induced the hottest period since at least the middle of the 16th century when records began. Temperatures averaged 2.3°C (4.1°F) above average that summer, which led to some 70,000 premature deaths. Glaciers in the Alps also melted at twice the rate of the previous record summer five years previously (Jolly 2014).

On August 6, 2003, London experienced its highest temperature since reliable records began to be kept (around 1770): 95.7°F (34°C). Four days later, the temperature at Heathrow International Airport peaked at 100.2°F, the first three-digit reading in the climatic history of the British Isles. In Gravesend, Kent, the same day, the temperature reached 100.8°F. British commuter railroads were slowed because of fear that tracks might buckle in the heat. England was hardly alone; during the same week, one German locale recorded an all-time record high of 40.8°C (105.4°F). A reporting station in Switzerland reached 41.5°C (106.7°F), and one in Portugal soared to 47.3°C (117.1°F) (McCarthy 2003). The same week, some areas in southwestern Spain hit 117°F ("Europe's Heat" 2003).

The year 2003 turned out to be the warmest in the entire Central England Temperature (CET) record, which dates to 1659. The year's final average temperature of 10.65°C beat the previous records of 1990 (10.63°C) and 1999 (10.62°C). The trend was not restricted to Europe. Albuquerque, New Mexico, for example, set a record-high minimum in 2003 at 4°F above long-term averages (Fleck 2003). Power supplies failed in many urban areas, and sun-seeking tourists from the likes of Ireland and Scandinavia returned home from Italy and Greece having spent their holidays in hotel rooms because outside temperatures near 100°F made it too hot to sunbathe.

In France during the same summer, an estimated 14,800 people died; many elderly citizens expired in apartments that had no air conditioning. During the two weeks when heat was most intense in France, the nation's mortality rate increased by 54 percent (Schar and Jendritzky 2004, 559). A politically charged debate enveloped the country as temperatures exceeded 100°F several times during August 2003. The nation's health minister resigned as morgues overflowed with bodies. A quarter of France's 58 nuclear power plants shut down at the height of the heat wave because the nation's rivers had become too warm to properly cool reactors. Paris recorded an overnight low of 78 one morning, its highest low temperature on record. During the same period, more than 1,300 people died of heat-related causes in Portugal. Another 7,000 heat-related deaths were recorded in Germany, and nearly 4,200 lives were lost in both Spain and Italy. More than 2,000 people died in Britain ("Summer Heat Wave" 2003, 2).

One analysis put Europe's death toll at 35,000 or more during the scorching summer of 2003. The Earth Policy Institute based in Washington, D.C., warned that such deaths are likely to increase because "even more extreme weather events lie ahead" ("Summer Heat Wave" 2003). Europe's 2003 summer heat wave was the climatic crown jewel of the hottest decade in the history of the continent, which reaches back at least to 1500 CE. The average temperature was 2°C (3.6°F) higher than the long-term average.

"Taking into account the uncertainties in our reconstructions, it appears that the summer of 2003 was very likely warmer than any summer back to 1500," said Jurg Luterbacher, who led a 500-year study of European temperatures. Results from regional climate model simulations suggest that about every second summer will be as hot or even hotter than 2003 by the end of the 21st century (Luterbacher et al. 2004, 1502). The summer of 2003 in Europe was by far the hottest single season in the 500-year record examined in this study. Luterbacher and colleagues stated with a 95-percent degree of confidence that "European climate is very likely warmer than that of any time during the past 500 years" (Luterbacher et al. 2004, 1499).

As described in *Science*, Luterbacher and colleagues used temperature data from sources such as ice cores and tree rings to reconstruct the climatic history of Europe in the preceding 500 years. Against this background, data from the past century, particularly its most recent decades, stand out as extraordinary and point strongly to human-induced global warming as the key influence (Henderson 2004). Although the summer 2003 European heat wave was a statistical anomaly according to current average temperatures, a team of Swiss scientists found that, given expected increases in greenhouse gas levels, such events appear increasingly likely in the future (Schar and Jendritzky 2004).

As Europe sizzled, U.S. President George W. Bush formed a "century club" comprised of people willing to jog with him at his Crawford, Texas, ranch when the temperature exceeded 100°F. After exercising, the joggers reclined in the air-conditioned comfort of Bush's ranch house—a comfort not available to most of the Europeans who had died in the heat. Skeptics of global warming asserted that the French needed more air conditioning, and environmentalists replied that most Europeans had never before required mechanically cooled air to survive. Air conditioning requires copious amounts of energy that is often generated by fossil fuels and eventually aggravates warming.

Heat Waves in India and Pakistan

Temperatures in India usually peak in May and then monsoon rains begin. In recent years, however, May heat waves have been more intense than usual, and monsoons have often failed, producing less rain than average. In 2015, an intense heat wave killed more than 2,500 people in one month as temperatures rose as high at 118°F in Andhra Pradesh, a state on the subcontinent's southeast coast, as well as in the neighboring states of Telangana and Odisha. On May 20, 2016, the temperature set an all-time record for India of 123.8°F (51°C) in Phalodi, in Rajasthan, the desert state in the country's northwest.

Most of those who died worked outdoors (such as those in construction trades) or were elderly or homeless. Temperatures hit 115°F in New Delhi. High humidity aggravated the heat wave as temperatures averaged 5.5°C (10°F) above average during the period. The government of India late in April 2016 banned cooking during the daylight hours after a desiccating heat wave killed more than 300 people during that month, 80 of them from out-of-control cooking fires. Temperatures rose to 113°F in Hyderabad; relative humidity was just 13 percent. By April 2016, 330 million people in India were lacking enough water to meet their daily needs ("India Orders" 2016). Ground–based measurements were 4 to 5°C (8–10°F) above average; in isolated areas, temperatures rose to 12°C (22°F) above seasonal averages ("Heat Wave" 2016). On May 18 and 19, however, the record heat and drought were broken suddenly by violent, deadly deluges in southern India. Days later, heat and drought returned.

As the NASA Earth Observatory reported, "The hot weather was not limited to India. Pakistan, southern Iran, the United Arab Emirates, and Oman also faced extremely hot temperatures. On some days, it was even hotter in parts of Pakistan than it was in India, according to news reports. On June 3, temperatures soared to 50.7 [°C or 123.3°F] in Sweihan, a town in the United Arab Emirates" ("India Faces" 2015). The 2014 monsoon had failed, and weather forecasters were anticipating below-average rains in 2015. Dr. Harsh Vardhan, India's minister for Earth sciences, called the extreme heat and below-average rainfall "manifestations of climate change."

"World leaders are already speaking out on the link between freaky weather and climate change. So, let us not fool ourselves that there is no connection between the unusual number of deaths from the ongoing heat wave and the certainty of another failed monsoon," said Vardhan ("India Blames" 2015). Atmospheric pollution, he said, also has been influencing the monsoon.

As heat killed thousands during India's scorching 2015 heat wave, temperatures as high as 45°C (113°F) killed more than 1,000 people in Karachi, Pakistan. The death toll was worst among the poor and those who labored physically outdoors. It also was aggravated by the Muslim feast of Ramadan, which requires that Muslims abstain from eating food and drinking water during daylight hours. Fifteen hours without water was killing elderly people, laborers, and street vendors, among others. Elsewhere in Pakistan, temperatures rose to as high as 122°F.

With a population of more than 20 million, Karachi also suffers frequent power outages that shut down electric fans and air conditioning and inhibit water pumping. "The first to die were the people on the streets—heroin addicts, beggars, the homeless," said Anwar Kazmi, a government spokesman. "Then it was the elderly, particularly those who didn't have anyone to take care of them" (Imtiaz and Walsh 2015). The heat-induced suffering also had become a political issue in Karachi when activists called on the rich to turn off their air conditioning and endure the heat alongside the poor. After a week of intense heat, Mufti Muhammad Naeem took the highly unusual action of issuing a fatwa that excused people who were ill from abstaining from food and water during daylight hours of Ramadan. Michael Oppenheimer, professor of Geosciences at Princeton University, said, "What we're seeing, really, is a window into what global warming looks like. It looks like heat. It looks like fire. It looks like this kind of environmental disaster" (Gaarder 2012, 8A).

Temperature records are also being set in the Arctic. On May 29, 2012. Narsarsuag, Greenland, on the island's southern coast, hit 24.8°C (76.6°F), the warmest temperature on record for Greenland and only 0.7°C (1.3°F) below the all-time record high, which was set July 26, 1990. The previous May record (22.4°C or 72.3°F) at Kangerlussuaq (Sondre Stormfjord in Danish) on May 31, 1991, was far surpassed. The exceptional warmth was the result of intense high pressure and a downslope (foehn) wind.

Elsewhere in the world, long-term evidence of warming was plentiful. According to the Mongolian ministry of nature and environment, temperatures in that nation have risen over 60 years (1942 to 2002) by 1.56°C annually, including by 3.61°C during winter. Rising temperatures have come with intensifying drought in many areas ("Climate Warms" 2003).

Eugene Robinson, a columnist with the *Washington Post*, commented:

Still don't believe in climate change? Then you're either deep in denial or delirious from the heat. As I write this, the nation's capital and its suburbs are in post-apocalyptic mode. About one-fourth of all households have no electricity, the legacy of an unprecedented assault by violent thunderstorms Friday night. Things are improving: At the height of the power outage, nearly half the region was dark.

The line of storms, which killed at least 17 people as it raced from the Midwest to the sea, culminated a punishing day when the official temperature here reached 104 degrees, a record for June. Hurricane-force winds of up to 80 miles per hour wreaked havoc with the lush tree canopy that is perhaps Washington's most glorious amenity. . . . According to scientists, climate change means not only that we will see higher temperatures but that there will be more extreme weather events like the one we just experienced. Welcome to the rest of our lives. (Robinson 2012)

Recent Warmth during Summers

Stoked by an intense El Niño as well as rising greenhouse gas levels, the calendar year 2015 was Earth's hottest by a wide margin. Around the globe, heat made history. Heat may not be unusual in the Persian Gulf area, but during midsummer 2015, for example, temperatures reached as high as 125°F, and the city government of Baghdad cautioned against outdoor activity, declaring a four-day emergency. Many people could not escape the heat because years of armed conflict have destroyed many homes and turned them into refugees. Many more citizens lack air conditioning because they cannot afford it or, even if they can, must endure chronic power cuts. During the hottest afternoons, many people wander shopping malls not to buy anything but to get some cool air. During the summer of 2015, with temperatures shattering record highs and air conditioning down to a few hours a day, several thousand people surged into the streets, protesting government corruption they held responsible for the lack of electricity. Several government ministers subsequently lost their jobs.

As water temperatures in the Persian Gulf rose into the 90s, evaporation raised humidity over adjacent land. On Thursday, July 30, 2015, a sea breeze pregnant with sopping humidity in midafternoon produced a dew point of 90 on the southwestern Iranian coast at the Bandar Mahshahr Airport, with a temperature of 115°F (45°C). That reading produced at heat index of 163. Further inland, with lower dew points, temperatures reach 124°F in Baghdad. In Mitribah, Kuwait, on the same day, the thermometer hit 127°F, within 1.5 degrees of the all-time record that was set in 2014. Heat also is not unusual in Phoenix, Arizona, but as the temperature reached a monthly record of 120°F there on June 19, 2016, Kim Leeds baked cookies in her car.

Nick Wiltgen of Weather.com reported on July 31, 2015 that a "Filipino migrant rights activist collapsed and later died of apparent heat stroke during a visit to his country's consulate in Dubai this week. Highs over the past week have hovered near 113 degrees (45 degrees Celsius) in Dubai, and the Dubai area has not reported a temperature below 90 degrees (32 degrees Celsius) even at night since July 24" (Wiltgen 2015).

On July 1, 2015, an all-time July high was set at London's Heathrow Airport at 98.6°F. It also was the hottest day in the annals of the Wimbledon tennis tournament, which dates back to 1877. A ball boy collapsed in the heat and was taken off the court on a stretcher. The same day the high hit 103.5°F in Paris. In Washington state the same week, temperatures peaked at 113°F. According to Timothy Egan, a columnist for *The New York Times* who grew up in that area, "Wildfires burn today in the Olympic Mountains west of Seattle, a forest zone that is typically one of the rainiest places on Earth" (Egan 2015).

At the same time, heat and drought in the U.S. West, Canada, and Alaska that produced record temperatures during late June and early July 2015 was enhanced by a "kink" in the jet stream and a ridge of high pressure. A similar pattern afflicted Western Europe at the same time, where, as the NASA Earth Observatory observed:

> On July 2, the air temperature in Maastricht reached 38.2 [°C or 100.8°F), the highest temperature ever recorded in the Netherlands. The mercury reached 40.3°C (104.5°F) in Kitzingen, Germany, on July 5, and 38.9°C (102.0°F) in Geneva, Switzerland, on July 7, the highest temperature ever recorded in either nation. Dozens of cities—including London, Paris, Berlin, Cordoba, Frankfurt, Madrid—easily surpassed records for the hottest June or July day on record. In some cities, both monthly records fell in a matter of days. ("Europe and Pacific Northwest" 2015)

The record heat of 2015 followed several years of intensifying warmth around the world. Paris roasted at 100°F as Oklahoma City hit 107 as parts of Interstate 44 were closed because they buckled in the heat. Omaha hit 106. On July 23, the minimum in Phoenix, Arizona, was 97 after a record-high maximum of 112°F. Sacramento's low on July 23 was 79 after a high of 107. On July 27, 2006, that city experienced the last of 11 consecutive 100-degree days, a record. Salt Lake City's low on July 23 was 77 after a high of 100 on July 22. In Sioux City, South Dakota, on July 31, the temperature stood at 85°F at 4 a.m. On August 2, the temperatures at 3 a.m. were 86 in Kansas City and Des Moines. The average nighttime temperature in Los Angeles was seven degrees warmer than it was a century earlier (Harden 2006).

Record Heat Roll Call

In California during the summer of 2006, a heat wave killed more than 140 people, the highest toll in that state since the widespread advent of air conditioning in the 1950s. This heat wave was exceptional for its length (almost two weeks), its record temperatures, and—most notably—the fact that temperatures did not cool at night as usual in California. Most of those who died were elderly who lived alone without air conditioning and fieldworkers.

One unidentified fieldworker around 40 years of age reached a hospital emergency room only after his body temperature reached 109.9°F. Unusually high humidity and nighttime high temperatures played a role in the death toll. Roughly 16,500 cows died of the heat, according to California Dairies, the state's largest milk cooperative. Panting, miserable cows, who lack the benefit of sweat glands, yielded 10 percent to 20 percent less milk than usual (Steinhauer 2006). The walnut harvest was damaged when intense sunlight and heat burned the nuts inside their shells.

Record heat during the scorching summer of 2006 caused power consumption to soar. On July 22, the temperature in Woodland Hills, a Los Angeles suburb, reached 119°F. Most startling were the minimum temperatures in many cities of the western United States, where low dew points usually cause temperatures to fall on summer nights, even after warm days. Seattle had a high of 97 followed by a record-high low of 72. Boise reached a low of 77 on July 24 after a high of 98 (Tehran, Iran, was 104 and 84 the same day). Advertisements for the documentary *An Inconvenient Truth* in *The New York Times* on July 21, 2006, ran alongside reports of record highs under the headline "How Hot Is It in Your City?" Record heat continued in 2007 as Phoenix registered 29 days at 110°F or higher.

Average temperatures in California increased almost two degrees Fahrenheit between 1950 and 2000, according to data from NASA and California State University–Los Angeles. Land-use patterns as well as generalized greenhouse warming played a role in the temperature rise. The state's urban areas, most notably Southern California and the San Francisco Bay Area, experienced the largest temperature increases. Rural nonagricultural regions warmed the least. The Central Valley warmed the most slowly, and coastal areas warmed faster, and the southeast desert warmed most quickly of all. Only one area was cooler: a small part of the state's rural northeast interior ("California's Temperature" 2007).

Bill Patzert of NASA's Jet Propulsion Laboratory in Pasadena said that increasing minimum temperatures in California has led to narrower daily temperature ranges throughout the state. "California nights are heating up, giving us a jump start on hotter days," Patzert said: "This is primarily due to increased urbanization, not increases in cloudiness or precipitation. Rainfall and snowfall didn't increase significantly for most California stations during the study period" ("California Temperature" 2007). "California's complex topography and large latitude range give it some of the most diverse microclimates in North America," said Steve LaDochy of California State University, also a coauthor of the report. "Climate change models and assessments often assume global warming's influence here would be uniform. That is not the case" ("California Temperature" 2007).

More Record Heat in the United States

On July 29, 2009, under a blazing sun, Seattle reached an all-time high of 103 degrees (with a record-high minimum of 75°F as well) at Seattle–Tacoma airport. Meanwhile, a sodden New York City was enduring its second-coolest July on record. While Seattle was clocking 103, Iowa was having its coolest July on record at 68°F, and Phoenix in Arizona was averaging 98.3 for the month of July, also a record high, all the result of low pressure over the East and high pressure over the West that stayed anchored in place all of July.

On September 27, 2010, Los Angeles set an all-time high of 113°F, breaking an old record of 112 set in 1990. Russia (at 111°F), Saudi Arabia and Iraq (both at 126), Niger (118), Sudan (121), and Pakistan (at 129) set all-time temperature records. On July 6, 2010, Baltimore hit 104°F, Philadelphia 104, and New York City 103. By 7 a.m. the next day, the temperature in New York City was 86. In

Philadelphia it was 87. The same week, Beijing reached 109°F. In August, 2007, Atlanta, Georgia, saw five of its 10 hottest days on record during one heat wave. The same summer, Phoenix, Arizona, set a record for number of days—32—above 110°F.

On July 24, 2010, a high of 105°F in Norfolk, Virginia, tied the city's all-time high temperature set on August 7, 1918, with a heat index of 116. The next day, Norfolk broke the record with 106. Washington, D.C., in June 2010 had the highest average temperature for that month on record (begun in 1871), with 18 days at 90°F or higher. In Kansas City, police tucked cold packs into bulletproof vests. Also in Washington, D.C., several dozen Boy Scouts collapsed from heat illness during a parade to celebrate the centennial of scouting (Dorrell 2010). Shortly after, the temperature dropped 30 degrees from 100°F in an hour during one of the area's worst thunderstorms on record, one that cut power to more than a half-million homes and businesses.

In New York City, the 2010 meteorological summer (June, July, and August), at 77.8°F was the hottest on record in New York City. In 1966, with an average of 77.3°F, a time before many people had access to air conditioning, more than 1,100 people died in heat that often broke 100 degrees.

Heat in Russia 2010

In midsummer 2010, Russia suffered a withering heat wave that raised temperatures to 100°F in Moscow (where the average midsummer high is 75°F) and scorched 24 million acres of wheatland in southern parts of the country, where no rain had fallen since April. The heat kindled peat bog fires around Moscow, "sending coils of smoke into the hazy air. Ignited by a casually flipped cigarette butt or an improperly placed campfire, these fires burrow into the dry peat and then creep along, dozens of feet underground. They pollute the air, ignite above-ground forest fires and are typically only fully extinguished after a heavy rain in the fall" (Kramer 2010). So much smoke was generated in some areas that airports shut down. On August 4, 2010, the Russian federal government banned wheat exports, causing a major rise of 10 percent in worldwide wheat futures in one day. The fires also contained radioactivity remaining from the Chernobyl nuclear accident in 1986. Large areas that were burned had been coated with fallout following the accident. Moscow's health minister urged residents to stay indoors or leave the city even if they missed work. By August 9, the damage of the smoky shroud to people's lungs was being measured in packs of cigarettes per day. Estimates ranged from three to eight. The death rate in Moscow rose by 50 percent from early July to early August.

As Moscow endured weeks of 90-plus highs, large numbers of Russians who lacked air conditioning drowned in local rivers and lakes. Peat and wildfires around the city threw an acrid eye-stinging blanket over the city, adding more carbon dioxide to the atmosphere. The bogs were drained in Soviet times to harvest the peat as fuel, making them dry enough to burn. Moscow's 10 million residents were advised to wear face masks. Moscow temperatures hit an all-time record high of 101°F on July 29, a reading matched August 2. St. Petersburg hit 96 on July 29, and Helsinki,

Finland, rose to 93 on July 28. Russia also suffered large areas of prolonged drought even as northern Pakistan mourned the deaths of more than 1,000 people in its worst monsoon deluge on record.

Heat Waves: 2011, 2012, and 2013

In the United States, many areas seared at 100-plus during 2011. The Tucson, Arizona, city morgue had to bring in refrigerator trucks when it ran out of space following the deaths of 150 undocumented workers crossing the border in the heat, many more than in previous years. In 2011, Texas, with an average temperature of 86.8°F, had the hottest meteorological summer of any state to that time, surpassing the previous state record set in Oklahoma during 1934 by more than 1°F. In 2011, Dallas experienced 70 days with highs above 100°F, 40 of them consecutively. Nearby Little Rock, Arkansas, hit 113 on August 3, 2011; and Tulsa, Oklahoma, reached 114 the same day.

The 2011 Texas heat wave brought record drought as well. The Texas Forest Service Association reported fires destroying 1,500 homes and razing an area the size of Connecticut. "No one on the face of this Earth has ever fought fires in these extreme conditions" (Friedman 2011).

The century mark was reached in some unusual places during 2011's summer, including Portland, Maine, at 102°F. Washington, D.C., reached its highest on record (105), as did Newark (108). July 2011 also was notable for its high nighttime temperatures, often in the 80s (Sternberg 2011). July 22, 2011 was the hottest day in memory for many people in the mid-Atlantic states—104 (heat index 121) in Washington, D.C.; 104 in New York City; and 108 in Newark, New Jersey.

Showing the fingerprint of global warming, the nights were more exceptionally warmer than the days. The thickening blanket of greenhouse gases held in the heat more effectively than in the past. For the month of July 2011, the United States had 2,722 daily record highs and 6,106 record warm minimum temperatures. Although parts of the West were unseasonably cool that year, nationwide the United States experienced only 443 daily record cool minimum temperatures, about a quarter of the record lows (Pollack 2011).

In Omaha, Nebraska, the monthly average temperature during July 2011 was 81.9 degrees, the warmest since August 1983. Omaha nights were unusually warm and extremely humid. The average daily low was a sultry 73.8 degrees, about average for Houston, tying the all-time record for Omaha from 1934. The temperature did not go below 80 degrees for four consecutive days from July 17th through the 20th. This was the longest since 1934—one of the hottest of the Dust Bowl summers. Low temperatures remained above normal for 34 consecutive days from June 29 through August 1, 2011 (Pollack 2011).

Excessive heat during the summer of 2012, even in sizzling Phoenix, Arizona, "seem[ed] to have stretched even the most elastic tolerance levels to their limits," according to Fernanda Santos in *The New York Times* (2012). "The temperature rises cruelly here as the day goes on—hot in the morning, very hot by midday and still hot late at night. While that is not uncommon for August, when the mercury breaches the triple digits practically every day, it has been particularly vicious of late as the same routine has played out day after day" (Santos 2012). In Phoenix, daily minimums above 90°F became routine. During the day, the high there reached 116°F on August 8. The average temperature (high and low) in Phoenix averaged 100.2°F for the first half of August.

Worldwide Heat

Globally, Australia experienced its warmest year on record and Japan's and South Korea's summers were the hottest, according to the National Oceanic and Atmospheric Administration (NOAA). As with 2014, 2013 was notable for its near-record warmth in the absence of an El Niño. As in 2014, much of the central and northern United States were cooler than average, a worldwide anomaly.

During July and early August 2013, eastern China endured a record-breaking heat wave. Shanghai set all-time record highs three times in three weeks, peaking at 40.8°C (105.4°F) on August 7. During January 2013, Australia was swept by record heat for several weeks, with temperatures reaching 45°C (113°F) in several locations. The Australian Bureau of Meteorology said that the Austral summer set record for the highest average temperature across the continent, with a high of 40.33°C (104.59°F) on January 7. The next day was almost as hot, with an average national temperature of 40.11°C (104.20°F) the following day. The minimum on January 7–8 also set a record at 32.36°C (90.25°F).

As the Australian Bureau of Meteorology analysis said,

> Australia has always experienced heat waves, and they are a normal part of most summers. However, the current event affecting much of inland Australia has definitely not been typical. The most significant thing about the recent heat has been its coverage across the continent, and its persistence. It is very unusual to have such widespread extreme temperatures—and have them persist for so long. On those two metrics alone, spatial extent and duration, the last two weeks surpasses the only previous analog in the historical record (since 1910) a two week country wide hot spell during the summer of 1972–1973. ("How Widespread?" 2013)

The entire year of 2014 was on pace to set an instrumental record. September 2014 was the warmest September, and April through September was the hottest in 130 years of worldwide records. This did not hold true in the central and northeastern United States, however, which continued to be in the grip of a persistent Artic Oscillation. Worldwide, however, according to NOAA, the average world temperature in August was 61.36°F (16.35°C), breaking the record set in 1998. The year 2014 broke the annual heat record set in 2010 and did so in the absence of El Niño conditions. Illustrating the long-term trend, the last monthly record for cold was set in 1916, according to NASA, but the oldest monthly heat record worldwide dates from 1997. December 2014 also was the 358th consecutive month that worldwide temperatures have been above the 20th-century average.

Further Reading

"Baked Alaska." NASA Earth Observatory, May 27, 2015. http://earthobservatory.nasa.gov /IOTD/view.php?id=85932&src=eoa-iotd.

"California's Temperature Rising." Environment News Service, April 9, 2007. http://www .ens-newswire.com/ens/apr2007/2007-04-09-09.asp#anchor6 (no longer available).

"Climate Warms Twice as Fast." British Broadcasting Company Monitoring Asia-Pacific, January 6, 2003. (LEXIS).

Davey, Monica. "Death, Havoc, and Heat Mar Chicago Race." *The New York Times*, October 8, 2007, A1, A14.

Dorell, Oren. "Scouts Ambushed by Heat, Storm." *USA Today*, July 26, 2010, A1.

Egan, Timothy. "Seattle on the Mediterranean." *The New York Times*, July 3, 2015. http:// www.nytimes.com/2015/07/04/opinion/seattle-on-the-mediterranean.html.

"Europe and Pacific Northwest Face Record Heat." NASA Earth Observatory, July 12, 2015. http://earthobservatory.nasa.gov/IOTD/view.php?id=86204.

"Europe's Heat Wave Toll Tops 19,000." *Omaha World-Herald*, September 26, 2003, 2.

Fleck, John. "Dry Days, Warm Nights." *Albuquerque Journal*, December 28, 2003, B1.

Freidman, Thomas L. "Is It Weird Enough Yet?" *The New York Times*, September 14, 2011, A29.

Gaarder, Nancy. "Experts Unravel Causes of Punishing Heat Wave." *Omaha World-Herald*, June 29, 2012, 1A, 8A.

Harden, Blaine. "Tree-Planting Drive Seeks to Bring a New Urban Cool; Lower Energy Costs Touted as Benefit." *Washington Post*, September 4, 2006; A1. http://www.washingtonpost .com/wp-dyn/content/article/2006/09/03/AR2006090300926_pf.html.

"Heat Wave Hits Thailand, India." NASA Earth Observatory, May 4, 2016. http:// earthobservatory.nasa.gov/IOTD/view.php?id=87981&src=eoa-iotd.

Henderson, Mark. "Past Ten Summers Were the Hottest in 500 Years." *Times* (London), March 5, 2004, 10.

"How Widespread Was the Australian Heatwave?" NASA Earth Observatory, January 23, 2013. http://earthobservatory.nasa.gov/IOTD/view.php?id=80232&src=eoa-iotd.

Imtiaz, Saba, and DeClan Walsh. "Karachi Heat Wave Death Toll Tops 650 during Ramadan Fast." *The New York Times*, June 23, 2015. http://www.nytimes.com/2015/06/24 /world/asia/pakistan-says-more-than-600-have-died-in-heat-wave.html.

"India Blames Climate Change for Heat Wave Deaths." Environment News Service. June 2, 2015. http://ens-newswire.com/2015/06/02/india-blames-climate-change-for-heat-wave -deaths/.

"India Faces Deadly Heat Wave." NASA Earth Observatory, June 5, 2015. http:// earthobservatory.nasa.gov/IOTD/view.php?id=85986&src=eoa-iotd.

"India Orders Some Areas to Cook at Night." *Omaha World-Herald*, April 30, 2016, 7A.

Jolly, David. "Heat Waves in Europe Will Increase, Study Finds." *The New York Times*, December 8, 2014. http://www.nytimes.com/2014/12/09/world/europe/global-warming-to-m ake-european-heat-waves-commonplace-by-2040s-study-finds.html.

Kramer, Andrew E. "Russians and Their Crops Wilt under Heat Wave." *The New York Times*, July 19, 2010. http://www.nytimes.com/2010/07/20/world/europe/20russia.html.

Luterbacher, Jurg, et al. "European Seasonal and Annual Temperature Variability, Trends, and Extremes Since 1500." *Science* 303 (March 5, 2004): 1499–1503.

McCarthy, Michael. "The Four Degrees: How Europe's Hottest Summer Shows Global Warming Is Transforming Our World." *The Independent* (London), December 8, 2003, 3.

Mooney, Chris, and Jason Samenow. "The North Pole Is an Insane 36 Degrees Warmer Than Normal as Winter Descends." *Washington Post*, November 17, 2016. https://www

.washingtonpost.com/news/energy-environment/wp/2016/11/17/the-north-pole-is-an
-insane-36-degrees-warmer-than-normal-as-winter-descends/?utm_term=.4c41bc
26bbbe&wpisrc=nl_headlines&wpmm=1.

Pollack John. U.S. Weather Service Meteorologist e-mail, August 15, 2011.

Robinson, Eugene. "Feeling the Heat." *Washington Post*, July 2, 2012. http://www.washing
tonpost.com/opinions/eugene-robinson-feeling-the-heat/2012/07/02/gJQANNZGJW
_print.html.

Santos, Fernanda. "Nine Straight Days of 110 or More: That's Hot, Even for Phoenix." *The
New York Times*, August 14, 2012. http://www.nytimes.com/2012/08/15/us/unrelenting
-heat-keeps-torrid-grip-on-phoenix.html.

Schar, Christoph, and Gerd Jendritzky. "Hot News from Summer 2003." *Nature* 432 (December 2, 2004): 559–561.

Steinhauer, Jennifer. "In California, Heat Is Blamed for 100 Deaths." *The New York Times*,
July 28, 2006. http://www.nytimes.com/2006/07/28/us/28heat.html.

Sternberg, Steve. "July Heat Left Records Smoldering." *USA Today*, August 1, 2011, A1.

"Summer Heat Wave in Europe Killed 35,000." United Press International, October 10, 2003
(LEXIS).

Thurman, Judith. "In Fashion: Broad Stripes and Bright Stars: The Spring-Summer Men's
Fashion Shows in Milan and Paris." *The New Yorker*, July 28, 2003, 78–82.

Wiltgen, Nick. "Iraq [and] Iran Heat 125 Degrees; Feels Like Temperature Reaches 159."
Weather.com. July 31, 2015. http://www.weather.com/news/news/iraq-iran-heat-middle
-east-125-degrees.

See also: Atmospheric Circulation; Australia, Heat and Drought in; Desertification;
Drought, United States; Drought, Worldwide; El Niño and La Niña; Heat Waves,
Forecasts of; Temperatures, Global; Temperatures, Greenhouse Gas Levels and;
Wildfires

HEAT WAVES, FORECASTS OF

Infrared forcing (the greenhouse effect or global warming) does not cause Earth to
heat evenly. Weather remains variable, so a generally warming climate expresses
itself in occasional bursts of heat in specific areas as well as occasional cold episodes. Heat often occurs earlier in spring and may last longer into the early fall.
Different areas of the globe experience record heat in various years as the general
warming trend coincides with conditions such as southerly winds (in the Northern Hemisphere) and persistent high pressure. Heat waves are often associated with
semistationary high pressure at the surface and aloft, which produces clear skies,
light winds, and warm-air advection. One model suggests that these conditions will
occur more frequently with increasing concentrations of greenhouse gases in the
atmosphere (Meehl and Tebaldi 2004, 996).

Studies in the United States anticipate more intense and frequent heat waves and
more extreme weather events such as floods, droughts, and wildfires with increasing atmospheric levels of greenhouse gases. "The general conclusion of the new
research is that many of the extremes being witnessed worldwide are consistent with
what scientists expect on a warming planet," wrote Justin Gillis in *The New York
Times* (2012). "Heat waves, in particular, are probably being worsened by global
warming, the scientists said. They also cited an intensification of the hydrological

cycle, reflected in an increase in both droughts and heavy downpours." The study was published during 2012 in the *Bulletin of the American Meteorological Society* ("BAMS" 2012). So much water is being removed from the oceans and precipitated on land "that the sea level actually dropped as storms moved billions of gallons of water onto land, they said, but by late 2011 the water had returned to the sea, which resumed a relentless long-term rise" (Gillis 2012).

Types of Heat Illness

Heat illness usually begins with the minor discomfort and inconvenience of heat cramps, which result when blood vessels widen to dissipate extra heat. These muscle spasms usually disappear with relief from the heat and a drink of cold water. Heat stress is cumulative, however, even over many days. Increasingly warm nights and lack of relief can lead to heat stress in various more serious forms, one of which is heat exhaustion. An affected person will feel dizzy and weak and then progress to nausea and vomiting. As uncomfortable as it may be, heat exhaustion is usually curable with fluids in a cool environment.

The truly dangerous stage of heat stress is heat stroke, which can set in with a deceptive euphoria. A person passes from heat exhaustion to heart stroke as body temperature rises to around 105°F. Once body temperature reaches the level of a medical emergency, lethargy sets in, along with severe headaches and disorientation, delirium, comas, and then sometimes death from damage to the brain and nervous system (Parker and Shapiro 2008, 32–34).

Heat Waves and Mortality

Historically, heat stress has been the foremost weather-related cause of death in the United States. During the second half of the 20th century, however, even as temperatures rose, the rate of heat-related deaths declined dramatically with the increased use of air conditioning, better medical care, and increased public awareness of heat stress effects. Robert Davis, associate professor of environmental sciences at the University of Virginia, and his colleagues studied heat-related mortality in 28 major U.S. cities from 1964 though 1998. They found that the heat-related death rate—41 per million people a year in the 1960s and 1970s—declined to 10.4 per million during the 1990s (Davis et al. 2003).

Laurence S. Kalkstein has estimated that a doubling of the carbon dioxide level in the atmosphere could increase heat-related mortality to seven times current levels if acclimatization is not factored in. With acclimatization—human adaptation to higher temperatures—the estimated increase in heat-wave mortality estimated by Kalkstein is four times the current rate (Kalkstein 1993, 1397). Kalkstein observed that each urban area has its own "temperature threshold" at which the death rate from heat illnesses rises rapidly. Seattle, for example, has a lower threshold than Dallas. "Mortality rates in warmer cities seemed to be less affected no matter how high the temperature rose," Kalkstein wrote (1993, 1398). He suggested that residents of urban areas in poor countries will find adaptation more difficult because of limited access to air conditioning.

Fatal Heat By 2100?

Levels of heat and humidity that are fatal to human beings may become routine in the Persian Gulf by 2100 as some population centers in the Middle East experience temperature levels intolerable to humans. A foretaste of such conditions (a dew point, or wet-bulb temperature, of 95, and a heat index of around 165) arrived in that area once for a short period in 2015. As Jeremy S. Pal and Elfatih A.B. Eltahir wrote in *Nature Climate Change* (2015), the date of such conditions as noted in scientific literature may advance from 200 years to less than a century:

> A human body may be able to adapt to extremes of dry-bulb temperature (commonly referred to as simply temperature) through perspiration and associated evaporative cooling provided that the wet-bulb temperature (a combined measure of temperature and humidity or degree of 'mugginess') remains below a threshold of 35°C. This threshold defines a limit of survivability for a fit human under well-ventilated outdoor conditions and is lower for most people. We project using an ensemble of high-resolution regional climate model simulations that extremes of wet-bulb temperature in the region around the Arabian Gulf are likely to approach and exceed this critical threshold under the business-as-usual scenario of future greenhouse gas concentrations. Our results expose a specific regional hot spot where climate change, in the absence of significant mitigation, is likely to severely impact human habitability in the future.

Victims of heat waves in our time usually include the very young, the elderly, and people who are ill or have compromised immune systems. The most extreme of heat waves in 2100, however, "would probably be intolerable even for the fittest of humans, resulting in hyperthermia" after six hours of exposure, according to this study (Schwartz 2015). Given such conditions, the Muslim hajj (pilgrimage) to Mecca in Saudi Arabia at summer's height could become lethal.

Could Europe's Heat of 2003 Become Typical?

In 2014, three scientists working with Great Britain's national weather agency, the Met Office (analogous to the National Weather Service in the United States), wrote in *Nature Climate Change* that earlier studies confirmed that Europe's temperatures of 2003 could become typical across the continent by the 2040s. By the end of the 21st century, the average summer temperatures in Europe during 2003 may become the new average. Their findings suggest that once every five years, Europe is likely to experience "a very hot summer" in which temperatures are approximately 1.6°C (2.9°F) above the 1961–1990 average. This is up from a probability a decade earlier that such events would occur only once every 52 years, a 10-fold increase.

Nikolaos Christidis and two colleagues, Gareth S. Jones and Peter A. Stott, used mathematical models to study historical data for much of Western Europe and the Mediterranean and found that the probability of summer heat waves had risen sharply in two decades (from 1990 to 2012). Between those two years, the probability of a heat wave on the magnitude of that in 2003 had risen from once in 1,000 years to once in 127 years (Jolly 2014).

According to British scientists, Europe's scorching heat wave of 2003 will be considered typical seasonal weather by the middle of the 21st century and may be below average in a century. The British team made a case that the summer of 2003 was Europe's hottest in southern, western, and central Europe in at least five centuries. From the eastern Atlantic to the Black Sea, the mercury was 2.3°C (4.14°F) above average. According to their models, by the 2040s at least one European summer in two will be hotter than in 2003. "By the end of this century, 2003 would be classed as an anomalously cold summer relative to the new climate," the scientists wrote in *Nature* in December 2004 ("Phew" 2004).

According to a team from the Swiss Federal Institute of Technology in Zurich, the European heat wave of summer 2003 was of a severity that could have been expected once every 46,000 years (at least estimating from past weather records). This indicates how truly paradigm breaking the heat wave was (McGuire 2005, 53).

According to their models, summers in 2100 will average about 6°C (10.8°F) hotter than 2003's averages. "We estimate it is very likely (confidence level more than 90 percent) that human influence has at least doubled the risk of a heat wave exceeding this threshold magnitude," wrote Peter A. Stott and colleagues in *Nature* (Stott et al. 2004, 610). They continued: "It seems likely that past human influence has more than doubled the risk of European summer temperature as hot as 2003, and with the likelihood of such events projected to increase 100-fold over the next four decades, it is difficult to avoid the conclusion that potentially dangerous anthropogenic interference in the climate system I already underway" (Stott et al. 2004, 613). Separately, scientists at the French meteorological agency Météo France said they expect summer temperatures in France to rise by between 4 and 7°C (7.2–12.6°F) by 2100. "By the end of the century, a summer with temperatures as we had in 2003 will be considered a cool summer," said researcher Michel Deque ("Phew" 2004).

Heat and drought during the summer of 2003 reduced grain harvests across Europe, forcing food prices to rise worldwide. During late August 2003, the International Grains Council, an intergovernmental body, reduced its forecast of the world's grain harvest for 2003 by 36 million tons as a result of "heat and drought, particularly in Europe" (Lean 2003). The damage was most severe in Eastern Europe, which harvested its worst wheat crop in three decades. In Ukraine, the harvest fell from 21 million tons in 2002 to 5 million in 2003, while Romania had its worst crop on record. Germany was the worst-hit European Union country; some farmers in the southeast regions of that country lost half their grain harvest (Lean 2003). World grain reserves fell sharply, continuing a four-year pattern.

Gerald A. Meehl and Claudia Tebaldi studied heat waves in Chicago during 1995 and Paris during 2003, forecasting that "future heat waves in these areas will become more intense, more frequent, and longer-lasting in the second half of the 21st century" (Meehl and Tebaldi 2004, 994). They anticipate that areas suffering severe heat waves will probably experience more intense heat than surrounding areas: "The model show[s] that present-day heat waves over Europe and North America coincide with a specific atmospheric circulation pattern that is intensified by ongoing increases in greenhouse gases" (Meehl and Tebaldi 2004, 994). They concluded that "areas already experiencing strong heat waves (e.g., Southwest, Midwest, and Southeast United States and the Mediterranean region) could experience even more intense heat waves in the future" (Meehl and Tebaldi 2004, 997).

James Hansen and colleagues, writing in 2012, put the string of episodic heat waves into climatic context:

> "Climate dice," describing the chance of unusually warm or cool seasons relative to climatology, have become progressively "loaded" in the past 30 years, coincident with rapid global warming. The distribution of seasonal mean temperature anomalies has shifted toward higher temperatures and the range of anomalies has increased. . . .
>
> We conclude that extreme heat waves, such as that in Texas and Oklahoma in 2011 and Moscow in 2010, were "caused" by global warming, because their likelihood was negligible prior to the recent rapid global warming. (Hansen et al. 2012)

Further Reading

"BAMS State of the Climate: 2011." Special Supplement to Bulletin of the American Meteorological Society, 93(7) (July 2012).

Davis, Robert E., et al. "Changing Heat-Related Mortality in the United States." *Environmental Health Perspectives*, July 23, 2003. doi: 10.1289/ehp.6336.

Gillis, Justin. "Global Warming Makes Heat Waves More Likely, Study Finds." *The New York Times*, July 10, 2012. http.//www.nytimes.com/2012/07/11/science/earth/global-warming-makes-heat-waves-more-likely-study-finds.html.

Hansen, James E., Makiko Sato, and Reto Ruedy. "Perceptions of Climate Change: The New Climate Dice." NASA Goddard Institute for Space Studies and Columbia University Earth Institute, 2012. http://globalwarming-sowhat.com/warm—cool-/loaded-temps-dice-since .pdf.

Jolly, David. "Heat Waves in Europe Will Increase, Study Finds." *The New York Times*, December 8, 2014. http://www.nytimes.com/2014/12/09/world/europe/global-warming-to-make-european-heat-waves-commonplace-by-2040s-study-finds.html.

Kalkstein, Laurence S. "Direct Impacts in Cities." *Lancet* 342 (December 4, 1993): 1397–1400.

Lean, Geoffrey. "Hot Summer Sparks Global Food Crisis." *The Independent* (London), August 31, 2003, 4.

McGuire, Bill. *Surviving Armageddon: Solutions for a Threatened Planet.* New York: Oxford University Press, 2005.

Meehl, Gerald A., and Claudia Tebaldi. "More Intense, More Frequent, and Longer Lasting Heat Waves in the 21st Century." *Science* 305 (August 13, 2004): 994–997.

Pal, Jeremy S., and Elfatih A.B. Eltahir. "Future Temperature in Southwest Asia Projected to Exceed a Threshold for Human Adaptability." *Nature Climate Change*, November 2015. http://www.nature.com/nclimate/journal/vaop/ncurrent/full/nclimate2833.html.

Parker, Cindy L., and Steven M. Shapiro. *Climate Chaos: Your Health at Risk: What You Can Do to Protect Yourself and Your Family*. Westport, CT: Praeger, 2008.

"Phew, What a Scorcher—and It's Going to Get Worse." Agence France Presse, December 1, 2004 (LEXIS).

Schwartz, John. "Deadly Heat Is Forecast in Persian Gulf by 2100." *The New York Times*, October 26, 2015. http://www.nytimes.com/2015/10/27/science/intolerable-heat-may-hit -the-middle-east-by-the-end-of-the-century.html.

Stott, Peter A., D.A. Stone, and M.R. Allen. "Human Contribution to the European Heatwave of 2003." *Nature* 432 (December 2, 2004): 610–613.

See also: Atmospheric Circulation; Australia, Heat and Drought in; Desertification; Drought, United States; Drought, Worldwide; El Niño and La Niña; Heat Waves, Evidence of; Temperatures, Global; Temperatures, Greenhouse Gas Levels and; Wildfires

HURRICANES, OBSERVATIONS OF

While climate scientists have been vigorously debating global warming's influence on the frequency and intensity of tropical cyclones (and thunderstorms and tornadoes), nature has been providing examples of growing ferocity that illustrate a geophysical truism: these storms draw their power from warm water. In November 2013, Typhoon Haiyan, among the most intense cyclonic events in recorded history with sustained winds of 200 miles an hour and gusts to 235, devastated parts of the Philippines, leaving corpses hanging in trees, "scattered on sidewalks or buried in flattened buildings—some of the 10,000 people believed killed in one Philippine city alone by ferocious Typhoon Haiyan that washed away homes and buildings with powerful winds and giant waves" ("Philippine Typhoon" 2013). The storm was the Philippines' worst natural disaster, with a storm surge of nearly 20 feet that "erased villages and towns. . . . [and] sent walls of water over half a mile inland along a crowded coastline" (Bradsher 2013).

"Even by the standards of the Philippines," the Associated Press reported, "Haiyan [was] a catastrophe of epic proportions and has shocked the impoverished and densely populated nation of 96 million people. . . . It appears to have killed many more people than the previous deadliest Philippine storm, Thelma, in which about 5,100 people died in the central Philippines in 1991" ("Philippine Typhoon" 2013). "The water was as high as a coconut tree," said 44-year-old Sandy Torotoro, a bicycle taxi driver who lives near the airport with his wife and 8-year-old daughter. "I got out of the Jeep and I was swept away by the rampaging water with logs, trees and our house, which was ripped off from its mooring. . . . When we were being swept by the water, many people were floating and raising their hands and yelling for help. But what can we do? We also needed to be helped," Torotoro said ("Philippine Typhoon" 2013).

The New York Times described the devastation of Tacloban, a city directly hit by the storm:

> Shattered buildings line every road of this once-thriving city of 220,000 and many of the streets are still so clogged with debris from nearby buildings that they are barely discernible. The civilian airport terminal here has shattered walls and gaping holes in the roof where steel beams protrude, twisted and torn by winds far more powerful than those of Hurricane Katrina when it made landfall near New Orleans in 2005. Decomposing bodies still lie along the roads, like the corpse in a pink, short-sleeved shirt and blue shorts facedown in a puddle 100 yards from the airport. . . .
>
> As a violet sunset melted on Monday evening into the nearly total darkness of a city without electricity, lighted only by a waxing half moon, dispirited residents walked home after another day of waiting at the airport in hope of freshwater, food or a flight out. Looters sacked groceries and pharmacies across the city over the weekend, leaving bare shelves for a population now quickly growing hungry and thirsty. (Bradsher 2013)

A study published in mid-2013 in the *Proceedings of the United States National Academy of Sciences* (PNAS) by hurricane scientist Kerry Emanuel of the Massachusetts Institute of Technology suggested that the annual increase in the number of tropical cyclones each year could rise by as many as 20 by 2100 because of a warming climate. Currently, the average number is 90 or so. This study also suggests an increase in intensity, which confirms earlier observations. Severe storms are a major by which the geophysical system redistributes heat, and higher temperatures provide them more energy.

In the eastern Pacific and Indian Oceans, the cyclone season has become nearly year round. In March 2015, an unusually early storm, Maysak, intensified to a category 5 typhoon. When Maysak was at near peak intensity, astronauts on the International Space Station could see the storm. "Looking down into the eye—by far the widest one I've seen—it seemed like a black hole from a sci-fi movie," said NASA astronaut Terry Virts ("Typhoon Maysak" 2015).

MIT's Emanuel said, "Our study suggests that the largest increases might occur in the western North Pacific region, but with noticeable increases in the South Indian Ocean and in the North Atlantic region. . . . It is important to emphasize that most studies suggest that the frequency of the highest category tropical cyclones (those of category 3 and higher on the Saffir–Simpson scale) should increase as the globe warms," Emanuel said. "There is less agreement about the frequency of the weakest category of storms" (Rice 2013).

Climatologist Judith Curry of Georgia Institute of Technology in Atlanta offered this criticism of Emanuel's study: "The conclusions from this study rely on a large number of assumptions, many of which only have limited support from theory and observations and hence are associated with substantial uncertainties. Personally,

I take studies that project future tropical cyclone activity from climate models with a grain of salt" (Rice 2013). Another researcher, Roger Pielke Jr., a professor of environmental studies at the University of Colorado, said "Kerry Emanuel is a smart scientist; I'll trust that he has done good work here." However, Pielke says his research as well as some of Emanuel's earlier research indicate that the association of rising temperatures with hurricane frequency and intensity is not a closed case (Rice 2013). Pielke also said that increases predicted in Emanuel's work that are projected over a century to come do not reflect history: "Over the past century, no such increase in frequency or intensity has been observed (in the United States or globally) and in fact, the United States is currently in the midst of the longest stretch with no category 3 or greater hurricane landfall since at least 1900" (Rice 2013).

Tropical Cyclones in Unaccustomed Places

Typhoons also have been hitting the Philippines at unaccustomed times and places. During December 2012, for example, Typhoon Bobha killed hundreds of people in eastern Mindanao, an area that has historically been too far south for typhoons. In December 2011, Tropical Storm Washi killed more than 1,200 people in the same area.

After 2000, the mid-Pacific islands of Hawaii found themselves in the path of tropical storms more often. Heretofore, such storms had nearly always moved south of the islands or lost strength when approaching waters slightly cooler than 80°F that are typically required to sustain them. Now Hawaii's Ocean Resources Management Plan anticipates much larger storm surges and higher wind velocities for Hawaii as Earth warms. "Global warming is predicted to cause an increase in frequency and power of both storm surge and hurricanes," the updated plan stated ("Tropical Storm" 2013). Hawaiians are now becoming accustomed to state-of-emergency declarations and power outages as the U.S. Coast Guard closes ports at Hilo and Kahului on Kawaihae, the "big island," and Maui.

Wind velocities are expected to increase by as much as 25 percent by 2100 because of warming oceans. In addition, according to updated emergency plans, "The Intergovernmental Panel on Climate Change states that a one-meter (39-inch) rise in sea level would enable a 15-year storm to flood areas that today are only flooded by a 100-year storm. . . . Changes in precipitation, with heavier rainfalls, are also expected, which impact the amount of runoff from Hawaii's watersheds" ("Tropical Storm" 2013). The Hawaiian study was supported by others that project a doubling in the number of such storms striking the island chain ("More Cyclones" 2013).

The number and intensity of tropical cyclones striking land adjacent to the Arabian Sea (with unprecedented loss of life and damage) rose after 2000, and the storms also became more intense. Anthropogenic air pollution, including black carbon and other aerosols, were suspected as being contributing factors (Sriver 2011; Evan et al. 2011).

Amato T. Evan and colleagues wrote in *Nature* (2011),

Throughout the year, average sea surface temperatures in the Arabian Sea are warm enough to support the development of tropical cyclones, but the atmospheric monsoon circulation and associated strong vertical wind shear limits cyclone development and intensification, only permitting a pre-monsoon and post-monsoon period for cyclogenesis. Thus a recent increase in the intensity of tropical cyclones over the northern Indian Ocean is thought to be related to the weakening of the climatological vertical wind shear. At the same time, anthropogenic emissions of aerosols have increased sixfold since the 1930s, leading to a weakening of the southwesterly lower-level and easterly upper-level winds that define the monsoonal circulation over the Arabian Sea.

During June 2007, for example, Tropical Cyclone Gonu, with sustained winds of more than 120 miles per hour, churned 35-foot-high waves and struck Oman on the Arabian peninsula, causing at least 13 deaths. The storm was the strongest on record in the northwestern Arabian Sea. The cyclone hit the Omani coastal towns of Sur and Ras al Hadd with sustained winds of more than 100 miles per hour (mph). Judith Curry, a hurricane expert at Georgia Tech, said the cyclone's strength was "really rather amazing" for the region and appeared to be amplified by sea temperatures hovering around 87°F. Even weakened, she said, the storm could prove disastrous in Oman or Iran. "Cyclones are very rare in this region and hence governments and people are unprepared," she said (Revkin 2007, June 6).

In January 2016, hurricane Alex became only the second such storm on record during that month in the Atlantic Ocean basin (the first formed in 1938). It was also the strongest on record, with sustained winds of 85 mph as it hit the Azores Islands. One other hurricane formed in the Atlantic—Alice—in late December 1954. Both Alice and the unnamed 1938 storm had maximum winds of 80 mph.

Wetter and Windier but Not More Frequent?

Global warming will not increase Atlantic hurricane intensity even as it feeds storms that are wetter, according to research by Tom Knuston, a meteorologist at the fluid dynamics lab in Princeton, New Jersey, run by the National Oceanic and Atmospheric Administration (NOAA). Knutson and several colleagues say his computer model study, published online May 18, 2008, in *Nature Geoscience*, argues "against the notion that we've already seen a really dramatic increase in Atlantic hurricane activity resulting from greenhouse warming" (Borenstein 2008).

Knutson and colleagues wrote,

Our regional climate model of the Atlantic basin reproduces the observed rise in hurricane counts between 1980 and 2006, along with much of the interannual variability when forced with observed sea surface temperatures and atmospheric conditions. . . . Here we assess, in our model system, the changes in large-scale climate that are projected to occur by the end of the 21st century by an ensemble of global climate models, and find that Atlantic hurricane and tropical storm frequencies are reduced. At the same time, near-storm rainfall rates increase substantially. Our results do not support the notion of large increasing trends in either tropical storm or hurricane frequency driven by increases in atmospheric greenhouse gas concentrations. (Knutson et al. 2008)

Knutson's data indicate that the number of such storms will decline 18 percent by the end of the 21st century. The number hitting land will decline 30 percent, Knutson argues. However, the greatest decrease in frequency will be among weaker storms, Knutson's research indicates. Rainfall within 30 miles of a hurricane should rise some 37 percent, and wind strength should increase by around 2 percent, according to his models.

Knutson's study stoked debate over global warming and hurricanes. Kerry Emanuel said that his conclusions are "demonstrably wrong" because they are based on a computer model that does not examine storms correctly. Kevin Trenberth, climate analysis chief at the National Center for Atmospheric Research in Boulder, Colorado, said that Knutson's computer model is poor at assessing tropical weather and "fails to replicate storms with any kind of fidelity." Kevin Trenberth of the National Center for Atmospheric Research said it is not just the number of hurricanes "that matter, it is also the intensity, duration and size, and this study falls short on these issues" (Borenstein 2008). Knutson acknowledged that his model is coarse and lacks an accurate analysis of the strength of individual storms.

NOAA hurricane meteorologist Chris Landsea said Knutson's work was "very consistent with what's being said all along. I think global warming is a big concern, but when it comes to hurricanes the evidence for changes is pretty darn tiny," Landsea said (Borenstein, 2008).

"The Lemming-Like March to the Sea"

Along beaches in many U.S. locales, ranks of condos have risen over the last few decades despite the debate over hurricane intensity. At the same time, a panel of scientists has warned of the "lemming-like march to the sea" (Revkin 2006, July 25). Even scientists who disagree over global warming's role in hurricane intensity say that the increased concentration of housing on and near the coasts will cause more damage even if hurricanes do not intensify over time.

At a meeting of the National Association of Insurance Commissioners, Andrew Logan, insurance director of the Ceres investor coalition (representing $4 trillion in market capital), said that "insurance as we know it is threatened by a perfect storm of rising weather losses, rising global temperatures and more Americans living in harm's way" (Morrison and Sink 2007). Ceres analysts estimate that losses related to catastrophic weather—the vast majority of insurance payouts—have increased 15-fold in the U.S. property-casualty industry in three decades.

A month after Hurricane Katrina hit the U.S. coast along the Gulf of Mexico in late summer 2005, many insurance executives expected the industry's cost to be around $35 billion or equal the inflation-adjusted private payout for the World Trade Center attacks in 2001. The previous U.S. industry record for a season of U.S. hurricanes had been $27 billion in 2004 when four powerful hurricanes hit Florida. The pre-Katrina hurricane record payout (also adjusted for inflation) was $23 million for Hurricane Andrew in 1992 (Treaster 2005). Total insurance losses in the United States from damage by Hurricanes Katrina, Rita, and Wilma amounted to $57.6 billion by the end of 2005, more than twice 2004's all-time record. During 2005, world insurance loses from weather-related natural disasters as estimated by the Munich Re Foundation, part of one of the world's largest reinsurance companies, rose to an all-time record of more than $200 billion U.S.

By the summer of 2006, the annual premium to insure some oil and gas platforms in the Gulf of Mexico had risen from $2 million to $25 million (Pleven et al. 2006, A1). Some businesses with exposure to hurricanes were canceling projects, and others were gambling without insurance. Wal-Mart dropped its coverage for severe windstorms and decided to pay for damage internally. Omaha investor Warren Buffett's firm Berkshire Hathaway stepped up its investments in reinsurance even as others dropped out. "If you like to watch football, you probably enjoy the game a little more if you have a bet on it," Buffett told the *Wall Street Journal*. "I like to watch the Weather Channel" (Pleven et al. 2006, A1).

During 2006, several major insurers (among them Allstate, the largest firm in the United States) were canceling or refusing to write new policies in some coastal areas of Florida, Louisiana, and New York that they deemed too risky because of anticipated hurricanes. Where insurance was available in coastal areas prone to hurricane landfalls, premiums had risen to several thousand dollars a month on some properties, rivaling the cost of a mortgage (Adams 2006). Insurance premiums on some 1,500-square-foot houses in Key West, Florida, doubled to $13,000 a year in 2006, $10,000 of which was being assessed for wind-storm insurance (Waddell 2006).

By 2007, some insurance companies were canceling homeowners' policies en masse along parts of the U.S. East Coast, citing potential hurricane threats. Between 2004 and 2007, more than 3 million homeowner policies along the East Coast were terminated for adverse storm risk. In many cases, exclusion from the insurance market was making many homes unsalable because coverage is required to obtain a mortgage. Some of the homes refused private insurance were several miles inland Some states (Florida was an example) were assuming the risk out of public funds. By 2007, Florida's Citizens Property Insurance Corporation became that state's largest homeowners' insurer with 1.3 million policies (Vitello 2007). Massachusetts also had a state high-risk pool with 200,000 policies by 2007, including 44 percent of homeowners on Cape Cod.

Further Reading

Adams, Marilyn, "Strapped Insurers Flee Coastal Areas." *USA Today*, April 26, 2006, B1.

Borenstein, Seth. "Jump in Atlantic Hurricanes Not Global Warming, Says Study That Predicts Fewer Future Storms." Associated Press, May 19, 2008 (LEXIS).

Bradsher, Keith. "Struggle for Survival in Philippine City Shattered by Typhoon." *The New York Times*, November 11, 2013. http://www.nytimes.com/2013/11/12/world/asia /philippines-storm-surge-leaves-scenes-of-devastation.html.

Emanuel, Kerry. "Downscaling CMIP5 Climate Models Shows Increased Tropical Cyclone Activity over the 21st Century." *Proceedings of the United States National Academy of Sciences* 110(30) (July 2013): 12219–12224.

Evan, Amato T., et al. "Arabian Sea Tropical Cyclones Intensified by Emissions of Black Carbon and Other Aerosols." *Nature* 479 (November 3, 2011): 94–97.

Knutson, Thomas R., et al. "Simulated Reduction in Atlantic Hurricane Frequency under Twenty-First-Century Warming Conditions." *Nature Geoscience* online, May 18, 2008. http://www.nature.com/ngeo/journal/vaop/ncurrent/abs/ngeo202.html. doi: 10.1038/ngeo202.

"More Cyclones for Hawaiian Islands." *Nature* 497 (May 16, 2013): 290.

Morrison, John, and Alex Sink. "The Climate Change Peril That Insurers See." *Washington Post*, September 27, 2007, A25. http://www.washingtonpost.com/wp-dyn/content/article/2007/09/26/AR2007092602070_pf.html.

"Philippine Typhoon Leaves up to 10,000 Dead in Tacloban City." Associated Press in Canadian Broadcasting Corp., November 11, 2013. http://www.cbc.ca/news/world/philippine-typhoon-leaves-up-to-10-000-dead-in-tacloban-city-1.2421580.

Pleven, Liam, Ian McDonald, and Karen Richardson. "As Hurricane Season Starts, Disaster Insurance Runs Short." *Wall Street Journal*, July 10, 2006, A1, A8.

Revkin, Andrew C. "Climate Experts Warn of More Coastal Building." *The New York Times*, July 25, 2006, D2.

Revkin, Andrew C. "Cyclone Nears Iran and Oman." *The New York Times*, June 6, 2007. http://www.nytimes.com/2007/06/06/world/middleeast/06storm.html.

Rice, Doyle. "Storm Warning: Climate Change to Spawn More Hurricanes." *USA Today*, July 8, 2013. http://www.usatoday.com/story/weather/2013/07/08/climate-change-global-warming-hurricanes/2498611/.

Sriver, Ryan L. "Climate Change: Man-Made Cyclones." *Nature* 479 (November 3, 2011): 50–51.

Treaster, Joseph B. "Gulf Coast Insurance Expected to Soar." *The New York Times*, September 24, 2005. http://www.nytimes.com/2005/09/24/business/24insure.html.

"Tropical Storm Nears Hawaii, New Ocean Plan Predicts Worse." Environment News Service, July 29, 2013. http://ens-newswire.com/2013/07/29/tropical-storm-nears-hawaii-new-ocean-plan-predicts-worse/.

"Typhoon Maysak Approaches the Philippines." NASA Earth Observatory, April 4, 2015. http://earthobservatory.nasa.gov/IOTD/view.php?id=85638&src=eoa-iotd.

Vitello, Paul. "Hurricane Fears Cost Homeowners Coverage." *The New York Times*, October 16, 2007. http://www.nytimes.com/2007/10/16/nyregion/16insurance.html.

Waddell, Lynn. "Rising Insurance Rates Push Florida Homeowners to Brink." *The New York Times*, June 29, 2006. http://www.nytimes.com/2006/06/29/us/29florida.html.

See also: Hurricanes, Scientific Debates; Hurricanes, Snow; Oceans' Absorption of Heat; Temperatures, Global; Temperatures, Greenhouse Gas Levels and

HURRICANES, SCIENTIFIC DEBATES OF

Hurricanes thrive in warm water and quickly decline in intensity when sea surface temperatures fall below 80°F. This and other facts have caused an intense debate about whether a warmer world will mean more frequent and more intense hurricanes.

Although the temperature of the ocean undoubtedly plays a role in hurricane frequency and intensity, many other factors are at work as well, making a simple linear association of hurricane intensity with global warming difficult to prove. As with other phenomena, global warming's effect on hurricanes plays out in a broader climatic context.

Benjamin D. Santer of the U.S. Department of Energy's Lawrence Livermore National Laboratory has compiled data indicating that sea surface temperatures are rising, based on several climate models, ranging from 0.32°C to 0.67°C during the 20th century. "For the period 1995–2005," Santer wrote in *The Proceedings of the National Academy of Sciences*, "we find an 84 percent chance that external forcing explains at least 67 percent of the observed SST [sea surface temperature] increases in the two tropical cyclogenesis regions" (Santer et al. 2006, 13905). "Even under modest scenarios for emissions, we're talking about sea surface temperature changes in these regions of a couple of degrees," Santer said. "That's much larger than anything we've already experienced, and that is worrying" (Revkin 2006, September 12).

The relationship (or lack thereof) between hurricane intensity and warming atmospheric temperatures is complicated by the fact that water temperatures (like air temperatures) sometimes vary over periods of several decades, with the long-term trend "signal" provoked by greenhouse gas levels. For example, water temperatures in the Atlantic Ocean, which produces nearly all the hurricanes that have an impact on the United States of America, have been rising steadily since the 1970s, paralleling a general global rise in air temperatures. The frequency and intensity of hurricanes (as well as the number hitting U.S. coastlines and inflicting major damage) also have been rising during the same period.

Any study that examines records back to the 1970s can find a tight relationship between ocean warming, hurricane intensity, and air temperatures. However, during the 1950s and 1960s, air temperatures were generally cooler than during the 1970s, but hurricane intensity was higher on average. By 2005, this divergence was fueling a testy debate between some hurricane experts regarding whether and to what degree hurricane intensity and frequency was related to the overall warming trend. This debate often spilled over into the public realm as Florida and surrounding areas were smacked by four major hurricanes in 2004 (Charley, Frances, Ivan, and Jeanne) and as the 2005 hurricane season set records for the number of named storms, including Katrina, Rita, and others that were exceptionally large and violent. Each of these hurricanes ranked in the top 10 of storms to hit the United States in terms of insurance losses.

Hurricanes during the Pliocene

During the early Pliocene Epoch 3 million to 5 million years ago, persistent warm El Niño conditions were more prevalent in the equatorial Pacific than during cooler periods. Alexey V. Fedorov and colleagues (2010) modeled tropical storm activity in the same area, in which waters of the upper ocean are vigorously mixed, heat uptake is affected, heat is transported toward the poles, and global temperatures rise. Distribution and frequency of tropical cyclones therefore exert an influence in climatic response to global warming. "A potential analog to modern greenhouse conditions," they wrote, "the climate of the early Pliocene epoch can provide important clues to this response."

> Here we describe a positive feedback between hurricanes and the upper-ocean circulation in the tropical Pacific Ocean that may have been essential for maintaining warm, El Niño-like conditions. during the early Pliocene. . . . More frequent and/or stronger hurricanes in the central Pacific imply greater heating of the parcels, warmer temperatures in the eastern equatorial Pacific, warmer tropics and, in turn, even more hurricanes. Using a downscaling hurricane model, we show dramatic shifts in the tropical cyclone distribution for the early Pliocene that favor this feedback. Further calculations with a coupled climate model support our conclusions. The proposed feedback should be relevant to past equable climates and potentially to contemporary climate change. (Fedorov et al 2010)

Several Variables Shape Hurricanes

Heat is important in the development and strength of hurricanes, but they are extremely sensitive to other influences. Even dust storms blowing off the Sahara desert in Africa can have important effects. After the devastating hurricane seasons of 2004 and 2005, the next year was quiet. Hurricane activity over the Atlantic Ocean was so much lower in part because of an increase in dust levels over the ocean from the Sahara. During the 2006 hurricane season, SSTs remained relatively cool and only five hurricanes formed, one-third of the total in 2005.

William Lau and Kyu-Myong Kim at NASA's Goddard Space Flight Center in Greenbelt, Maryland, showed that airborne Saharan dust probably caused 30 percent to 40 percent of the water-temperature drop in the Atlantic in areas where hurricanes usually form. The dust blocked sunlight in tropical regions between June 2005 and June 2006. The team's research was reported in December 2007 in the American Geophysical Union's *Geophysical Research Letters* ("Saharan Dust" 2007).

"Previous studies have looked at how hot, dry air associated with a Saharan dust outbreak affects an individual storm, but our study is the first to focus on dust's radiative effect on sea surface temperatures, which may affect storms for the entire season. Nobody had suggested that link before," Lau said ("Saharan Dust" 2007).

Still other factors influence hurricane numbers and intensity, including El Niño conditions in the tropical eastern Pacific Ocean and the strength of the West African monsoon. If other conditions are right, a hurricane season can be intense even if surface waters are cooler than usual, and vice versa.

A study of the eastern Caribbean (near the Puerto Rican island of Vieques) covering 5,000 years and published in *Nature* (May 24, 2007) revealed centuries-long periods in which all of these variables played important roles in the number and intensity of hurricanes. The study's authors, scientists from the Woods Hole Oceanographic Institution, said their findings did not conflict with other scientists' assertions that increasing hurricane activity can be linked to human-caused warming of the climate and seas.

Lead author Jeffrey P. Donnelly said the findings pointed to the importance of solving an unresolved puzzle: whether and how global warming will affect the El Niño–La Niña cycle. More intense or longer Pacific warmups could stifle Atlantic and Caribbean hurricanes even with warmer seas, Dr. Donnelly said (Revkin 2007, May 24).

"Warm sea surface temperatures are clearly the fuel for intense hurricanes," he said. "What our work says is that without sea temperatures varying a lot, the climate system can flip back and forth between active and inactive regimes." A disturbing possibility, he added, was a warming of waters while conditions in the Pacific and Africa are in their hurricane-nurturing mode. "If you flip that knob and also have warming seas," Dr. Donnelly said, "oh boy, who knows what could happen?" (Revkin 2007, May 24).

Donnelly and Woodruff (2007) compiled a record of intense hurricane activity in the western North Atlantic Ocean for 5,000 years based on sediment cores from a Caribbean lagoon that contains coarse-grained deposits associated with intense hurricane landfalls.

> The record indicates that the frequency of intense hurricane landfalls has varied on centennial to millennial scales over this interval. Comparison of the sediment record with paleoclimate records indicates that this variability was probably modulated by atmospheric dynamics associated with variations in the El Niño [and] Southern Oscillation and the strength of the West African monsoon, and suggests that sea surface temperatures as high as at present are not necessary to support intervals of frequent intense hurricanes. To accurately predict changes in intense hurricane activity, it is therefore important to understand how the El Niño [and] Southern Oscillation and the West African monsoon will respond to future climate change. (Donnelly and Woodruff 2007, 465)

More Studies, More Debate

Tracing any specific storm to warming is a tenuous exercise, but hurricanes may intensify generally as oceans warm. Hurricanes are essentially heat engines, so storms that approach their upper limits of intensity are expected to be slightly stronger—and produce more rainfall—in a warmer climate because of higher sea surface temperatures. According to a simulation study by a group of scientists at NOAA's Geophysical Fluid Dynamics Laboratory (GFDL), a 5 percent to 12 percent increase in wind speeds for the strongest hurricanes (typhoons) in the northwest tropical Pacific is projected if tropical sea surfaces warm by a little more than 2°C.

More than a decade previously, such an increase in the upper-limit intensity of hurricanes with global warming had been suggested on theoretical grounds by Kerry Emanuel, a hurricane specialist at the Massachusetts Institute of Technology. Holland and Webster, however, were the first to examine the question using a hurricane prediction model that was being used operationally to simulate realistic hurricane structures. Other models also indicate increases in precipitation of as much as 20 percent (Knutson et al. 1998, 1018; Knutson et al. 2001, 2458).

Another study reached similar conclusions. By the 2080s, warmer seas could cause the average hurricane to intensify about an extra half step on the Saffir–Simpson scale, according to a study conducted on supercomputers at the GFDL in Princeton, New Jersey. The same study anticipates that rainfall as far away as 60 miles from the core of a hurricane could be nearly 20 percent heavier. This study is significant because it used half a dozen computer simulations of global climate devised by separate groups at institutions around the world.

Thomas R. Knutson and Robert E. Tuleya's models indicate that given sea surface temperature increases of 0.8 to 2.4°C, hurricanes would become 14 percent more intense (based on central pressure), with a 6-percent increase in maximum wind speeds and an 18-percent rise in average precipitation rates within 100 kilometers of storm centers. "One implication of the results," they wrote, "is that if the frequency of tropical cyclones remains the same over the coming century, a greenhouse-gas induced warming may lead to a gradually increasing risk in the occurrence of highly destructive category-5 storms" (Knutson and Tukeya 2004, 3477).

This study of hurricanes and warming was "by far and away the most comprehensive effort" to assess the question using powerful computer simulations, said Emanuel. "This clinches the issue" (Revkin 2004). The study added that rising sea levels caused by global warming would lead to more flooding from hurricanes. With almost every combination of greenhouse-warmed oceans and atmosphere and formulas for storm dynamics, the results were the same, said Tuleya: more powerful storms and more rainfall (Revkin 2004).

Mark A. Saunders and Adam S. Lea of the Benfield University College London Hazard Research Centre, Department of Space and Climate Physics, isolated factors affecting hurricane strength and frequency in the Atlantic Ocean and Caribbean between 1965 and 2005, removing such influences as wind shear. According to Saunders and Lea, "the sensitivity of tropical Atlantic hurricane activity to August–September sea surface temperature over the period we consider is such that a 0.5 [°C] increase in sea surface temperature is associated with a roughly 40 percent increase in hurricane frequency and activity. The results also indicate that local sea surface warming was responsible for approximately 40 percent of the increase in hurricane activity relative to the 1950–2000 average between 1996 and 2005" (Saunders and Lea 2008, 557).

Other scientists lend support to Emanuel's case. S. T. O'Brien et al. (1992) assert that doubling the level of atmospheric carbon dioxide (increasing tropical SSTs between 1 and 4°C) will double the number of hurricanes and increase their strength by 40 percent to 60 percent. O'Brien et al. also believe that warmer average temperatures will extend the hurricane season, because the season effectively ends for any given area when water temperature falls below 26°C (80°F). R. J. Haarsma et al. (1993) also estimate that a doubling of greenhouse gases levels will increase the frequency of hurricanes by 50 percent and the average intensity of storms by 20 percent.

A study by NOAA's Geophysical Fluid Dynamics Laboratory (Princeton, New Jersey) published early in 2010 supported earlier models that conclude that fewer hurricanes will develop in the Atlantic basin under warmer conditions but that those that do form will likely be more intense. This study by Morris A. Bender et al. supports earlier findings but has employed models with better resolution; they can project conditions over smaller areas than the previous studies. Projecting expected warming to the end of the 21st century, this study projects that categories 4 and 5 hurricanes may double in frequency. The strongest storms—winds 234 kilometers (145 miles) per hour or more—may triple. As water temperatures rise, an increase in water vapor in the atmosphere is expected to intensify stronger storms and inhibit formation of weaker ones (Kerr 2010, 399; Bender et al. 2010, 454).

Hurricane Frequency and Intensity: The Critics

A report by Christopher W. Landsea of the Miami, Florida, office of the National Oceanic and Atmospheric Administration asserts that an increase in carbon dioxide levels may raise the threshold temperature in which hurricanes thrive (currently above 80°F) in proportion to the rise in temperature (Landsea 1993). This change might nullify the increase in strength that is implied by warmer water surface temperatures. Landsea also speculates that increased frequency of El Niño–type weather patterns in the Pacific tends to dampen hurricane development in the Atlantic, although storm frequency and intensity often rises under El Niño conditions in the eastern Pacific.

Landsea contends that the intensity of Atlantic hurricanes has decreased since the middle of the 20th century. Landsea et al. (1996) also assert that hurricane frequency and intensity had not increased during the past half-century. They wrote that "a long-term (five decade) downward trend continues to be evident primarily in the frequency of intense hurricanes. In addition, the mean maximum intensity (i.e., averaged over all cyclones in a season) has decreased" (Landsea et al. 1996, 1700).

Landsea believes that any warming-induced change in hurricane frequency and intensity will probably be lost in the "noise" of year-to-year hurricane variability. He concludes, "Overall, these suggested changes are quite small compared to the observed large natural variability of hurricanes, typhoons and tropical cyclones. However, more study is needed to better understand the complex interaction between these storms and the tropical atmosphere and ocean" (Landsea 1999).

Three greenhouse skeptics—Sherwood Idso, Robert Balling, and R. S. Cerveny—have challenged the theories of Kerry Emanuel, a hurricane specialist at the Massachusetts Institute of Technology, as related to warming temperatures to hurricane intensity (Idso et al. 1990). They assembled hurricane data for the central Atlantic, the East Coast of the United States, the Gulf of Mexico, and the Caribbean Sea between 1947 to 1987, comparing their information to estimates of the sea surface temperatures in the Northern Hemisphere. Idso, Balling, and Cerveny concluded,

"There is basically no trend of any sort in the number of hurricanes experienced in any of the four regions with respect to variations in temperature" (Idso et al. 1990, 261).

Doubts also have been raised about the relationship of warming seas and the intensity of tropical cyclones by scientists whose views are less politicized than those of Idso, Balling, and Cerveny. A. J. Broccoli and S. Manabe (1990) included cloud-related feedbacks in their models and found a 15-percent reduction in the number of days with hurricanes, even assuming a temperature rise caused by a doubling of atmospheric carbon dioxide. Broccoli and Manabe noted that their results were easily skewed by (and highly dependent on) how they decided to represent cloud processes within their models.

William M. Gray, professor emeritus of Atmospheric Sciences at Colorado State University, is a long-standing opponent of the idea that warming temperatures have anything to do with hurricanes. According to his tally, between 1957 and 2006, 83 hurricanes hit the United States, 34 of them major. Between 1900 and 1949, 101 hurricanes hit the same area, 39 of which were major, with wind speeds above 110 miles an hour. From 1966 to 2006, said Gray, only 22 major hurricanes hit the United States, whereas between 1925 and 1965, 39 such storms hit the same area. "Even though global mean temperatures have risen by an estimated 0.4 [°C] and CO_2 by 20 percent, the number of major hurricanes hitting the United States declined," Gray wrote (Gray 2006). Since 1995, however, the number of major storms hitting the U.S. Atlantic and Gulf of Mexico coasts has risen sharply. Gray associates the increase with strengthening circulation in the Atlantic Ocean.

The Temperature-Intensity Debate

In the mid-1980s, Emanuel published an article in the *Journal of the Atmospheric Sciences* (Emanuel 1986) in which he argued that hurricane intensity is governed in part by the degree of thermodynamic disequilibrium between the atmosphere and the underlying ocean. Therefore, Emanuel reasoned, warmer ocean waters would breed more intense tropical cyclones. By the year 2000, Emanuel's forecast of "hypercanes" had become gist for at least one best-selling book, *The Coming Superstorm* by Art Bell and Whitley Strieber, which rose to number 15 on *The New York Times* best-seller list for nonfiction in January 2000. A major motion picture, *The Day After Tomorrow*, was loosely based on that book.

In 1987, Emanuel published an article in *Nature* in which he asserted that a 3°C increase in sea surface temperatures could increase the potential destructive power (as measured by the square of the wind speed) of storms by 40 percent to 50 percent (Emanuel 1987). Emanuel asserted in 1988 (1988, April 1; 1988, July–August) that a warming of the sea surface by 6 to 10°C (assuming no temperature change in the lower stratosphere) would make a supersized, ultrapowerful "hypercane" theoretically possible.

Emanuel's theory has helped spark intense debate among hurricane specialists. Critics of his theory responded that even the most pessimistic climate models do not project such a large amount of warming in the tropics. William Gray, perhaps the best-known hurricane specialist in the United States (partly because of his broadcasts on the Weather Channel) has assembled statistics indicating that Atlantic hurricane activity between 1970 and 1987 was been less than half the level observed between 1947 and 1969, an indication that temperature rises already experienced have not caused more intense tropical storms.

Hurricanes increase in strength with the warmth of the water over which they travel. If warm surface water is shallow, the storm's own turbulence may mix enough underlying water to drive its surface temperature below the 80°F required to sustain the storm. Under current conditions, the top wind speed of a hurricane is probably around 200 miles an hour, and the lowest possible central pressure is probably about 885 millibars (McKibben 1989, 95).

A 2005 study published in *Nature* (Emanuel 2005) indicated that the "dissipation of power" of Atlantic hurricanes had more than doubled during the previous 30 years with a dramatic spike since 1995 with global warming and other variations in ocean temperatures working together. In this study, Emanuel was the first scientist to directly indicate a statistical relationship between warming and storm intensity (Merzer 2005, n.p.). The trend reflects longer storm lifetimes and greater intensities, both of which Emanuel associates with increasing sea surface temperatures. "The large upswing in the last decade is unprecedented and probably reflects the effect of global warming. . . . My results suggest that future warming may lead to an upward trend in tropical cyclone destructive potential and—taking into account an increasing coastal population—a substantial increase in hurricane-related losses in the 21st century," Emanuel wrote (2005, 686).

Can an increasing numbers of Atlantic hurricanes since the 1990s be ascribed to more acute observation and analysis rather than climate change? Some research released in 2009 suggests that might be the case, including one study from NOAA. That study, led by Christopher W. Landsea, asserted in the *Journal of Climate* that several tropical storms that formed and dissipated over the ocean would not have been detected in presatellite days (Dean 2009, July 2; Dean 2009, August 12; Landsea et al. 2009).

The Debate Continues

Studies of increasing hurricane intensity released during September 2008 again swung the debate to global warming's causal role. Research published in *Nature* (Schiermeier 2008) indicated that maximum wind speeds of the strongest tropical cyclones have increased significantly since 1981 because of rising ocean temperatures. "It'll be pretty hard now for anyone to claim that cyclone activity has not increased," said Judith Curry of Georgia Tech (Atlanta), who was not involved in this study but whose work has reached similar conclusions (Schiermeier 2008).

With his colleagues, James Elsner, a climatologist at Florida State University in Tallahassee, found that the strongest tropical storms have intensified, especially in the North Atlantic and northern Indian Oceans. The scientists analyzed satellite data of cyclone wind speeds, finding only a slight increase in the average number or intensity of all storms. The number of storms at categories 4 and 5 on the

Saffir–Simpson scale increased, and these are the ones most likely to cause catastrophic damage (Elsner et al. 2008, 92).

The team estimated that a 1°C increase in sea surface temperature could contribute to a 31 percent increase in the global frequency of categories 4 and 5 storms per year—that is, from 13 to 17. Since 1970, the tropical oceans have warmed an average of 0.5°C. Computer models suggest they may warm by a further 2°C by 2100 (Schiermeier 2008).

Criticism of the Elsner study fell out along familiar lines. Christopher Landsea said the data rely on inaccurate information for storms in the Indian Ocean and skew the record in the North Atlantic by beginning their measurements in 1981 during a cyclical lull in hurricane activity and ending in 2006 during a relatively active period that began circa 1995. The study's statistics are elegant, said Landsea, but they rely on suspect information. Thomas R. Knutson, also a frequent skeptic of studies linking ocean warming and hurricane intensity, told *The New York Times*, that the Elsner study covered too short a period to be reliable as a predictor of a trend. Thus, he said, "It's not a definite smoking gun for a greenhouse warming signal on hurricanes" (Chang 2008).

J. Lighthill et al. (1994) argued that while global warming may exert some influence on cyclone formation in the tropics, natural variability is more important. This position drew a reply from Emanuel (1995). L. Bengtsson, a member of the German Max Planck Institut für Meteorologie, and colleagues (1996) contend that global warming will strengthen the upper-level westerlies in areas where hurricanes usually develop, inhibiting storm development and intensity, negating any boost the storms may get from warmer sea surface temperatures. (This is also typical of the El Niño pattern that seems to be occurring more frequently as temperatures rise.)

In its second assessment in 1995, the International Panel on Climate Change restated its earlier position that the state of science on the subject does not permit a conclusion regarding whether global warming will affect the number and intensity of tropical cyclones. Tom Karl et al. wrote two years later in *Scientific American*, "Overall, it seems unlikely that tropical cyclones will increase significantly on a global scale. In some regions, activity may escalate; in others, it will lessen" (Karl et al. 1997, 83).

In another study, Johan Nyberg and colleagues reconstructed hurricane activity in the North Atlantic Ocean as far back as 270 years using proxy records for vertical wind shear and sea surface temperature from corals and a marine sediment core. In an exercise of what scientists call "paleotempestology," samples are taken from lagoons into which storm tides wash, an event associated with strong winds and storm surges that occur only during the strongest tropical storms. Like all proxies, these are far from perfect. They do not, for example, account for changes in the paths of dominant hurricanes because they sample only a minute fraction of the area over which the storms move (Elsner 2007, 648). Records would be required over a much larger area to provide these statistics valid meaning.

Nyberg and colleagues found that the average frequency of major hurricanes decreased gradually from the 1960s until the early 1990s, reaching a long-term low cycle during the 1970s and 1980s. After 1995, frequency increased to levels similar to other periods of high intensity in their record "and thus appears to represent a recovery to normal hurricane activity, rather than a direct response to increasing sea surface temperatures" (Nyberg et al. 2007, 698). The upshot of this and other research is that although hurricanes are sustained by warm water, vertical wind shear (winds blowing from different directions at various heights that disturb hurricanes' circulation) can tear them apart and disperse storm-sustaining heat. This research raises other questions. El Niño conditions may be fostered by warming oceans, but El Niño conditions in the Pacific tend to cause above-average wind shear in the Atlantic, which seems to tear up hurricane circulation. The picture is not as simple, therefore, as equating warmer water with more frequent and intense hurricanes.

Is Tropical Storm Activity Too Variable to Detect a Global Warming Signal?

Short-term tropical storm activity varies so greatly in any one ocean basin that scientists often have trouble detecting long-term climate "signals" (or patterns) on a worldwide basis. When one area (the Atlantic Ocean, or example) is active, another (the Northern Pacific, perhaps) may be quieter than usual. Variation year to year may be much greater than multiyear trends, confounding any attempt to generalize trends whether with respect to global warming or to any other trend. Thus, some scientists find a rise in storm numbers and intensity and attribute it at least in part to a warming climate, and others find the opposite. The debate brings to mind the old aphorism about quoting scripture—the outcome may say more about who is doing the quoting than about the record being examined.

For example, in the context in which some scientists found rising tropical storm intensity came a 2008 study by Florida State University researchers that asserted that the 2007 and 2008 tropical storm seasons worldwide were among the quietest in the previous 30 years (even as the Atlantic Ocean experienced an active season). The study, "Global Tropical Storm Activity," authored by Ryan Maue (2008), agreed with research by Stan Goldenberg of NOAA that tropical storm activity is inherently variable and all but unconnected to a warming climate. Goldenberg called the idea associating global warming with increasing tropical storm intensity "a simplistic notion" (Dorell 2008).

Further Reading

Bender, Morris A., et al. "Modeled Impact of Anthropogenic Warming on the Frequency of Intense Atlantic Hurricanes." *Science* 327 (January 22, 2010): 454–458.

Bengtsson, L., M. Botzet, and M. Esch. "Will Greenhouse Gas-Induced Warming over the Next 50 Years Lead to a Higher Frequency and Greater Intensity of Hurricanes?" *Tellus* 48A (1996): 57–73.

Broccoli, A. J., and S. Manabe. "Can Existing Climate Models Be Used to Study Anthropogenic Changes in Tropical Cyclone Intensity?" *Geophysical Research Letters* 17 (1990): 1917–1920.

Chang, Kenneth. "Strongest Storms Grow Stronger Yet, Study Says." *The New York Times*, October 4, 2008, A18.

Dean, Cornelia. "El Niño Variant Is Linked to Hurricanes in Atlantic." *The New York Times*, July 2, 2009. http://www.nytimes.com/2009/07/03/science/earth/03hurricane.html.

Dean, Cornelia. "An 'Increase' in Big Storms May Just Be Better Detection." *The New York Times*, August 12, 2009. http://www.nytimes.com/2009/08/13/science/earth/13atlantic.html.

Donnelly, Jeffrey P., and Jonathan D. Woodruff. "Intense Hurricane Activity over the Past 5,000 Years Controlled by El Niño and the West African Monsoon." *Nature* 447 (May 24, 2007): 465–468.

Dorell, Oren. "Report: Tropical Storm Activity at 30-Year Low." *USA Today*, November 12, 2008, 3A.

Elsner, James B. "Tempests in Time." *Nature* 447 (June 7, 2007): 647–648.

Elsner, James B., James P. Kossin, and Thomas H. Jagger. "The Increasing Intensity of the Strongest Tropical Cyclones." *Nature* 455 (September 4, 2008): 92–95.

Emanuel, K. A. "An Air-Sea Interaction Theory for Tropical Cyclones. Part I: Steady-State Maintenance." *Journal of the Atmospheric Sciences* 43 (1986): 585–604.

Emanuel, K. A. "The Dependence of Hurricane Intensity on Climate." *Nature* 326 (2) (April 1987): 483–485.

Emanuel, K. A. "The Maximum Intensity of Hurricanes." *Journal of the Atmospheric Sciences* 45 (April 1, 1988): 1143–1156. http://www.aos.wisc.edu/~aos718/basic%20dynamics /mpi.emanuel.pdf.

Emanuel, K. A. "Toward a General Theory of Hurricanes." *American Scientist* 76 (July–August 1988): 370–379. ftp://texmex.mit.edu/pub/emanuel/PAPERS/Amer_sci.pdf.

Emanuel, K. A. "Comments on 'Global Climate Change and Tropical Cyclones.' Part I." *Bulletin of the American Meteorological Society* 76 (1995): 2241–2243.

Emanuel, Kerry A. "Increasing Destructiveness of Tropical Storms over the Past 30 Years." *Nature* 436 (August 4, 2005): 686–688.

Fedorov Alexey V., Christopher M. Brierley, and Kerry Emanuel. "Tropical Cyclones and Permanent El Niño in the Early Pliocene Epoch." *Nature* 463 (February 25, 2010): 1066–1070.

Gray, William M. "Hurricanes and Hot Air." *Wall Street Journal*, July 26, 2006, A12.

Haarsma, R. J., J. F. B. Mitchell, and C. A. Senior. "Tropical Disturbances in a G[lobal] C[limate] M[odel]." *Climate Dynamics* 8 (1993): 247–257.

Idso, S. B., R. C. Balling Jr., and R .S. Cerveny. "Carbon Dioxide and Hurricanes: Implications of Northern Hemispheric Warming for Atlantic/Caribbean Storms." *Meteorology and Atmospheric Physics* 42 (1990): 259–263.

Karl, T. R., N. Nicholls, and J. Gregory. "The Coming Climate." *Scientific American* 276 (1997): 79–83.

Kerr, Richard A. "Models Foresee More-Intense Hurricanes in the Greenhouse." *Science* 327 (January 22, 2010): 399.

Knutson, Thomas R., and Robert E. Tukeya. "Impact of CO_2-Induced Warming on Simulated Hurricane Intensity and Precipitation: Sensitivity to the Choice of Climate Model and Convective Parameterization." *Journal of Climate* 17(18) (September 15, 2004): 3477–3495.

Knutson, T. R., R. E. Tuleya, and Y. Kurihara. "Simulated Increase in Hurricane Intensities in a CO_2-Warmed Climate." *Science* 279 (February 13, 1998), 1018–1020.

Knutson, Thomas R., et al. "Impact of CO_2-Induced Warming on Hurricane Intensities as Simulated in a Hurricane Model with Ocean Coupling." *Journal of Climate* 14 (2001): 2458–2469.

Landsea, Christopher W. "A Climatology of Intense (or Major) Atlantic Hurricanes." *Monthly Weather Review* 121 (1993): 1703–1713.

Landsea, Christopher W. *NOAA: Report on Intensity of Tropical Cyclones*. Miami, Florida, August 12, 1999. http://www.aoml.noaa.gov/hrd/tcfaq/tcfaqG.html#G3.

Landsea, Christopher W., et al. "Downward Trends Atlantic Hurricanes During the Past Five Decades." *Geophysical Research Letters* 23 (1996): 1697–1700.

Landsea, Christopher W., et al. "Impact of Duration Thresholds on Atlantic Tropical Cyclone Counts." *Journal of Climate*, August, 2009. http://ams.allenpress.com/perlserv/?request=get-abstract&doi=10.1175%2F2009JCLI3034.1. doi: 10.1175/2009JCLI3034.1.

Lighthill, J., et al. "Global Climate Change and Tropical Cyclones." *Bulletin of the American Meteorological Society* 75 (1994): 2147–2157.

Maue, Ryan. "Global Tropical Storm Activity." 2008. http://policlimate.com/tropical/.

McKibben, Bill. *The End of Nature*. New York: Random House, 1989.

Merzer, Martin. "Study: Global Warming Likely Making Hurricanes Stronger." *Miami Herald*, August 1, 2005, n.p. (LEXIS).

Nyberg, Johan, et al. "Low Atlantic Hurricane Activity in the 1970s and 1980s Compared to the Past 270 years," *Nature* 447 (June 7, 2007): 698–701.

O'Brien, S. T., B. P. Hayden, and H. H. Shugart. "Global Climatic Change, Hurricanes, and a Tropical Forest." *Climatic Change* 22 (1992): 175–190.

Revkin, Andrew. "Global Warming Is Expected to Raise Hurricane Intensity." *The New York Times*, September 30, 2004, A20.

Revkin, Andrew. "Study Links Tropical Ocean Warming to Greenhouse Gases." *The New York Times*, September 12, 2006. http://www.nytimes.com/2006/09/12/science/12ocean.html.

Revkin, Andrew C. "Study Finds Hurricanes Frequent in Some Cooler Periods." *The New York Times*, May 24, 2007. http://www.nytimes.com/2007/05/24/science/earth/24storm.html.

"Saharan Dust Has Chilling Effect on North Atlantic." NASA Earth Observatory Press Release, December 14, 2007. http://earthobservatory.nasa.gov/Newsroom/NasaNews/2007/2007121425986.html.

Santer, B. D., et al. "Forced and Unforced Ocean Temperature Changes in Atlantic and Pacific Tropical Cyclogenesis Regions." *Proceedings of the National Academy of Sciences* 103(38) (September 19, 2006): 13905–13910.

Saunders, Mark A., and Adam S. Lea. "Large Contribution of Sea Surface Warming to Recent Increase in Atlantic Hurricane Activity." *Nature* 451 (January 31, 2008): 557–560.

Schiermeier, Quirin. "Global Warming Blamed for Growth in Storm Intensity." *Nature* 455, September 3, 2008. http://www.nature.com/news/2008/080903/full/news.2008.1079.html.

See also: Hurricanes, Observations of; Hurricanes, Snow; Oceans' Absorption of Heat; Temperatures, Global; Temperatures, Greenhouse Gas Levels and

HURRICANES, SNOW

U.S. East Coast cities can get caught between cold air inland and intense storms moving northeastward along a warming Gulf Stream; these confluences have always produced heavy, explosive snowstorms. In late January and February 2015, for example, Boston received more than 40 inches of snow in one week and eight feet in a month. It was a year for explosive snow events. Thousands of miles away, Istanbul, Turkey, received two feet in a February blizzard at about the same time, "catching off guard a city of 14 million that rarely gets more than a short-lived dusting" (Yeginsu 2015).

While the rest of the world experienced its warmest winter on record (according to NASA), by March 15, 2015, Boston had more than 110 inches of snow for the season, some of which had fallen with wind chills at minus 20 to minus 30, including the biggest three snowstorms since records began after the Civil War. Two of these were "snow hurricanes," intense blizzards with winds as high as 70 miles an hour. Eastport, Maine, more than 250 miles northeast of Boston, was buried under 109 inches of snow in 23 days by February 19.

More evidence of change in the hydrological cycle arrived during the first week of March 2015, as Boston was digging out from its eight feet of snow. On March 5, Capracotta, a town in Italy's Appenine Mountains, received 100.8 inches of snow (256 centimeters) in 18 hours, breaking the world record for snowfall in one diurnal cycle. The town, 90 miles east of Rome, sits 4,662 feet (1,421 meters) above sea level. The old record, 90.6 inches, was set at Mount Ibuki, Japan, on February 14, 1927. The U.S. record is 75.8 inches at Silver Lake, Colorado, on April 14–15, 1927.

Intense Snow and Global Warming

What happened? Boston and other parts of New England had been caught on the tongue of the Arctic Oscillation that delivered intensely cold air that combined with warmer-than- usual Gulf Stream to produce "ocean effect" snowstorms that amplified the usual nor'easter pattern. This was not unlike the seven-foot snow blast that hit south of Buffalo, New York, earlier in the season, which combined the same arctic air with relatively warm air over Lake Erie. "The environment in which all storms form is now different than it was just 30 or 40 years ago because of global warming," said Kevin Trenberth, a senior scientist at the National Center for Atmospheric Research in Boulder, Colorado. Higher temperatures warm the oceans and allow the atmosphere to hold a greater amount of water vapor, said Brad Johnson, a meteorologist with the University of Georgia. "Both of these factors, among others, contribute to stronger storms in general." he said (Rice 2015).

During massive, early-season lake-effect snows downwind of the Great Lakes in November 2014, the Arctic Oscillation threw a massive early-season cold wave over most of the central and eastern United States, producing the coldest November since 1976. It was warmer in Kivalina, Alaska, and Iqaluit, on Baffin Island, than in Chicago. The lake water was still around 50°F, and the record-setting cold triggered some of the heaviest lake-effect snows on record, worse than the legendary blizzard of 1977, which brought far less moisture. For a week in the middle of the month, as much as eight feet fell in some locations southwest of Buffalo, killing at least 13 people. Less than a week after 2014's record snow, temperatures rose to almost 50°F, provoking widespread flooding.

During the winters of 2009–2010 and 2010–2011, the eastern half of the United States (as well as parts of Western Europe) experienced snowfall that was far above average and some periods of record cold. This had some climate-change deniers contending not only that global warming had ended but also that a new ice age had begun. The jet stream had plunged out of the Arctic across the central United States, turned northward over south Texas, and headed up the East Coast, a perfect pattern for nor'easters that suck warm air and moisture from the Gulf Stream and combine it with cold blasts from the continent. The jet stream then continued north–northeastward to Greenland, giving that area record warmth. It plunged again over Europe ("Global Warming Blamed" 2011).

For a week in late November 2014, extremely cold air from the Arctic flowing over relatively warm water in Lake Erie from the west–southwest blasted the area south of Buffalo with as much as seven feet of snow—a dramatic illustration of a moisture-primed lake effect partly the result of a warming atmosphere. Cowlesville received 88 inches, Orchard Park 71 inches, and Lancaster 74 inches. The snow fell as intensely as six inches an hour and sometimes was accompanied by thunder and lightning. A 132-mile section of the New York State Thruway shut down for several days (Lam et al. 2014). Within days, a gigantic thaw produced floods.

According to the National Weather Service, 22.6 inches of snow fell at Buffalo Niagara International Airport during one storm in October 2006, a record for the month. "Our crews were wrapping up their workdays yesterday and the weather forecasters were still talking about rain," said Steve Brady, a spokesman for National Grid, which provides electricity to Buffalo as well as its largest suburbs. "We've had windstorms in the fall and ice storms in the fall, but never a snowstorm like this, this early" (Staba 2006). Similar "thundersnow" struck Omaha at the end of October 1997 as a foot of snow destroyed many of the city's old maple and ash trees, knocking out power in some parts of the city for a week.

Fewer Storms, Greater Intensity

Fed by greater evaporation from the Great Lakes, overall precipitation in the region could increase 10 percent to 20 percent, according to the U.S. Environmental Protection Agency. "A lot of scientists mention getting fewer storms but of greater intensity," said Helen Domske, a researcher at the University at Buffalo's Great Lakes Program (Zremski 2002). During the first few decades of warming, lake-effect snowstorms could be more frequent in Buffalo, thanks to Lake Erie's lack of ice. Easterling said, however, that temperatures eventually probably will warm to a point where lake-effect rain may become more common than snow. As a result, lake-effect snow could decrease by half within a century (Zremski 2002).

A comparative study of snowfall records in and outside the Great Lakes region indicated a statistically significant increase in lake-effect snowfall in the region since the 1930s. In areas where the lake effect is small, little change has been observed. Warmer lake waters and decreased ice cover were cited by the researchers as the major reasons for the snowfall increases in lake-effect areas. A team of researchers led by Adam W. Burnett, associate professor of geography at Colgate University, authored the study, "Increasing Great Lake-Effect Snowfall During the Twentieth Century: A Regional Response to Global Warming?" in the November issue of the *Journal of Climate* (Burnett et al. 2003).

Syracuse, New York, one of the snowiest cities in the United States, experienced four of its heaviest annual snowfalls on record during the 1990s, the warmest decade in the 20th century. "Recent increases in the water temperature of the Great Lakes are consistent with global warming," said Burnett. "This widens the gap between water temperature and air temperature—the ideal condition for [lake-effect] snowfall" ("Global Warming Means" 2003).

The research team compared snowfall records from 15 weather stations in the Great Lakes region with 10 stations outside the region. Records dating to 1931 were examined for eight of the lake-effect and six of the non–lake-effect areas. Records for the rest of the sample dated back to 1950. "We found a statistically significant increase in snowfall in the lake-effect region since 1931, but no such increase in the non–lake-effect area during the same period," said Burnett. "This leads us to believe that recent increases in lake-effect snowfall are not the result of changes in regional weather disturbances" ("Global Warming Means" 2003).

Great Lakes water levels have dropped before, of course, and the low levels of the years following 1997 followed a period of relatively high water levels in the lakes, which make up 18 percent of the world's freshwater supply. Recent low levels were notable, according to climatic experts, because of the "amount of lowering and the rapidity with which it occurred," as well as the fact that "the primary hydroclimatological driver was high air temperatures [increasing evaporation], not extremely low precipitation" (Assel et al. 2004, 1150).

Great Lakes and New York Lakes Fail to Freeze

The Great Lakes failed to freeze during the winter of 2001–2002. In addition, several lakes in upstate New York also remained liquid for the first time in at least three decades. In his 30 years of studying freeze–thaw cycles of lakes in New York State, Kenton Stewart had never before seen some of these lakes remain unfrozen for an entire winter. "The majority of the lakes in the state still froze, but a surprising number that developed ice covers in previous winters, had only a partial skim of ice that winter, or did not freeze at all," said Stewart, professor emeritus of biological sciences at the University at Buffalo ("New York Lakes" 2002). In subsequent winters, however, temperatures cooled and the lakes froze again. Shortening freezing seasons are not restricted to New York lakes, of course. The waters of Lake Mendota, near Madison, Wisconsin, for example, freeze about 40 fewer days today than during 1860 (Glick 2004, 32). By 2016, some lakes in far-north New York state such as Lower St. Regis Lake failed to freeze until after January 1, the latest on record.

According to a report by the Environment News Service, Kenton Stewart has studied the freeze–thaw cycles of more than 250 lakes in New York State. "Lakes that did not freeze this winter include some that did so during an El Niño year. Those that did freeze did so one to three weeks later than usual" ("New York Lakes" 2002). "One surprising thing about the unusually mild winter is that while it was as mild as some of the strong El Niño events that we've seen, it was not associated with an El Niño event in the Pacific Ocean that can have an atmospheric influence," said Stewart. "It also was not foreseen by the Climate Prediction Center of the National Oceanographic and Atmospheric Agency" ("New York Lakes" 2002).

Among the New York lakes that failed to freeze during 2001–2002 were Irondequoit Bay in Rochester; Hemlock and Canadice Lakes, south of Rochester; Cross Lake located west of Syracuse; Onondaga Lake in Syracuse; Otisco Lake located west of Syracuse; Big Green Lake in Green Lake State Park east of Syracuse; and Ashokan and other water supply reservoirs north of New York City ("New York Lakes" 2002).

Further Reading

Assel, Raymond A., Frank H. Quinn, and Cynthia E. Sellinger. "Hydroclimatic Factors of the Recent Record Drop in Laurentian Great Lakes Water Levels." *Bulletin of the American Meteorological Society* 85(8) (August 2004): 1143–1150.

Burnett, Adam W., et al. "Increasing Great Lake-Effect Snowfall during the Twentieth Century: A Regional Response to Global Warming?" *Journal of Climate* 16(21) (November 1, 2003): 3535–3542.

Glick, Daniel. "The Heat Is On: Geosigns." *National Geographic*, September 2004, 12–33.

"Global Warming Blamed for Heavy Snowstorms, Record Floods." Environment News Service, March 2, 2011. http://www.ens-newswire.com/ens/mar2011/2011-03-02-02.html.

"Global Warming Means More Snow for Great Lakes Region." Ascribe Newswire, November 4, 2003 (LEXIS).

Lam, Linda, Nick Wiltgen, and Jon Erdman Wilt. "Lake-Effect Snow Recap: Up to 88 Inches of Snow Buries Parts of Western New York, Including the Buffalo Southtowns." Weather.com, November 21, 2014. http://www.weather.com/storms/winter/news/lake-effect-snow-significant-lake-erie-lake-ontario-20141115.

"New York Lakes Fail to Freeze." Environment News Service, March 21, 2002. http://ens-news.com/ens/mar2002/2002L-03-21-09.html#anchor1.

Rice, Doyle. "Buried in Boston? Blame It on Climate Change—Maybe." *USA Today*, February 10, 2015. http://www.usatoday.com/story/weather/2015/02/09/northeast-snowstorms-climate-change-global-warming/23133913/.

Staba, David. "Snowstorm Blankets Buffalo, Killing at Least 3." *The New York Times*, October 14, 2006. http://www.nytimes.com/2006/10/14/nyregion/14storm.html.

Yeginsu, Ceylan. "In Turkey, Even Snow Can Be Tainted by Politics." *The New York Times*, February 20, 2015, A4.

Zremski, Jeremy. "A Chilling Forecast on Global Warming." *Buffalo News*, August 8, 2002, A1.

See also: Atmospheric Circulation; Hurricanes, Observations of; Hurricanes, Scientific Debates; Oceans' Absorption of Heat; Temperatures, Global; Temperatures, Greenhouse Gas Levels and

MONSOONS

The southwestern Asian monsoon affects nearly half the world's population, and its effects range from the Sahara desert to Japan. In addition to greenhouse gases, the strength of the monsoon is governed by lengthy cycles—10,000 to 100,000 years (or more)—in solar insolation related to Earth's orbit around the sun. The strength of the monsoon reached a 10,000-year low around the beginning of the 17th century and coinciding with the Maunder Minimum, a period of reduced solar activity, as well as cold temperatures during the Little Ice Age of the 16th to 19th centuries (Black 2002, 528). Evaluating the work of Anderson et al. (2002), David M. Anderson wrote, "This argument has critical implications in the face of global warming" (Black 2002, 528). Monsoon intensity was most notable during the 20th century because the pace of temperature increases accelerated.

The Indian Ocean monsoon affects the daily lives of hundreds of millions of people on the Indian subcontinent. In the past, a switch in upper-level winds that caused abnormally heavy rain or drought played a role in floods, droughts, and famines that killed hundreds of thousands of people. With half of India's 1.2 billion people working in agriculture, a drought during monsoon season can have a major effect on million of peoples' lives. "A large number of people don't get employment. There are acute drinking water problems," said Yoginder K. Alagh, chairman of the Institute of Rural Management.

In 2014, atmospheric scientist Wenju Cai and colleagues explained in *Nature* how wind patterns can change monsoon precipitation:

> In its positive phase, sea surface temperatures are lower than normal off the Sumatra–Java coast, but higher in the western tropical Indian Ocean. During the extreme positive-IOD [Indian Ocean Dipole] events of 1961, 1994 and 1997, the eastern cooling strengthened and extended westward along the equatorial Indian Ocean through strong reversal of both the mean westerly winds and the associated eastward-flowing upper ocean currents. This created anomalously dry conditions from the eastern to the central Indian Ocean along the Equator and atmospheric convergence farther west, leading to catastrophic floods in eastern tropical African countries, but devastating droughts in eastern Indian Ocean rim countries. (Cai et al. 2014, 254)

A Debatable Question

Is greenhouse warming exerting an effect on these patterns? As of now, this question is debatable. However, Cai and colleagues' research has allowed them to assert that wind reversals will become more frequent as the atmosphere warms and that "the frequency of extreme [positive IOD] events will increase by almost a factor of three, from one event every 17.3 years over the 20th century to one event every 6.3 years over the 21st century" (Cai et al. 2014, 254).

Anecdotal evidence has been supporting these observations. Monsoons seem to have become more erratic, although variability is normal as with most weather patterns, so establishing greenhouse-induced patterns (the difference between short-term weather and long-term climate) is difficult. In 2012 and the years immediately following, the annual monsoon rains were sparse over much of the Indian subcontinent some years and superabundant in others, leading to widespread rural disaster. Variability was not unknown in the past, but the frequent nature of unusual weather seemed new. The monsoon rains that generally fall between June and October have become less reliable.

A study of the monsoon's history by S. Sharmila and colleagues (2014) detected growing changes in the precipitation patterns of recent decades. B. N. Goswami, a coauthor of the paper (and former director of the Indian Institute of Tropical Meteorology, an agency of the Indian government), concluded, "Climate models suggest that while overall rainfall should increase in the coming decades, the region can expect longer dry spells and more intense downpours—forces that would seem to cancel each other out but in fact pose new threats" (Bajaj 2012). "Heavy rains are normally short duration, and therefore the water runs off," said Goswami, who added that more research was needed to fully understand the impact of climate change on monsoons. "Weak rains are important for recharging groundwater" (Bajaj 2012).

However, on the other side of this debate, Timothy DelSole and Jagadish Shukla at George Mason University in Fairfax, Virginia, "found no significant statistical relationship between summer monsoon rainfall in India and sea surface temperatures surrounding the country in May of each year from 1960 to 2005. They analyzed predictions made by five coupled ocean–atmosphere climate models based on the same temperature data and period, and found that the models were better than the statistical methods at predicting rainfall" ("Predicting" 2012).

South Asia also has been subject to erratic summertime droughts between 1950 and 2000. Reporting in *Science*, Massimo A Bollasina and colleagues investigated the South Asian monsoon's response to natural and anthropogenic forcings, and found that "the observed precipitation decrease can be attributed mainly to human-influenced aerosol emissions. The drying is a robust outcome of a slowdown of the tropical meridional overturning circulation, which compensates for the aerosol-induced energy imbalance between the Northern and Southern Hemispheres. These results provide compelling evidence of the prominent role of aerosols in shaping regional climate change over South Asia" (Bollasina et al. 2011, 502)

David M. Anderson and colleagues reconstructed wind speeds of the Asian southwest monsoon for the last 1,000 years using fossil *Globigerina bulloides* (sediment) abundance in box cores from the Arabian Sea. They found that "monsoon wind speed increased during the past four centuries as the Northern Hemisphere warmed" (Anderson et al. 2002, 596). They infer that "the observed link between Eurasian snow cover and the southwest monsoon persists on a centennial scale" (Anderson et al. 2002, 596). Alternately, they wrote in *Science*, the "forcing implicated in the warming trend (volcanic aerosols, solar output, and greenhouse gases) may directly affect the monsoon. Either interpretation is consistent with the hypothesis that the southwest monsoon strength will increase during the coming century as greenhouse gases continue to rise and northern latitudes continue to warm" (Anderson et al. 2002, 596).

Indian Ocean Temperatures and African Drought

Warming in the Indian Ocean is probably linked to drought in parts of eastern and southern Africa, where rainfall has declined some 15 percent since the 1980s during the monsoon season, which is March through May. The reduction is caused by irregularities in moisture transport between the ocean and land, which is caused by warming in the ocean, according to research published in August 2008 in the *Proceedings of the National Academy of Sciences* ("NASA Data" 2008). Irregularities in the African monsoon affect marginal farmers' ability to feed themselves, according to U.S. Agency for International Development's Famine Early Warning Systems Network.

"The last 10 to 15 years have seen particularly dangerous declines in rainfall in sensitive ecosystems in East Africa, such as Somalia and eastern Ethiopia," said Molly Brown of NASA's Goddard Space Flight Center in Greenbelt, Maryland, a coauthor of the study. "We wanted to know if the trend would continue or if it would start getting wetter" ("NASA Data" 2008).

Lead author Chris Funk of the University of California–Santa Barbara and colleagues showed that, paradoxically, a warmer ocean caused more precipitation offshore while drying the atmosphere over the land. The team then ran 11 models to forecast future trends, and 10 of them agreed that the trend will continue as the ocean warms, depriving Africa's eastern seaboard of rainfall. "We can be quite certain that the decline in rainfall has been substantial and will continue to be," Funk said. "This 15 percent decrease every 20 to 25 years is likely to continue" ("NASA Data" 2008).

Further Reading

Anderson, David M., Jonathan T. Overpeck, and Anil K. Gupta. "Increase in the Asian Southwest Monsoon during the Past Four Centuries." *Science* 297 (July 26, 2002): 596–599.

Bajaj, Vikas. "Crops in India Wilt in a Weak Monsoon Season." *The New York Times*, September 3, 2012. http://www.nytimes.com/2012/09/04/business/global/drought-in-india-devastates-crops-and-farmers.html.

Black, David E. "The Rains May Be A-Comin'." *Science* 297 (July 26, 2002): 528–529.

Bollasina, Massimo A., Yi Ming, and V. Ramaswamy. "Anthropogenic Aerosols and the Weakening of the South Asian Summer Monsoon." *Science* 334 (October 28, 2011): 502–505.

Cai, Wenju, et al. "Increased Frequency of Extreme Indian Ocean Dipole Events Due to Greenhouse Warming." *Nature* 510 (June 12, 2014): 254–258.

"NASA Data Show Some African Drought Linked to Warmer Indian Ocean." NASA Earth Observatory, August 5, 2008. http://earthobservatory.nasa.gov/Newsroom/NasaNews/2008/2008080527314.html (no longer available).

"Predicting the Indian Monsoon." *Nature* 484 (April 19, 2012): 291.

Sharmila, S., et al. "Asymmetry in Space-Time Characteristics of Indian Summer Monsoon Intraseasonal Oscillations during Extreme Years: Role of Seasonal Mean State." *International Journal of Climate*, June 2014, 1–16.

See also: Atmospheric Circulation; Oceans' Absorption of Heat; Temperatures, Global; Temperatures, Greenhouse Gas Levels and; Thunderstorms and Tornadoes

THUNDERSTORMS AND TORNADOES

Does a warmer atmosphere tend to augment severe small-scale, warm-weather storms such as tornadoes and thunderstorms? The idea that storms dissipate heat seems to indicate that warming, accompanied by the air's increased capacity to hold water vapor, will provoke larger and more violent storms. Anecdotal evidence, including several seasons of severe storms in the United States and elsewhere, indicates reason for caution. Tornadoes and thunderstorms form and strengthen for many reasons other than heat, however, and scientific studies and opinion remain divided.

Bigger Storms in Unusual Places

Violent weather has been hitting unusual areas with increasing frequency. Major tornadoes, for example, have been sighted in Michigan and Japan. An F3 Tornado killed several people in Michigan on October 18, 2007, an extremely unusual event so far north in the third week of October. The storm killed two people in Ingham County's Locke Township near Lansing when their home was ripped from its foundation and dumped into a pond 100 yards away. At least nine people were killed, at least 20 injured, and others trapped under rubble when a tornado hit northern Japan on November 6, 2006, a location where such storms had been all but unknown. Officials said the tornado demolished two prefabricated buildings at a construction site, eight or so private houses, and a factory, according to an official at the fire department in Engaru on the northeastern part of Japan's northern island of Hokkaido. Fire engines and ambulances were at the site, where crushed cars could be seen along with lumber and other scattered debris. Several houses were reduced to rubble.

Intense thunderstorms once were rare in Norway, but in Friday, August 26, 2016, on Hardangervidda, a plateau in southern Norway, at least 323 reindeer, an entire herd, were killed by lightning as they huddled in torrential rain that conducted a massive amount of electricity. "We are not familiar with any previous happening on such a scale," said Kjartan Knutsen of the Norwegian Environment Agency. "Reindeer often huddle together in groups during thunderstorms," Knutsen said. "It is a strategy they have to survive, but in this case their survival strategy might have cost them their [lives]. The corpses are all lying in one big group, piled together" (Libellaug 2016).

High-altitude clouds in Earth's tropics associated with severe storms and heavy rainfall have been increasing in volume and height as the atmosphere warms, according to work by scientists at NASA's Jet Propulsion Laboratory in Pasadena, California. Senior research scientist Hartmut Aumann summarized a study linking severe storms, torrential rain, and hail with seasonal variations in average sea surface temperatures of the tropical oceans. For every 1°C increase in average ocean surface temperature, this research observed a 45-percent rise in the frequency of high-level clouds. At the current rate of global warming, 0.13°C per decade, this study inferred that the frequency of such storms may increase 6 percent per decade ("NASA Study Links" 2008).

A record number of tornadoes—more than 600—ravaged the southern and eastern United States during April 2011 (the former record had been 542 in May 1973), products of a semistationary boundary between extremely warm air to the south and cooler air to the north that was traversed by a powerful jet stream southwest to northeast. Some tornadoes developed in groups of three or four, rotating around each other. Raleigh, North Carolina, was raked by several storms. An Enhanced Fujita (EF) scale tornado estimated at EF4 had winds near 200 miles an hour and blew through St. Louis's major airport, closing it for two days. Other areas of the city also were ripped apart. A few days later, on April 27, an EF4 or EF5 "wedge" tornado ripped through Tuscaloosa, Alabama, leaving a 1.5-mile path of ruin in its wake. The same tornado plowed through the northern suburbs of Birmingham an hour later. This was one of several twisters that killed more than 200 people across five states that evening in one of the deadliest outbreaks in U.S. history. Two of the four largest tornado outbreaks recorded in the United States occurred during three weeks in April 2011.

The New York Times reported:

At least 300 people across six states died in the storms, with more than half—195 people—in Alabama [which] in some places has been shorn to the slab. . . . Thousands have been injured, and untold more have been left homeless, hauling their belongings in garbage bags or rooting through disgorged piles of wood and siding to find anything salvageable. (Robertson and Severson 2011, A1)

The sky exploded with a fury without recorded precedent. Across northern Alabama on the evening of April 27, no "single storm took a linear path. Rather, an untold number of tornadoes hit an untold number of homes and buildings in a chaotic flurry, touching down and picking up power at different places at different times" (Brown and Seelye 2011).

Tornadoes have been striking areas where they once were extremely rare and even unknown. For example, on December 5, 2012, three people were killed near Auckland, New Zealand, by a tornado, which damaged more than 150 homes ("Deadly Tornado" 2012). On average, the country has seven tornados a year, most of them small, mild, and producing little damage. On May 5, 2012, a tornado devastated a suburb northeast of Tokyo, Japan, killing one person, injuring 30, and substantially damaging 200 houses. According to one report, "Public broadcaster NHK showed rows of houses without roofs, apartment complexes with smashed balconies and shattered windows, and tilting telephone poles that could barely stand" ("Tornado Rips" 2012).

A tornado that destroyed large parts of Aurora, Nebraska, produced hailstones that were among the largest ever reported in the United States, as well as a tornado that stood virtually in one place for half an hour, devastating the town of Deshler. In 2010, the size record for hailstones was broken twice. The former record, seven inches in diameter (near Auburn, Nebraska, in 2003), was exceeded by an eight-inch stone in South Dakota on July 23 and a 7.75-inch stone near Wichita, Kansas, on September 15.

Two tornadoes ravaged the Virginia town of Waverly and Appomattox County, near Richmond, on February 24, 2016, killing four people, the first February tornado fatalities in that state's recorded history. These were part of an unusually strong tornado outbreak from Texas throughout the South that killed several other people as well. Heretofore, the only February tornadoes in Virginia had been weak (EF0 to EF1).

Tornadoes Out of Season

On December 13, 2010, a tornado destroyed several homes near Salem, Oregon. Eighteen days later, several tornadoes were reported in eastern Missouri and western Illinois (some near St. Louis), the first such storms on December 31 in these states on the observational record. The same day. a tornado in northwest Arkansas killed three people; four more people were killed in Missouri. One tornado that ravaged Cincinnati, Arkansas, and killed three of its 100 people, was rated as a probable EF3.

Rising temperatures across parts of the United States by the end of the 21st century may double the number of days with conditions likely to produce severe thunderstorms, according to a study by Robert J. Trapp and colleagues in the *Proceedings of the National Academy of Sciences* (2007). The Atlantic and Gulf coastal regions—roughly New York City to Atlanta to New Orleans—stand the greatest chance of increasingly severe storms, including tornadoes, hail, strong winds, and downpours. A warmer atmosphere pumps more water vapor into the air and holds a larger amount before its release as precipitation.

Street-level evidence supports the study. To cite one of several examples, out-of-season tornadoes killed three people and injured several others they strafed several homes from central Wisconsin (Wheatland) to southwest Missouri on January 7 and 8, 2008. The same outbreak also affected Arkansas, Illinois, Missouri, and Oklahoma. On February 5 and 6, 2008, an outbreak of more than 70 tornadoes that would have been severe in May killed 50 people in a wide swath from Arkansas to Tennessee. Police in several towns and suburbs of cities (including Memphis) described devastation resembling war zones. Many witnesses expressed awe at the ferocity of the storms, which ripped some houses entirely off their foundations.

Such storms are rare, but they have occurred occasionally before. During February 1971, severe tornadoes killed 134 people in Mississippi and Louisiana. The 2008 outbreak appeared to have been the country's second worst for February. On January 21 and 22, 1999, 134 tornadoes scoured the South, killing nine people. Still, the storms in 2008 were notable for their violence and size.

A "multivortex" that hit Joplin, Missouri, on May 22, 2011, was the single deadliest tornado since officials began keeping records in 1950. The tornadoes killed 116 people, injured more than 500, and damaged or destroyed more than 2,000 buildings, including the city's major hospital. "Patients were sucked out of the operating room and dumped into a parking lot," according to according to reporters for the *Washington Post* (Vastag and O'Keefe 2011). "We have had more F4s and F5s than in past years," said Jack Hayes, director of the National Weather Service.

Storm Violence a Figment of Imagination?

Not everyone believes that storms have become more numerous or increasingly violent. For example, Richard A. Muller, a professor of physics at the University of California–Berkeley, asserted in *The New York Times* (2013), that "scientific evidence shows that strong to violent tornadoes have actually been *decreasing* for the previous 58 years, and it is possible that the explanation lies with global warming." Muller admitted that his assertion did not agree with popular impression.

Muller said that he is not opposed to the idea of global warming per se, "which I am convinced is real and caused by man-made emissions of greenhouse gases." He argued instead that the observational record does not support a conclusion that warmer temperatures raise the number or intensity of tornadoes. Using National Oceanic and Atmospheric Administration (NOAA) data of tornadoes reported in the United States between 1950 and 2011, he found, "At first glance, the increase looks dramatic: The number of tornadoes in 1950 was only 200, and by 2011 it had shot up to 1,700." However, Muller wrote:

> But the 200 number is ridiculously small. If it had truly been that low in the past, *The Wizard of Oz* never would have been written. And indeed NOAA accounts for those early small numbers by explaining that originally only the most violent tornadoes were recorded, leading to lower reported totals. By contrast, these days even little storms are logged; backyard dust devils are reported for insurance claims, and Doppler radar makes it hard for any tornado to escape notice. (Muller 2013)

In 1976, NOAA recorded roughly 800 tornadoes in the United States. "But here, too," Muller asserted, "the increase is a function of enhanced reporting. The scientists at NOAA note explicitly that the rise in the numbers has come almost exclusively from the improved documentation of so-called category EF0 tornadoes—the ones that are so weak that their legacy consists only of broken branches and damaged billboards. As another NOAA Web site warns, 'This can create a misleading appearance of an increasing trend in tornado frequency'" (Muller 2013).

Muller said that if EF0 tornadoes are excluded, "the number has remained fairly steady from 1954 to now." Tornadoes in categories EF3 to EF5, the most severe, have declined since the mid-20th century. "Global warming does not obviously lead to increased or more violent tornadoes," said Muller. "It is possible, for instance, that the increased energy brought by the higher temperatures is less significant than global warming's reduction in the north–south temperature difference (the poles warm more than the Equator). The latter could reduce the kind of hot-cold weather fronts that generate severe storms" (Muller 2013).

Warming's Influence on Severe Thunderstorms

Severe thunderstorms (with winds stronger than 58 miles an hour and large hail) are formed by *convective available potential energy* (CAPE) and strong wind shear. CAPE, according to Harold Brooks (2013), a meteorologist at NASA's Earth Observatory, "is a measure of how much raw energy is available for storms; it relates to how warm, moist, and buoyant air is in a given area. Wind shear is a measure of how the speed and direction of winds change with altitude. CAPE can provide storms with the raw fuel to produce rain and hail, and vertical wind shear can pull and twist weak storms into strong, windy ones" ("Severe Thunderstorms" 2013). The most severe thunderstorms may develop into supercells and produce large, destructive complexes with hurricane-force winds, some of them running in straight lines (*derechos*), and tornadoes.

Warmer air holds more moisture and thus produces more explosive storms with more lightning. For each degree Celsius increase in temperature, lightning strikes increase by around 12 percent, according to calculations by University of California–Berkeley scientist David Romps and colleagues. Given temperature increases projected for the end of the 21st century, strikes may increase 50 percent worldwide, sparking more wildfires and causing other damage (Smith 2015).

Scientists at NASA's Earth Observatory have developed evidence that a warming atmosphere should increase CAPE "by warming the surface and putting more moisture in the air through evaporation. On the other hand, disproportionate warming in the Arctic should lead to less wind shear in mid-latitude areas prone to severe thunderstorms. So one factor makes severe storms more likely, while the other makes them less so" ("Severe Thunderstorms" 2013). "Disproportionate warming" in the Arctic reduces the difference in temperatures between the tropics and polar regions. This contrast, drawn together over regions such as the U.S. Midwest, is a primary provocation of violent weather, including tornadoes and severe thunderstorms. Temperatures have been rising much more rapidly in the Arctic than in the tropics.

One model provided by a team of scientists led by Robert Trapp of Purdue University (2007, 19719) indicates that doubling greenhouse gas levels in the atmosphere may "increase the number of days that severe thunderstorms could occur in the southern and eastern United States. Cities such as Atlanta and New York could see a doubling of the number of days that severe thunderstorms could occur" ("Severe Thunderstorms" 2013). Models indicate that this effect may be strongest by the end of the 21st century in areas that already experience severe storms (the U.S. Midwest and Southeast) and least in areas such as the U.S. West that currently have the lowest risk of such storms. Storms also may decrease in number but become stronger on average.

Further Reading

Brooks, H. "Severe Thunderstorms and Climate Change." *Atmospheric Research* 123 (April 1, 2013): 129–138.

Brown, Robbie, and Katherine Q. Seelye. "A Chaotic Flurry of Twisters That Spread the Devastation Fast and Wide." *The New York Times*, April 29, 2011, A19.

"Deadly Tornado Hits New Zealand." *Washington Post*, December 6, 2012. http://www
.washingtonpost.com/world/asia_pacific/tornado-rips-through-northeast-of-tokyo
-dozens-of-people-injured/2012/05/06/gIQAV4Jw4T_print.html (no longer available).

Libellaug, Henrik Pryser. "Lightning Strike Kills More Than 300 Reindeer in Norway." *The
New York Times*, August 29, 2016. http://www.nytimes.com/2016/08/30/world/europe
/hardangervidda-norway-lightning-reindeer.html.

Muller, Richard A. "The Truth about Tornadoes." *The New York Times*, November 20, 2013.
http://www.nytimes.com/2013/11/21/opinion/the-truth-about-tornadoes.html.

"NASA Study Links Severe Storm Increases, Global Warming." NASA Earth Observatory,
December 19, 2008. http://earthobservatory.nasa.gov/Newsroom/view.php?id=36309.

Robertson, Campbell, and Kim Severson. "Thousands Hurt—Tornadoes Raze Towns." *The
New York Times*, April 29, 2011, A1, A18.

"Severe Thunderstorms and Climate Change." NASA Earth Observatory April 7, 2013.
http://earthobservatory.nasa.gov/IOTD/view.php?id=80825&src=eoa-iotd.

Smith, Lindsay N. "Storm Surge." *National Geographic*, September 2015, 6.

"Tornado Rips Through City Northeast of Tokyo, 1 Person Dead, Dozens Injured." Associated Press, May 5, 2012. http://news.yahoo.com/tornado-rips-city-northeast-tokyo-1
-person-dead-103010197.html.

Trapp, Robert J., et al. "Changes in Severe Thunderstorm Frequency during the 21st Century
Caused by Anthropogenically Enhanced Global Radiative Forcing." *Proceedings of the
National Academy of Sciences* 104(50) (December 4, 2007): 19719–19723.

Vastag, Brian, and Ed O'Keefe. "Obama to Visit Missouri Sunday; Deadly Tornadoes on the
Rise." *Washington Post*, May 24, 2011. http://www.washingtonpost.com/national
/environment/researchers-see-a-pattern-in-rise-of-deadly-tornadoes/2011/05/23
/AFinz49G_print.html.

See also: Atmospheric Circulation; Heat Waves, Evidence of; Heat Waves, Forecasts of; Temperatures, Global; Temperatures, Greenhouse Gas Levels and

WILDFIRES, UNITED STATES

By the end of summer 2015, one of worst wildfire seasons in the the United States
was beginning to wind down, except in California, which was still smoldering under
drought and seasonally dry, offshore winds. Eight million acres had burned in the
United States by mid-September, 5 million in Alaska alone. Many years of widespread fires were changing the landscape in many parts of the U.S. West from verdant forests to scrubland as trees failed to take root.

By 2015, for the first time in its 110-year history, the U.S. Forest Service was
spending more than half of its entire budget on firefighting, up from 16 percent in
1995. "This is a five-alarm fire," said Secretary of Agriculture Tom Vilsack. "You're
no longer the Forest Service; you're a fire department" (Rice 2015). Between the
1970s and the year 2000, the western U.S. firefighting season had expanded by 92
days a year (Fried et al. 2004). Increasing heat and drought now spur fires that add
extra carbon dioxide to the atmosphere and increase future warming in a feedback
loop.

Intentional Fires Add Carbon Dioxide

Even fires set intentionally to clear land contribute substantially to human-caused emissions—as much as 20 percent. Authors of a study published in 2009 called on the Intergovernmental Panel on Climate Change to integrate fire into its assessments of global warming. "The tragic fires in Victoria, Australia, emphasize the ubiquity of recent large wildfires and potentially changing fire regimes that are concomitant with anthropogenic climate change," said first author David Bowman, professor at the University of Tasmania. "Our review is both timely and of great relevance globally" ("Fire" 2009; Bowman et al. 2009, 481).

"Fire is a worldwide phenomenon that appears in the geological record soon after the appearance of terrestrial plants. Fire influences global ecosystem patterns and processes, including vegetation distribution and structure, the carbon cycle, and climate," the authors wrote in *Science*. They continued:

> Although humans and fire have always coexisted, our capacity to manage fire remains imperfect and may become more difficult in the future as climate change alters fire regimes. This risk is difficult to assess, however, because fires are still poorly represented in global models. Here, we discuss some of the most important issues involved in developing a better understanding of the role of fire in the Earth system. (Bowman et al. 2009, 481)

The authors asserted that "Earth is intrinsically a flammable planet due to its cover of carbon-rich vegetation, seasonally dry climates, atmospheric oxygen, widespread lightning, and volcano ignitions. Yet, despite the human species' long-held appreciation of this flammability, the global scope of fire has been revealed only recently by satellite observations, available beginning in the 1980s" ("Fire" 2009).

Further Reading

Bowman, David M. J. S., et al. "Fire in the Earth System." *Science* (April 24, 2009): 481–484.

"Fire an Integral, but Unappreciated, Part of Global Warming." Environment News Service, April 23, 2009. http://www.ens-newswire.com/ens/apr2009/2009-04-23 -095.asp.

A reporter for *The New York Times* described Cochiti Canyon, New Mexico:

The hills here are beautiful, a rolling, green landscape of grasses and shrubs under a late-summer sky. But it is starkly different from what was here before: vast forests of ponderosa pine. The repeated blazes that devastated the trees were caused by simple things: an improperly extinguished campfire in 1996, a tree falling on a power line in 2011. . . . Mother pines are nowhere in sight. Nature's script has been disrupted by a series of unusually intense, unusually large fires—a product of many factors that include government firefighting policies, climate change, and bad luck. (Schwartz 2015)

"The future in a lot of places is looking shrubbier," said Craig D. Allen, a research ecologist with the U.S. Geological Survey (Schwartz 2015). Droughts are not new, nor are large fires or even intense fires, said Thomas W. Swetnam, a professor emeritus at the University of Arizona and an expert in tree-ring analysis. "But the greater number of intense and large fires, and the repeated 'burns on top of burns' like the ones that cleared the landscape around Cochiti canyon, are part of a pattern of worsening conditions exacerbated by hotter droughts," wrote reporter John Schwartz (2015).

As climate warms, the area of forest burned by wildfires in the United States may increase more than 50 percent by 2050, with the greatest impact in the Pacific Northwest and Rocky Mountains, where burned areas may rise by between 78 percent and 175 percent. "Wildfires, such as those in California earlier this year, are a serious problem in the United States and this research shows that climate change is going to make things significantly worse," said Dr. Dominick Spracklen, from the School of Earth and Environment at the University of Leeds, lead author of the research. "Our research shows that wildfires are strongly influenced by

Pines Scorched on Pine Ridge

Wildfires stoked by heat and drought change landscapes, sometimes for an extremely long time. Nebraska's Pine Ridge, for example, lost its namesake signature ponderosa pines to fires during the summer of 2012. The area has become prairie, Pine Ridge without evergreen trees (Gaarder 2013, A1). Doak Nickerson, district forester for the Nebraska Forest Service, has seen much of the state's evergreen cover disappear during his lifetime. In a state that is less than 2 percent forest, the once-thick cover of greenery that dominated Pine Ridge had been shrinking for 50 years because of rising temperatures and decreasing precipitation when the fires of 2012 reduced the greenery to scattered patches. In 2012, Nebraska experienced its most extensive drought since the Dust Bowl of the 1930s. "The size of these megafires is symptomatic of what is changing in front of us," Nickerson said. "Increasingly hotter, drier summers, coupled with milder winters are part of it" (Gaarder 2013, A2).

Further Reading

Gaarder, Nancy. "Pines Burned Out of Pine Ridge." *Omaha World-Herald*, January 7, 2013, A1, A2.

temperature. Hotter temperatures lead to dryer forests resulting in larger and more serious fires," said Spracklen ("Wildfires Set" 2009).

A report from the U.S. National Research Council in 2010 estimated that each one degree Celsius rise in temperatures will increase the area burned by wildfires in the western United States by 200 percent to 400 percent (Hoag 2010).

A Year-Round Fire Season

The fire season in California and nearby states has grown in length an average of 86 days since the 1970s, according to a study at the University of California–Merced. "Southern California has a 12-month fire season now," said Scott L. Stephens, a professor of fire science at the University of California–Berkeley. "You can have a fire there at any time" (Park et al. 2015). In Alaska, the 2016 season began in February with two fires that ignited on bare, dry ground, very unusual so far north at that time of year.

Wildfires in the western United States probably will increase in the coming decades, according to a study that links episodic fire outbreaks in the past five centuries with periods of warming sea surface temperatures in the North Atlantic Ocean. Warming surface waters there correspond with episodes of drought and fires in the West as recorded in tree rings studied by the researchers, according to lead researcher Thomas Kitzberger of the University of Comahue in Argentina ("Western Wildfires" 2007).

The team analyzed nearly 34,000 individual fire scar dates from tree rings, primarily ponderosa pine and Douglas fir, at 241 sites, the largest record of tree rings linked to past wildfires ever assembled. Wildfire frequency in Washington, Oregon, California, Colorado, New Mexico, Arizona, and South Dakota was found to be affected. "This study underscores the value of building large networks of high-resolution fire history data to better understand how climate may affect fire regimes over large areas of the globe," said Kitzberger ("Western Wildfires" 2007).

Warmer temperatures and changing atmospheric circulation patterns provoking drought in the western United States have dramatically increased both the number of large forest wildfires and the length of the season during which they occur since the early 1980s. Specifically, according to one study published in 2006, the number of fires from 1986 to 2005 increased fourfold and the area burned increased sixfold compared to 1979 through 1985 (Westerling et al. 2006). The same study found that the average fire season (the period of time when major fires were taking place) had increased by 78 days, and the average length of major fires had risen from 7.5 days to 37.1 days (Running 2006, 927).

The authors wrote that increases in spring and summer temperatures by 0.9°C and the melting of mountain snowpacks one to four weeks earlier had contributed to the increased number and duration of wildfires. These fires also add carbon dioxide to the atmosphere. Compounding all of this, according to Westerling and colleagues, warmer and longer growing seasons at high elevations reduce the amount of carbon removed from the atmosphere by living trees, especially during droughts (Westerling et al. 2006, 943). Adding it all up, the forests of the western United

States are likely to become a net source of carbon to the atmosphere rather than a sink.

A. L. Westerling and colleagues (2006) compiled a comprehensive time series of large forest wildfires in the western United States for the period from 1970 to 2003 and compared those data with corresponding observations of climate, hydrology, and land surface conditions. Wildfire activity increased suddenly in the mid-1980s. Hydroclimate and fires are closely related, and climate variation has been the primary cause of the increase in fires during the period of their study, although land-use changes can also be important. Longer springs and summers that could result as the world warms will continue to lengthen the fire season and continue to cause more large wildfires.

Apocalyptic Reports

Reports from fires in Southern California during October 2007 were apocalyptic—more than a million people routed from their homes, many people huddled in stadiums and fairgrounds, highways choked with fleeing multitudes, hundreds of homes burned to the ground, firefighters completely overwhelmed as hot hurricane-force Santa Ana winds descended from the mountains to drive flames through brush dried by a record drought and burning 500,000 acres within a week in 100-degree afternoon temperatures and air nearly devoid of humidity. The phrase "environmental refugees" is being used more often, and the fires and late October highs in the 80s along the East Coast were tied into global warming on the NBC evening news. In some cases, the fires moved so quickly that firefighters found themselves trapped as they wrapped themselves in fireproof aluminum fabric to survive.

As is so often the case, climatic extremes have combined with other factors to produce disaster on an epic scale. An extremely strong Santa Ana fanned flames into suburbs that had expanded into land prone to wildfires. Winds gusted as strong as 111 miles an hour. An inventory by University of Wisconsin researchers found that about two-thirds of new building in Southern California over the past decade was on land susceptible to wildfires, said Mike Davis, a historian at the University of California–Irvine and author of environmental and social histories of the region that have anticipated the toll of wildfires. "It gives you some parameters for understanding the current situation," Davis said. "Another way to look at it is you simply drive out the San Gorgonio Pass, where the winds blow over 50 mph over a hundred days a year and you have new houses standing next to 50-year-old chaparral. . . . You might as well be building next to leaking gasoline cans" (Vick and Geis 2007).

The Fires Next Time

Even before the fires of 2007, scientists had warned they were coming. Fires that charred nearly three-quarters of a million acres in California during the fall of 2003 could presage increasingly severe fire danger as global warming weakens more forests through disease and drought. Warmer, windier weather and longer, drier summers could result in higher firefighting costs and greater loss of lives and property, according to researchers at the Lawrence Berkeley National Laboratory and the U.S. Forest Service. Both the number of out-of-control fires and the acreage burned are likely to increase, more than doubling losses in some regions, according to a study published in the scientific journal *Climatic Change*. The study examined Northern California, but "the concern for Southern California would be much higher" because that region is drier for longer periods, said researcher Evan Mills of the Lawrence Berkeley lab (Thompson 2003).

According to this study, a doubling of atmospheric carbon dioxide will provoke fires that "burn more intensely and spread faster in most locations." According to models developed by Jeremy S. Fried, Margaret S. Torn, and Evan Mills, the number of "escapes" (fires that exceed initial containment efforts) doubles current frequencies, according to their models. Contained fires also burn 50 percent more land under the warmer and windier conditions anticipated with a doubling of carbon dioxide concentration in the atmosphere. The researchers stressed that their projections "represent a *minimum* expected change, or best-case forecast. In addition to the increased suppression costs and economic damages, changes in fire severity of this magnitude would have widespread impacts on vegetation distribution, forest condition, and carbon storage, and greatly increase the risk to property, natural resources, and human life" (Fried et al. 2004, 169). The researchers projected at least a 50-percent increase in out-of-control fires in the south San Francisco Bay Area and a 125 percent increase in the Sierra Nevada foothills, with a more than 40-percent increase in the area burned. The state's northern coast saw no significant change under the computer model and conditions used in the study (Thompson 2003).

The study's projections use conservative forecasts that do not consider increased lightning strikes and the spread of volatile grasslands into areas now dominated by less flammable vegetation. Even potentially wetter winters simply mean more growth, providing additional fuel for summer fires, according to the study. "Fires may be hotter, move faster, and be more difficult to contain under future climate conditions," said Robert Wilkinson of the University of California–Santa Barbara's School of Environmental Science and Management in a federal report on the impact of climate change on California. "Extreme temperatures compound the fire risk when other conditions, such as dry fuel and wind, are present" (Thompson 2003). Damage may be aggravated by constriction of homes in brushlands that are vulnerable to fires.

Further Reading

Fried, Jeremy S., Margaret S. Torn, and Evan Mills. "The Impact of Climate Change on Wildfire Severity." *Climatic Change* 64 (May 2004): 169–191.

Hoag, Hannah. "Degree-by-Degree Breakdown of Climate Effects Published." *Nature* 466 (July 22, 2010): 425.

Park, Haeyoun, Damien Cave, and Wilson Andrews. "After Years of Drought, Wildfires Rage in California." *The New York Times*, July 15, 2015. http://www.nytimes.com/interactive /2015/07/15/us/california-fire-season-drought.html?hp&action=click&pgtype =Homepage&module=second-column-region®ion=top-news&WT.nav=top-news.

Rice, Doyle. "Wildfires Decimating Forest Service's Budget." *USA Today*, August 6, 2015, 3A.

Running, Steven W. "Is Global Warming Causing More, Larger Wildfires?" *Science* 313 (August 18, 2006): 927–928.

Schwartz, John. "As Fires Grow, a New Landscape Appears in the West." *The New York Times*, September 21, 2015. http://www.nytimes.com/2015/09/22/science/as-fires-grow-a-new -landscape-appears-in-the-west.html.

Thompson, Dan. "Experts Say California Wildfires Could Worsen with Global Warming." Associated Press, November 12, 2003. (LEXIS).

Vick, Karl, and Sonya Geis. "California Fires Continue to Rage." *Washington Post*, October 24, 2007, A1. http://www.washingtonpost.com/wp-dyn/content/article/2007/10/23/AR20 07102300347_pf.html.

Westerling, A. L., et al. "Warming and Earlier Spring Increase Western U.S. Forest Wildfire Activity." *Science* 313 (August 18, 2006): 940–943.

"Western Wildfires Linked to Atlantic Ocean Temperatures." Environmental News Service, January 3, 2007. http://www.ens-newswire.com/ens/jan2007/2007-01-03-09.asp#anc hor2 (no longer available).

"Wildfires Set to Increase 50 Percent by 2050." NASA Earth Observatory, July 28, 2009. http://earthobservatory.nasa.gov/Newsroom/view.php?id=39650&src=eoa-manews (no longer available).

See also: Australia, Heat and Drought in; Desertification; Drought, United States; Drought, Worldwide; Extreme Weather; Heat Waves, Evidence of; Heat Waves, Forecasts of; Pine Beetles; Wildfires, Worldwide

WILDFIRES, WORLDWIDE

Russia is ablaze, just like the western United States and northern Canada. On one day—July 11, 2012—more than 97 square miles (some 25,000 hectares) of forests were burning in Siberia, according to the Russian Federal Forestry Agency. Wildfires raged uncontrolled in boreal forests in central and eastern Siberia. Smoke from some of the wildfires in Siberia was being pushed across the Pacific Ocean. "On July 8 and 9, 2012," reported NASA's Earth Observatory, "smoke from Siberia arrived in British Columbia, Canada, and caused ground-level ozone to reach record high levels" ("Wildfires in Siberia" 2012).

The summers of 2012 and 2013 brought Russia two of its most severe wildfire seasons after intense heat waves across parts of Siberia. A persistent high-pressure ridge (a "blocking high" in meteorological parlance) pushed temperatures to 90°F as far north as Norilsk near the Arctic Circle. The average daily high in Norilsk in July is 61°F. Although most of Siberia's wildfires had occurred south of 57° north latitude and along the southern edge of the taiga (a broad swath of continental forest) in 2012 and 2013, the heat waves allowed ignition and continuous burning as far north as 65° north.

The higher the temperature, the easier fires can burn "because less energy is required to raise their temperature to the point of ignition. With temperatures soaring in northern Russia, it was easier for previously active fires to continue burning and for lightning to spark new ones" ("Heat Intensifies" 2013). Russia's north latitudes have warmed more rapidly since the 1970s than the world average—approximately 0.51°C per decade compared to the global average per decade of 0.17°C, or three times as fast. A study by Anatoly Shvidenko of the International Institute for Applied Systems Analysis "expect[s] a doubling in the number of forest fires in Russia's taiga forests by the end of the century, as well as increases in the intensity of those fires" ("Heat Intensifies" 2013).

Jon Ranson, a NASA Goddard Space Flight scientist, on a field expedition in Siberia, witnessed the fires firsthand on a flight between St. Petersburg and Krasnoyarsk:

Worldwide Fires and Carbon Emissions

A spectacular number of fires have scorched western North America, Siberia, and other locations in recent years, but fire behavior averaged worldwide has not changed much, according to a global inventory gathered by NASA satellites. On a typical day in August, NASA's Aqua and Terra satellites detect around 10,000 active fires.

"In 2014, fires released about 2,030 teragrams of carbon into the atmosphere," NASA reported.

That's just slightly below the 2001–2013 average of 2,034 teragrams per year. While global emissions were average in 2014, North America and the Indonesian archipelago saw a very active fire season. South America and northern Africa had an unusually quiet season. In North America, the burning was centered on Canada's Northwest Territories, where low winter precipitation, high summer temperatures, and low summer rainfall combined to produce a fierce fire season. . . . Meanwhile, fire emissions in South America were 41 percent below the 2000–2013 average. Emissions have been declining since 2005 as deforestation rates in Brazil have declined as well. The decline in emissions from equatorial Africa is part of a downward trend driven by the conversion of savanna to cropland, which reduces the amount of fuel available for fires. ("Taking Stock" 2015)

Further Reading

"Taking Stock of 2014 Fire Emissions." NASA Earth Observatory, August 11, 2015. http://earthobservatory.nasa.gov/IOTD/view.php?id=86388&src=eoa-iotd.

We left St. Petersburg near sunset, with the sun low on the horizon. Smoky sunsets create very red skies, and the colors were pretty spectacular. Dozens, or maybe hundreds [of fires], colored the sky on July 6. . . . It is sobering to realize that in two years so close together that the taiga has suffered such extreme fires. Is this the result of climate change? Or a freak occurrence? What I know, for sure, is that these fires appear to be consuming a lot of forest. They must be releasing a whole lot of carbon into the atmosphere, and what happens here does affect the rest of our world. ("Wildfires in Siberia" 2012)

Natural wildfires and those started by people (usually to clear farmland) in Malaysia, Indonesia, Borneo, and Papua New Guinea between 2000 and 2006 contributed as much carbon dioxide to the atmosphere as all other human sources combined ("NASA Study" 2009), according to data from a carbon-detecting NASA satellite and computer models. The results were presented in December 2008 in *Proceedings of the National Academy of Sciences*.

Wildfires and Boreal (Northern) Forests

A report by Greenpeace International suggests that between 50 percent and 90 percent of Earth's existing boreal forests are likely to disappear if atmospheric levels of carbon dioxide and other greenhouse gases double. These forests make up one-third of Earth's remaining tree cover, approximately 15 million square kilometers across Russia, Canada, the United States, Scandinavia, China, Mongolia, Japan, and parts of the Korean peninsula. Large forests also cover many mountain ranges outside of these areas. In total, boreal forests cover some 10 percent of the world's land area.

The Greenpeace report indicates that global warming's toll on boreal forests had begun by the early 1990s. The report warns that decaying forests may provide an extra boost to rising carbon dioxide levels, causing warming to feed on itself. The decline of boreal forests also endangers more than 1 million indigenous people who live in them, including the Dene and Cree of Canada; the Sami (Lapplanders) of Norway, Sweden, and Finland; the Ainu of northern Japan; and the Nenets, Yakut, Udege, and Altaisk of Siberia (Jardine 1994). Some larger forest animals such as the Siberian tiger are already near extinction. The Greenpeace report concludes that rapid logging of the boreal forests is intensifying pressure on animal life and accelerating the release of even more carbon dioxide and other greenhouse gases into the atmosphere.

If boreal forests continue to decline, their burning and rotting could contribute to the release of as much as 225 billion tons of extra carbon dioxide into the atmosphere, raising current levels by one-third and accelerating the pace of warming. Trees could have difficulty colonizing the thawing tundra to the north of their current ranges because they simply cannot migrate quickly enough and because the treeless tundra cannot evolve quickly enough to sustain them.

Wildfires Inhibit Rainfall

Wildfires produce small particles called *aerosols* that inhibit the formation of rain-producing cumulous clouds and intensify droughts that accelerate fire. Michael Tosca and colleagues (2015) used data from three satellites and ground-based weather stations, along with an atmospheric model, to study how aerosols in smoke affect the development of low-level cumulus clouds over land. In tropical regions, these clouds form in midmorning, peak in late afternoon, and dissipate by evening, often producing rain in the process.

The researchers found that the presence of smoke coincided with sharp decreases in the amount of cloud cover (Tosca et al. 2015). "Fire-emitted particles crippled the atmosphere's ability to build clouds and thunderstorms, and that ultimately caused a decrease in rainfall during what's already a seasonal drought," Tosca said. "While this 'smoky' image just shows one day, remember that these conditions persist essentially unabated from December through February. Taken together, it represents a huge climate perturbation."

Tosca attributes the reduction in cloud cover and rainfall to a phenomenon known as the *semidirect aerosol effect*. The dark soot particles in smoke absorb incoming sunlight and warm the atmosphere. This warming reduces the upward movement of moisture, which is crucial to the development of cumulus clouds.

Tosca and colleagues wrote in *Geophysical Research Letters* (2015):

> It is well established that smoke particles modify clouds, which in turn affects climate. . . . Here for the first time, we use temporally offset satellite observations from northern Africa between 2006 and 2010 to quantitatively measure the effect of fire aerosols on convective cloud dynamics. We attribute a reduction in cloud fraction [formation] during periods of high aerosol optical depths to a smoke-driven inhibition of convection. We find that higher smoke burdens limit upward vertical motion, increase surface pressure, and increase low-level divergence— meteorological indicators of convective suppression.

Further Reading

Tosca, M. G., et al. "Human-caused Fires Limit Convection in Tropical Africa: First Temporal Observations and Attribution." *Geophysical Research Letters* 42(15) (August 5, 2015). http://onlinelibrary.wiley.com/doi/10.1002/2015GL065063 /abstract#footer-support-info.

Kevin Jardine sketched the role of insect outbreaks in the anticipated destruction of boreal forests. "Pests [that] may invade boreal forests under warming conditions include the western spruce budworm, the Douglas-fir tussock moth, and the mountain pine beetle" (Kurz 1995, 127). Kurz et al. concluded that "biospheric feedbacks from temperate and boreal forest ecosystems will be positive feedbacks that further enhance the carbon content of the global atmosphere" (Kurz 1995, 129). By 1998, spruce budworms had devoured 50 million acres (20 million hectares) of Alaskan woodland.

> In moderation, the forest usually benefits from insect outbreaks because they reduce the likelihood of catastrophic fire by helping to eliminate older stands and diseased trees. However, populations of insects such as bark beetles, the Siberian silkworm and the spruce budworm can explode and devastate millions of acres of forest. The life cycles of the spruce budworm (*Choristoneura fumiferana*) and the spruce bark beetle (*Ips typographus*) are strongly influenced by climate, with both species likely to increase in numbers during the kind of warmer, drier weather predicted in a global warming world. . . . A Canadian government study released in 1987 showed that conifers were growing on average up to 65 percent slower than in the 1940s and 1950s, because of spruce budworm outbreaks, and possibly acid rain. (Jardine 1994)

In 2007, fires north of Alaska's Brooks Range singed 1,000 squares kilometers, a larger area than all recorded fires on Alaska's North Slope during the previous half-century, adding 1.3 million tons of carbon dioxide to the atmosphere. The number of tundra fires is accelerating, according to Adrian Rocha of the Marine Biological Laboratory in Woods Hole, Massachusetts. Because "projected changes in climate over the next century—increased aridity, thunderstorms, and warming in the Arctic—will increase the likelihood that these thresholds will be crossed and thus result in more larger and frequent fires" (Logan 2009).

A year after such a fire, "the most severely burned tundra emitted twice as much carbon as undamaged tundra normally stores away" (Logan 2009). William Bowden, a wetland ecologist of the University of Vermont, said that fire-blackened surfaces increase the amount of water in soil. "Both factors should promote soil warming, permafrost thaw, and possible thermokarst formation," he said (Logan 2009). (Thermokasts are areas of collapsed terrain where structurally important permafrost has thawed—a process that can damage the foundations of homes, roads, and pipelines. Permafrost melt will also increase the amount of greenhouse gases such as methane entering the atmosphere.) (Logan 2009).

"The amount of area burning now in Siberia is just startling—individual years with 30 million acres burned," said Thomas W. Swetnam, an expert on forest history at the University of Arizona. Thirty million acres is an area the size of Pennsylvania. "The big fires that are occurring in the American Southwest are extraordinary in terms of their severity, on time scales of thousands of years. If we were to continue at this rate through the century, you're looking at the loss of at least half the forest landscape of the Southwest" said Swetnam (Gillis 2011).

Justin Gillois of *The New York Times* described devastation that extended worldwide in 2011 as forests burned:

The great euphorbia trees of southern Africa are succumbing to heat and water stress. So are the Atlas cedars of northern Algeria. Fires fed by hot, dry weather are killing enormous stretches of Siberian forest. Eucalyptus trees are succumbing on a large scale to a heat blast in Australia, and the Amazon recently suffered two "once a century" droughts just five years apart, killing many large trees.

Trees are roughly 50 percent carbon and consume CO_2 when alive; a net absorber of carbon dioxide is lost each time a tree dies. Forests that burn and decay transform from carbon sinks to sources. Trees that survive have been weakened by drought, consume less carbon dioxide, and become prey to insect pests such as bark beetles, which swarm in increasing numbers and vigor as warming temperatures shorten their reproductive cycle. By 2011, in British Columbia alone, an area of pine forest the size of Wisconsin was infested with bark beetles (Gillis 2011).

Heat and drought feed wildfires and result in deforestation. In 2011, from the U.S. Southwest downslope into Texas, wildfires scorched dry land, burning millions of acres. By the end of the summer, 15 percent of Colorado's aspen forests were dead or dying (Gillis 2011).

Further Reading

Gillis, Justin. "The Threats to a Crucial Canopy." *The New York Times*, October 1, 2011. http://www.nytimes.com/2011/10/01/science/earth/01forest.html.

"Heat Intensifies Siberian Wildfires." NASA Earth Observatory, August 2, 2013. http://earthobservatory.nasa.gov/IOTD/view.php?id=81736&src=eoa-iotd.

Jardine, Kevin. "The Carbon Bomb: Climate Change and the Fate of the Northern Boreal Forests." Ontario, Canada: Greenpeace International, 1994. http://dieoff.org/page129.htm.

Kurz, Werner A., et al. "Global Climate Change: Disturbance Regimes and Biospheric Feedbacks of Temperate and Boreal Forests." Pp. 119–133 in George M. Woodwell and Fred T. MacKenzie, eds., *Biotic Feedbacks in the Global Climate System: Will the Warming Feed the Warming?* New York: Oxford University Press, 1995.

Logan, Tracey. "Alaska's Biggest Tundra Fire Sparks Climate Warning." *New Scientist*, July 30, 2009. http://www.newscientist.com/article/dn17537-alaskas-biggest-tundra-fire-sparks-climate-warning.html?DCMP=OTC-rss&nsref=environment.

"NASA Study Says Climate Adds Fuel to Asian Wildfire Emissions." NASA Earth Observatory, April 30, 2009. http://earthobservatory.nasa.gov/Newsroom/view.php?id=38433&src=eoa-nnews.

"Wildfires in Siberia." NASA Earth Observatory, July 12, 2012. http://earthobservatory.nasa.gov/IOTD/view.php?id=78515&src=eoa-iotd.

See also: Australia, Heat and Drought in; Desertification; Drought, United States; Drought, Worldwide; Extreme Weather; Heat Waves, Evidence of; Heat Waves, Forecasts of; Pine Beetles; Wildfires, U.S.

GLOBAL WARMING

OVERVIEW

What a show Hollywood could make of global warming. If only the West Antarctic ice sheet (covering an area roughly the size of Mexico) would disintegrate all at once, producing an overnight rise in sea levels of perhaps 18 to 20 feet. Such an event would be catastrophic for many millions of people around the world who live in large coastal urban areas from New York City to Shanghai. The rising oceans would inundate parts of intensely farmed river deltas such as the Nile and the Ganges. In the real world, however, the melting of the West Antarctic ice sheet is likely to take several centuries to unfold and thus reach its climax long after the last members of any present-day audience have left the theater. Ice turns to water one prosaic drop at a time, although scientists by 2014 were telling us that this script has already been written in the theatre of thermal inertia.

As the 21st century opened, ice was melting around the world. The Arctic ice cap was thinning, and scientists were speculating about how many years would pass before a summer during which all Arctic ice melts. Models developed by the U.N. Intergovernmental Panel on Climate Change project that between one-third and one-half of existing mountain glacial mass could disappear over the next 100 years. Sometime during this century, the last glacier may melt in Glacier National Park.

The melting of the world's ice lacks any Hollywood-style flair, but the eventual result may be climate change on a scale that Earth has not experienced since the Pliocene Epoch some 2 million to 3 million years ago. Human-induced climate change may be ending the cycle of glaciation that has been the norm on Earth for millions of years. Viewed on a timescale of the last 540 million years, however, glacial cycles have been relatively rare events. For three-quarters of the last 540 million years, Earth's ice caps have been negligible or nonexistent (Huber et al. 2000, xi).

Kilimanjaro, Snows of

Since the 1930s, Mount Kilimanjaro has lost 90 percent of its glacial mass. A third of the ice cap that Ernest Hemingway celebrated melted during the 12 years ending in 2006. By 2002, the mountain had lost 82 percent of the volume of its ice cap since it was first carefully measured in 1912, according to glaciologist Lonnie Thompson. Kilimanjaro's ice field shrank from 12 square kilometers in 1912 to only 2.6 square kilometers in 2000, reducing the height of the mountain by several meters. The ice covering the 19,330-foot peak "will be gone by about 2020," said Thompson (Arthur 2002).

By 2009, the ice cap on top of Kilimanjaro was nearly gone, having shrunk 26 percent since 2000 and 85 percent since 1912, long before Ernest Hemmingway penned the short story published in 1938 that made the mountain an icon of tourism and literature. A study in the *Proceedings of the National Academy of Sciences* (PNAS), published online in November 2009 discussed several causes for the erosion of the mountain's ice, one of which is general global warming. Summit ice cover (areal extent) decreased approximately 1 percent per year from 1912 to 1953 and around 2.5 percent per year from 1989 to 2007 (Thompson et al. 2009). The study results indicate that rising temperatures as well as sublimation—evaporation of ice directly into the atmosphere—are playing a role. The remnants of four off-white glaciers is all that remains of a landscape that one character in Hemingway's story said was "as wide as all the world, great, high, and unbelievingly white" (Rice 2009).

The authors of the *PNAS* paper commented:

The relative importance of different climatological drivers remains an area of active inquiry, yet several points bear consideration. Kilimanjaro's ice loss is contemporaneous with widespread glacier retreat in mid to low latitudes. The Northern Ice Field has persisted at least 11,700 years and survived a widespread drought about 4,200 years ago that lasted about 300 years. We present additional evidence that the combination of processes driving the current shrinking and thinning of Kilimanjaro's ice fields is unique within an 11,700-year perspective. If current climatological conditions are sustained, the ice fields atop Kilimanjaro and on its flanks will likely disappear within several decades. (Thompson et al. 2009)

Further Reading

Arthur, Charles. "Snows of Kilimanjaro Will Disappear by 2020, Threatening World-Wide Drought." *The Independent* (London), October 18, 2002, 7.

Rice, Doyle. "Hemingway's Famous Icy Peaks Are Thawing Fast." *USA Today*, November 3, 2009, 7D.

Thompson, Lonnie G., et al. "Glacier Loss on Kilimanjaro Continues Unabated." *Proceedings of the National Academy of Sciences.* Published online before print November 2, 2009. http://www.pnas.org/content/early/2009/10/30/0906029106.abstract. doi: 10.1073/pnas.0906029106.

The speed of ice melt is important not only because it harms habitat but also because its erosion changes Earth's carbon cycle, usually in ways that accelerate the rise in atmospheric greenhouse gases worldwide. For example, melting Arctic sea ice exposes darker water, which absorbs more heat. One-third of Earth's carbon is stored in far northern latitudes (mainly in tundra and boreal forests) (Mack et al. 2004, 440), so the speed at which ecosystem warming may release this carbon to the atmosphere is vitally important in forecasting global warming's pace and effects. The amount of carbon stored in Arctic ecosystems equals two-thirds of the amount currently found in the atmosphere (Loya and Grogan 2004, 406). Its release into the atmosphere will depend on the pace of temperature rise—and the Arctic, according to several sources, has been the most rapidly warming regions on Earth.

The ice world of the Inuit is melting. Ice is fundamental to the entire Arctic food chain, including to human beings who cannot afford to live on expensive food flown in from the lower latitudes:

> Melt the ice and everything changes. The food chain is disrupted as food sources change at all levels. The lives of anyone who lives in this environment change as well, as hunters are unable to rely on once-familiar sources of sustenance and income. The entire landscape (which is shaped by ice, especially in the winter) changes as well. The thickness of ice changes, creating risk of drowning in frigid waters. Hunters can no longer "read" the sky and become disoriented. The open-water fishing season grows longer, and the maritime catch includes creatures heretofore unknown in Arctic waters.

"We know the ice can change fast," said Eric Rignot, professor of earth sciences at the University of California–Irvine. "We've never seen it. No human has ever seen it." Rignot is fairly confident, however, that we are seeing it now—a conclusion borne out by the ice-sheet data he scrutinizes every week. A few decades from now, he said, we may look back with regret, wondering why more of us did not acknowledge the signs all around us, why we did not see "that the collapse had already started." Greenland by 2015 was losing 303 billion tons of ice a year.

Describing the tendency of most scientists to avoid speaking bluntly about ice-sheet collapse, Rignot told Jon Gertner of *The New York Times Sunday Magazine* (2015),

> You can fiddle around and say, "It's going to take a long time" or "We don't know." But even the most conservative people in our community will tell you: "We warm the climate by two or three degrees C [and] Greenland's ice is gone." He was suggesting to me, in other words, that . . . two degrees Celsius would not halt the glacial decline. . . . And five degrees? There's no way Antarctica is going to stay if we warm up by five degrees.

The effects of melting ice often play out far from mass audiences in the midlatitudes. For example, with ice receding hundreds of miles offshore of Alaska and Russia during several recent summers, walruses gathered by the thousands onshore in Alaska and Siberia. Joel Garlich-Miller, a walrus expert with the U.S. Fish and

Wildlife Service, said that walruses began to gather onshore late in July 2007, a month earlier than usual. A month later, their numbers had reached record levels from Barrow to Cape Lisburne, 300 miles southwest, on the Chukchi Sea. This gathering has recurred several summers in the years since. By 2015, television cameras had gone elsewhere and the now-annual walrus migration to shore received a one-inch mention in *The New York Times*. The gathering of 35,000 animals by then was being called a "September phenomenon" by the U.S. Fish and Wildlife Service ("Alaska" 2015).

The effects of melting ice are not restricted to polar regions. A good look at rapid mountain glacier erosion in temperate regions worldwide has been provided by researchers from Turkey's Ege University and NASA's Goddard Space Flight Center. Their analysis of 41 years of Landsat data documented a decline of 14 glaciers in Turkey from 25 square kilometers (10 square miles) to 11 square kilometers (4.19 square miles) (Yavasli et al. 2015). Five of the glaciers have melted completely. The researchers attributed glacier recession to higher minimum summer temperatures with no changes in precipitation or cloud cover over this period. By 2015, only two substantial glaciers remained in Turkey ("Turkish Glaciers" 2015). The glacier on Mount Süphan, a dormant volcano, has lost 75 percent of its ice, and what remains is now covered by debris. And so it is around the world as ice melts at the poles and mountains, usually quietly and out of human sight.

Although warmer air holds more moisture, not everyone will see more precipitation in a globally warmed world. Many deserts already are expanding in a worldwide pattern influenced by atmospheric circulation patterns that meteorologists call *Hadley cells*.

Most deserts range between 20° and 40° north and south latitude. Whereas precipitation patterns are also influenced by other factors (such as ready access or lack thereof to ocean-borne moisture), rainfall is strongly influenced by Hadley cells, which determine whether air generally rises or falls at certain latitudes. Rising air portends instability, low pressure, and storminess; descending air generally provides high pressure and clear skies. In a warmer world, Hadley cells expand, which causes deserts to expand, a process that is already evident from news reports around the world.

As surface temperatures warm, thunderstorms are becoming larger and more violent, and thunderheads are building higher into the atmosphere. Thunderstorms are one way that the atmosphere discharges heat, and more heat makes them larger and more intense. Where previously thunderheads usually stopped growing within the layer of the atmosphere closest to the surface (the troposphere), they now sometimes build into a boundary level (the tropopause) and then inject moisture into the stratosphere, which accelerates depletion of ozone by speeding up reactions based on the presence of chlorofluorocarbons (CFCs) used before the ban by the Montreal Protocol in the late 1980s.

Many climate scientists believe that the middle of the 21st century will witness dramatic acceleration in global warming. In part, this acceleration will result from exhaustion of various natural "sinks" that have been absorbing greenhouse gases. At about the same time, various feedback loops are also expected to accelerate natural increases in atmospheric greenhouse gas levels and, consequently, worldwide temperatures. These include several natural processes that add greenhouse gases to the atmosphere such as melting permafrost in the Arctic and, in the far future, possible gasification of solid methane deposits (clathrates) in the oceans, as well as an increasingly dark Arctic Ocean that will absorb more heat as the ice cap melts.

The El Niño–La Niña (El Niño Southern Oscillation) climate cycle is Earth's most prominent source of short-term climate change. It alters upper-air storm tracks, bringing droughts to areas that usually receive abundant rainfall, and deluges to some of the driest deserts in the world. El Niño, which warms the Pacific Ocean near the equator west of South America and then extends westward, generally raises worldwide temperatures. La Niña, which occurs in the same area, usually has the opposite effect. These conditions often set in precisely within days of Christmas, thus their name—a reference in Spanish to the Christ child.

In 2015 and 2016, global temperatures surged and had a powerful assist from a strong El Niño pattern in the equatorial Pacific Ocean, which has raised ocean temperatures in that area as much as 3°C (5.4°F). above 20th-century averages, an all-time record. Global air temperatures set unprecedented highs on the instrumental record (since 1880) in 2015, far above previous El Niño years such as 1997 and 1998, indicating an underlying warming trend from rising greenhouse gas levels in the atmosphere.

These are some of the ways in which a rise in temperatures spurred by fossil fuel use has been and will influence the lives of every living thing on planet Earth.

Further Reading

"Alaska: Walrus Again Crowd onto Shore." *The New York Times*, September 11, 2015, A15.

Gertner, Jon. "The Secrets in Greenland's Ice Sheet." *The New York Times Sunday Magazine,* November 15, 2015. http://www.nytimes.com/2015/11/15/magazine/the-secrets-in -greenlands-ice-sheets.html.

Huber, Brian T., Kenneth G. MacLeod, and Scott L. Wing. *Warm Climates in Earth History.* Cambridge, UK: Cambridge University Press, 2000.

Loya, Wendy M., and Paul Grogan. "Carbon Conundrum on the Tundra." *Nature* 431 (September 23, 2004): 406–407.

Mack, Michelle C., et al. "Ecosystem Carbon Storage in Arctic Tundra Reduced by Long-Term Nutrient Fertilization." *Nature* 432 (September 23, 2004): 440–443.

"Turkish Glaciers Shrink by Half." NASA Earth Observatory, July 2, 2015. http://earthobservatory.nasa.gov/IOTD/view.php?id=86140&src=eoa-iotd.

Yavasli, D. D., C. J. Tucker, and K. A. Melocik. "Change in the Glacier Extent in Turkey during the Landsat Era." *Remote Sensing of Environment* 163 (June 15, 2015): 32–41. http://www.sciencedirect.com/science/article/pii/S003442571500098X.

ATMOSPHERIC CIRCULATION

Although warmer air holds more moisture, not everyone will see more precipitation in a globally warmed world. Many deserts already are expanding in a worldwide pattern influenced by atmospheric circulation patterns that meteorologists call *Hadley cells*.

Most deserts range between 20° and 40° north and south latitude. Although precipitation patterns are also influenced by other factors (such as ready access or lack thereof to ocean-borne moisture), rainfall is strongly influenced by Hadley cells, which determine whether air generally rises or falls at certain latitudes. Rising air portends instability, low pressure, and storminess; descending air generally provides high pressure and clear skies. In a warmer world, Hadley cells expand, which causes deserts to expand, a process that is already evident from news reports around the world.

Droughts around the World

Droughts in regions where Hadley cells favor descending air now span the globe from Australia to Spain, Iraq, Afghanistan, parts of China, and the American Southwest, including California. In June 2008, California declared a drought and warned that water rationing would follow its driest spring in 88 years. Los Angeles announced plans to use cleansed sewage water to augment supplies ("Hunger" 2008).

In Iraq, drought caused partially by changes in Hadley cells has been aggravated by overpumping of the Euphrates River upstream in Turkey and Syria, local mismanagement of the river's flow by farmers, and increasing evaporation because of a warming climate (Robertson 2009). In China, the Gobi desert is also within the northern reaches of Hadley cells and has been expanding, sending occasional dust storms into Beijing and aggravating air pollution from coal-fired power plants that give that and other cities in China the dirtiest air on Earth.

Drought-stricken northern China usually grows three-fifths of China's crops. The current drought, probably the worst in northern China in at least 50 years, is crippling the country's best wheat farmland and wells that supply water to industry and millions of people. China also has been wasting water for decades, as it depletes aquifers, some of which have now receded to half a mile below the surface. Between October 2008 and late February 2009, the area went 100 days without rain or snow. Beijing's winter snows have become rare in the last 25 years. Many people in rural areas haul water from distant taps in plastic bags (Wines 2009).

By 2009, the agricultural breadbasket of Argentina, which, like most of Australia, lies within Hadley cell updrafts, was suffering its worst drought since the 1930s. Farmers of corn, wheat, and soybeans lost $5 billion in one year, and cattle starved in fields that had turned to dust. "Nothing edible grows," Hilda Schneider, a rancher who lost 500 head to cattle ("Cow Skulls" 2009). In mid-2008, several thousands people were forced out of their homes because of food and water shortages in northern Afghanistan. Large parts of the country received much less than normal rain and snow. Nearly 2,000 families departed the drought-stricken Chemtal district of Balkh province in one week late in May 2008 even after the government donated 37 tons of food and water. Water and food shortages have forced several families to eat grass, and some people died of starvation. Many farm fields were ravaged by locusts as well.

Drought intensified in the Fertile Crescent region of northern Iraq and eastern Syria during the winter of 2007–2008 as sufficient winter rains and snows failed to arrive in the mountains of Turkey, which have fed rivers in the area since the first beginnings of human urban civilization. Lack of moisture also limited irrigation, which is crucial for agriculture in the desert.

Barcelona, Spain, was so dry by May 2008 that water was being imported for the first time ever by ship. A desalinization plant and pipeline from the Ebro River (to the west) was being planned. Parts of southeastern Spain were turning to desert by 2008, even amid new developments of vacation homes and golf courses. Water has become a valuable commodity and source of copious conflict. This area has experienced cyclical droughts in the past, but this one may be long-lasting. Local aquifers are retreating below the range of pumps as well as the area's climate comes to resemble that of northern Africa.

Local aquifers in Spain are retreating below the range of pumps as well as the climate comes to resemble that of northern Africa. The Spanish environment ministry warned that a third of the country may turn to desert in coming years (Rosenthal 2008). A desalinization plant and pipeline from the Ebro River (to the west) were being planned ("Drought Forces" 2008).

Drought Spreads in Brazil

By 2014, southeastern Brazil, which lies within an expanding Hadley cell south of the equator, was suffering its worst drought in a century centered near the rapidly growing city of São Paulo, with almost 20 million people, whose reservoirs were running out of water, some of them at 3 percent to 5 percent of capacity. Rainfall for the year was 12 to 16 inches below average by late October. NASA's Earth Observatory ("Drought Shrinking" 2014) reported that "[across] southeastern Brazil, production of key crops like coffee and sugar are in steep decline, and citizens are facing periodic outages in the water supply—even as news agencies report that local water authorities have not instituted conservation measures." The usual rainy season (which begins in late September on average) had not begun a month later.

By February 2015, São Paulo's largest reservoir system was nearly depleted, as *The New York Times* described an increasingly desperate situation:

> Many residents are already enduring sporadic water cutoffs, some going days without it. Officials say that drastic rationing may be needed, with water service provided only two days a week. Behind closed doors, the views are grimmer. In a meeting recorded secretly and leaked to the local news media, Paulo Massato, a senior official at São Paulo's water utility, said that residents might have to be warned to flee because there's not enough water, there won't be water to bathe, to clean homes.

By this time, southeastern and eastern Brazil was enduring its third year of intense drought as the Cantareira reservoir system, the main water supply for half of São Paulo, had shrunk to 10 percent of capacity, even at the end of the usual rainy season (it had been 5 percent in October 2014) ("Water Levels" 2015). As with many water shortages, the one in São Paulo had several causes—population growth, polluted rivers in the city, deforestation in the watershed, and massive leaks in the water-supply system. The Tietê and Pinheiros Rivers stink so badly from pollution that some passersby vomit. Deforestation tends to reduce rainfall. The main underlying problem, however, is lack of water provoked by changing atmospheric circulation patterns related to global warming. "Climate change has arrived to stay," Geraldo Alckmin, the governor of São Paulo state, said in February 2015. "When it rains, it rains too much, and when there's drought, it's way too dry" (Romero 2015).

Pakistan and Iran: Water-Starved Countries

Consider the plight of another rapidly growing urban area half a world away. On the outskirts of Karachi, Pakistan, with 18 million people, according to the Associated Press, "protesters burning tires and throwing stones have what sounds like a simple demand: They want water at least once a week. But that's anything but in Karachi, where people go days without getting water from city trucks, sometimes forcing them to use groundwater contaminated with salt" (Jawad 2014).

The recurring drought made the problem worse. And as the city rapidly grows, the water shortages have intensified. Karachi's major water source, the Indus River, was shriveling because of drought. By 2014, the river was able to supply less than half the city's needs, forcing imports from underground aquifers that are being rapidly depleted. Increasingly, criminals are tapping the aquifers and profiting from the scarce water. Some had switched from illegal drugs to water. Many poor people reported walking several miles a day to receive only as much water as they could carry. Areas that depend on monsoon rains such as Pakistan and India are finding that the rains have become irregular and less predictable, with droughts sometimes alternating with devastating floods. Meanwhile, growing populations and industrial production drive the rising demand for scarce water.

Nearly 80 percent of India's water is used in agriculture. Increasing affluence also increases water demand per person as more people buy water-intensive animal protein. The manufacturing of consumer goods and the generation of electric power, both of which are increasing, also require more water. "The Himalayan glaciers are receding, agricultural yields are stagnating, dry days have increased, patterns of monsoon have become more unpredictable," Jairam Ramesh, then minister of environment and forests, told the Mint newspaper in 2009. "So, we are seeing the effects" of climate change (Asokan 2012).

Except for occasional, irregular floods during the monsoon season, Pakistan is becoming "a water-starved country," its government warned in 2015 (Masood 2015). Local deforestation, waste, and mismanagement was compounding the problem, along with drought—all problems familiar to residents of São Paulo. In addition, glaciers that supply Pakistan's Indus River have been melting, portending more intense water shortages in the future. Pakistan also stores little water and has not built a new dam in almost 50 years.

"It is a very serious situation," said Pervaiz Amir, country director for the Pakistan Water Partnership. "The frequency of monsoon rains has decreased but their intensity has increased," said Amir. "That means more water stress, particularly in winters." Water has become a political issue with jihadi leaders accusing India of constricting the headwaters of the Indus, which originate in its territory. Hafiz Saeed, the leader of Lashkar-e-Taiba, a militant group that carried out deadly attacks in Mumbai, India, in 2008, criticized Indian "water terrorism" at public rallies (Masood 2015).

Drought in Iran

By 2015, a seven-year drought in Iran (within the Northern Hemisphere's Hadley zone) had scorched areas in the country's southwestern region where pistachio crops were being sustained on rapidly depleting groundwater because rainfall has ebbed and temperatures have risen. By the end of 2015, Iran had depleted 70 percent of its groundwater in half a century. In areas where groundwater had been tapped out, "[seemingly] endless fields . . . are filled with dead trees, their bone-colored branches a deathly contrast to the turquoise sky. Farms have been abandoned and squatters have moved in. Wind-driven dust is polluting many of Iran's cities, including Tehran" (Erdbrink 2014).

As reporter Thomas Erdbrink reported, in Zanjan Province northwest of Tehran, the historic Mir Baha-eddin Bridge

> crosses a riverbed of sand, stones and weeds. In Gomishan, on the shores of the Caspian Sea, the fishermen who once built houses on poles surrounded by freshwater now have to drive for miles to reach the receding shoreline. In Urmia, close to the Turkish border, residents have held protests to demand that the government return water to a once-huge lake that is now the source only of dust storms. In Tehran, reservoirs have reached a level that soon may force many people to live without water. The changing landscape is all too visible in Kerman Province. (Erdbrink 2014)

"In a not-so-distant past, the area was a beltway of green stretching for hundreds of square miles, using groundwater to produce grain and pistachios. Now, the sun bakes treeless plains that are increasingly giving way to deserts," Bridge said (Erdbrink 2014). Kerman Province remains one of the largest producers of pistachios in the world, but no other option remains once groundwater has been exhausted.

In Iran, only 5 percent of Lake Urmia's water remains, where "as recently as a decade ago, cruise ships filled with tourists plied the lake's waters in search of flocks of migrating flamingos. Now, the ships are rusting in the mud and the flamingos fly over the remains of the lake on their way to more hospitable locales" (Erdbrink 2014). The lake once was 90 miles long and 35 miles wide, the size of Utah's Great Salt Lake. Iran is also under an expanding Hadley cell and has been suffering a drought serious enough to stimulate plans for water rationing in Tehran, also now a megacity with more than 20 million people, and other Iranian cities. As is usually the case, climate change's effects (rising temperatures as well as drought) are being compounded by leakage, dam construction, political corruption, and wasteful water usage, including inefficient irrigation that contribute to shortages even with average rainfall.

"Major rivers near Isfahan, in central Iran, and Ahvaz, near the Persian Gulf, have gone dry, as has Hamoun Lake, in the Afghanistan border region. Dust from the dry riverbeds has added to already dangerously high air pollution levels in Iran, home to four of the 10 most polluted cities in the world," wrote *New York Times* reporter Thomas Erdbrink (2014).

Hadley Cells and Epic Drought in the U.S. Southwest

Hadley cell expansion also plays a role in intensifying long-term drought in California and other parts of the U.S. Southwest. Many reservoirs were down to a fraction of their capacity by 2014, including the Central Valley's Folsom Lake, which was down to 17 percent. According to one report, "Cattle ranchers have had to sell portions of their herd for lack of water. Sacramento and other municipalities have imposed severe water restrictions. Wildfires broke out this week in forests that are usually too wet to ignite. Ski resorts that normally open in December are still closed; at one here in the Sierra Nevada that is open, a bear wandered onto a slope full of skiers last week, apparently not hibernating because of the balmy weather" (Onishi and Wollanjan, 2014). The drought was the most intense on record.

Snowpack in the Sierra Nevada was less than 20 percent of average and wrecking havoc with water supplies for more than two-thirds of the state's 36 million people and an agricultural industry worth $445 billion a year. "We've got a ridge of high pressure sitting overhead, and it is not budging," said Diana Henderson, a forecaster with the National Weather Service (Onishi and Wollanjan 2014).

In Las Vegas, Nevada, with its grand outdoor pools, ersatz waterfalls, and air-conditioners humming in 120-degree heat, always a symbol of American indulgence and damn-nature ecological arrogance, lack of water has become a problem. Water planners now realize that within a few decades this grand cultural artifact may run out of water. The Colorado River, the principal water source for a large part of the American Southwest, is slowing to a muddy trickle in a drought that may be the worst in 1,250 years—since the two dry centuries that felled the Anasazi and (farther south) the Maya. The same warm spell lured the Vikings to Greenland, only to trap them in the Little Ice Age.

Lake Mead, which was created behind Hoover Dam after its construction in the 1930s, supplies more than 90 percent of Las Vegas's water. It was last full in 1998. Since then, Lake Mead had lost 60 percent of its water, 130 vertical feet, as of 2015 (Owen 2015, 53). Las Vegas has been pumping Lake Mead dry with a series of pipes, each one lower than the last. The latest pipe array, completed in 2016, reaches under the lake, raising the possibility that Las Vegas may someday drain Lake Mead entirely dry, acting as if a plug has been removed in a gigantic bathtub.

As *New York Times* reporter Michael Wines noted in 2014, "Reservoirs have shrunk to less than half their capacities, the canyon walls around them ringed with white mineral deposits where water once lapped. Seeking to stretch their allotments of the river, regional water agencies are recycling sewage effluent, offering rebates to tear up grass lawns and subsidizing less thirsty appliances from dishwashers to shower heads." By 2014, for the first time, federal authorities restricted the amount of water entering Lake Mead from Lake Powell, some 180 miles upstream on the Colorado River. Lake Mead supplies water to 40 million people—not only to Las Vegas but also to Los Angeles—and several million acres of irrigated farm and ranchland that produce 15 percent of the U.S. food supply.

Water is already becoming precious, as Wines reported: "Virtually all water used indoors, from home dishwashers to the toilets and bathtubs used by the 40 million tourists who visit Las Vegas each year, is treated and returned to Lake Mead. Officials here boast that everyone could take a 20-minute shower every day without increasing the city's water consumption by a drop." The Las Vegas area's water consumption did not increase between 2000 and 2012, even as it added 400,000 new residents.

Water Vapor and Global Warming

Naturally present in the atmosphere, water vapor is a powerful greenhouse gas that is not measured in many models. Although some climate contrarians assert that the ubiquity of water vapor makes changes in carbon dioxide levels irrelevant—thus turning science on its head—developing science has found that increasing temperatures and humidity levels work together to aggravate the effects of global warming. "Although there continues to be some uncertainty about its exact magnitude," wrote Andrew C. Dessler and Steven C. Sherwood in *Science*, "the water vapor feedback is virtually certain to be strongly positive, with most evidence supporting a magnitude of 1.5 to 2.0 w/m2/K [watts per square meter per degree Kelvin]. [This is sufficient] to roughly double the warming that would otherwise occur. To date, observational records are too short to pin down the exact size of the water vapor feedback in response to long-term warning from anthropogenic greenhouse gases. . . . But evidence for the water-vapor feedback—and the large future warming it implies—is now strong" (Dessler and Sherwood 2009, 1021).

Increases in water vapor levels at Earth's surface also can influence levels in the stratosphere and further aggravate ozone depletion there. Tropical wildfires and slash-and-burn agriculture have helped double the moisture content in the stratosphere over the last 50 years, a Yale University researcher concluded after examining satellite weather data ("Biomass Burning" 2002). "In the stratosphere, there has been a cooling trend that is now believed to be contributing to milder winters in parts of the Northern Hemisphere," said Steven Sherwood, assistant professor of geology and geophysics. "The cooling is caused as much by the increased humidity as by carbon dioxide" ("Biomass Burning" 2002).

Water-vapor assessment by ground, balloon, aircraft, and satellite measurements shows a global stratospheric water-vapor increase of as much as 2 parts per million by volume during the last 45 years, a 75 percent rise. Modeling studies by the University of Reading in England indicate that since 1980 the stratospheric water vapor increase has produced a surface temperature rise that is about half of that attributable to increased carbon dioxide alone.

"Higher humidity also helps catalyze the destruction of the ozone layer," added Sherwood ("Biomass Burning" 2002). Cooling in the stratosphere causes changes to the jet stream that produce milder winters in North America and Europe. By contrast, harsher winters may result in the high latitudes of the Arctic. Sherwood said that about half of the increased humidity in the stratosphere has been attributed to methane oxidation. No one seems to know, however, what has caused the rest of the additional moisture.

"More aerosols lead to smaller ice crystals and more water vapor entering the stratosphere," Sherwood explained. "Aerosols are smoke from burning. They fluctuate seasonally and geographically. Over decades there have been increases linked to population growth" ("Biomass Burning" 2002). Ozone experts in Canada and the United States said findings in Australia are a serious warning about the risks in trying to manipulate parts of the atmosphere in isolation. "You have to look at all these chemicals and see how they interact and evolve over time," said Tom McElroy, an ozone specialist with the Meteorological Service of Canada (Calamai 2002).

Other causes not directly related to human activity also may be increasing stratospheric moisture levels, according to Philip Mote, a University of Washington research scientist. "Half the increase [of water vapor] in the stratosphere can be traced to human-induced increases in methane, which turns into water vapor at high altitudes, but the other half is a mystery," said Mote. "Part of the increase must have occurred as a result of changes in the tropical tropopause, a region about 10 miles above the equator, that acts as a valve that allows air into the stratosphere" ("Most Serious" 2001). "A wetter and colder stratosphere means more polar stratospheric clouds, which contribute to the seasonal appearance of the ozone hole," said James Holton, the late chair of the University of Washington's Department of Atmospheric Sciences and expert on stratospheric water vapor. "These trends, if they continue, would extend the period when we have to be concerned about rapid ozone depletion" ("Most Serious" 2001).

Taller Storms, More Ozone Loss

As time passes, scientists are finding that ozone loss is more complex and that simply eradicating chlorofluorocarbons (CFCs) will not heal the ozone holes over the poles. They have discovered that global warming may accelerate stratospheric ozone loss over the middle latitudes through a process known as *convective injection* that causes moisture from the largest storms to reach the stratosphere.

As surface temperatures warm, thunderstorms are becoming larger and more violent, and thunderheads are building higher into the atmosphere. Thunderstorms are one way by which the atmosphere discharges heat, and more heat makes them larger and more intense. Where previously thunderheads usually stopped growing within the layer of the atmosphere closest to the surface (the troposphere), they now sometimes build into a boundary level (the tropopause) and then inject moisture into the stratosphere, which accelerates ozone depletion by speeding up reactions based on the presence of CFCs remaining from before the ban established by the Montreal Protocol in the late 1980s.

Scientists do not know the degree of damage or how soon we will pay with increased melanoma. Studies of such things soon will be underway. What we do know is that, for the first time, we may be exposed to more ultraviolet radiation—which plays a role in the development of skin cancers—over middle latitudes where hundreds of millions of people live. When James G. Anderson, lead author of the recent study, produced work in 2007 demonstrating this effect, many other scientists were skeptical. As his work was refined, they came to support it.

These supersized thunderstorms carry water vapor into a level of the atmosphere that is usually extremely dry, accelerating the destruction of ozone, which protects land animals, including humans, from skin cancer. An increasing level of carbon dioxide near Earth's surface also acts as a blanket by trapping heat. Deprived of emitted warmth, the stratosphere cools, aggravating the depletion of ozone. Chemical reactions that drive ozone depletion tend to accelerate as the stratosphere cools, retarding the restoration of ozone anticipated after the ban of CFCs under the Montreal Protocol. As levels of greenhouse gases rise, the cooling of the middle and upper atmosphere is expected to continue, with attendant consequences for ozone depletion. Because of this relationship, problems with ozone depletion depend, in a fundamental way, on mitigating greenhouse warming. In 2006, the Antarctic ozone hole was the largest on record, despite the two-decade ban of CFCs.

"The rate of these reactions was shocking to us," James Anderson told Henry Fountain of *The New York Times*. "It's irreversible." This is also one major reason why the ozone did not simply itself after CFCs were banned. "The world said: 'Oh, we've controlled the source of CFCs; we can move onto something else," Anderson said. "But the destruction of ozone is far more sensitive [than previously believed] to water vapor and temperature" (Anderson et al. 2012, 835; Ravishankara 2012, 809).

Mario Molina, whose work during the 1970s linking CFCs to ozone depletion led to a Nobel Prize in chemistry and the CFC ban, told *The New York Times* that the new findings add "one more worry to the changes that society's making to the chemical composition of the atmosphere" of concern to many people worldwide in areas that experience increasingly severe thunderstorms, including a large part of the United States (Fountain 2012, A14).

Further Reading

Anderson, James G., et al. "UV Dosage Levels in Summer: Increased Risk of Ozone Loss from Convectively Injected Water Vapor." *Science* 337 (August 17, 2012): 835–839.

Asokan, Shyamantha. "Indian States Fight over River Usage." *Washington Post*, April 1, 2012. http://www.washingtonpost.com/world/asia_pacific/indian-states-fight-over-river-usage/2013/04/01/73026ae0-9895-11e2-b68f-dc5c4b47e519_print.html.

"Biomass Burning Boosts Stratospheric Moisture." Environment News Service, February 20, 2002. http://ens-news.com/ens/feb2002/2002L-02-20-09.html (no longer available).

Calamai, Peter. "Alert over Shrinking Ozone Layer." *Toronto Star*, March 18, 2002, A8.

"Cow Skulls and Dust: Drought Grips Argentina." *Omaha World-Herald*, January 26, 2009, 3A.

Dessler, Andrew C., and Steven C. Sherwood. "A Matter of Humidity." *Science* 323 (February 20, 2009): 1020–1021.

"Drought Forces Barcelona to Ship in Drinking Water." *Wall Street Journal*, May 14, 2008, A13.

"Drought Shrinking São Paulo Reservoirs." NASA Earth Observatory, October 23, 2014. http://earthobservatory.nasa.gov/IOTD/view.php?id=84564&src=eoa-iotd.

Erdbrink, Thomas. "Its Great Lake Shriveled, Iran Confronts Crisis of Water Supply." *The New York Times*, January 30, 2014. http://www.nytimes.com/2014/01/31/world/middleeast/its-great-lake-shriveled-iran-confronts-crisis-of-water-supply.html.

Fountain, Henry. "A Storm Study Ties Warming to Ozone Loss." *The New York Times*, July 27, 2012, A1, A14.

"Hunger, Water Scarcity Displaces Thousands of Afghans." *The New York Times*, June 4, 2008. http://www.nytimes.com/reuters/world/international-afghan-displacement.html (no longer available).

Jawad, Adil. "Pakistan's Largest City Thirsts for a Water Supply." Associated Press. August 24, 2014. http://www.benningtonbanner.com/ci_26396810/pakistans-largest-city-thirsts-water-supply (no longer available).

Masood, Salman. "Starved for Energy, Pakistan Braces for a Water Crisis." *The New York Times*, February 13, 2015. http://www.nytimes.com/2015/02/13/world/asia/pakistan-braces-for-major-water-shortages.html.

"Most Serious Greenhouse Gas Is Increasing, International Study Finds." *Science Daily*, April 27, 2001. http://www.sciencedaily.com/releases/2001/04/010427071254.htm.

Onishi, Normimitsu, and Malia Wollanjan. "Severe Drought Grows Worse in California." *The New York Times*, January 17, 2014. http://www.nytimes.com/2014/01/18/us/as-californias-drought-deepens-a-sense-of-dread-grows.html.

Owen, David. "Where the River Runs Dry." *The New Yorker*, May 25, 2015, 52–63.

Ravishankara, A. R. "Water Vapor in the Lower Stratosphere." *Science* 337 (August 17, 2012): 809–810.

Robertson, Campbell. "Iraq, a Land Between Two Rivers, Suffers as One of Them Dwindles." *The New York Times*, July 14, 2009, A1.

Romero, Simon. "Taps Start to Run Dry in Brazil's Largest City; São Paulo Water Crisis Linked to Growth, Pollution and Deforestation." *The New York Times*, February 17, 2015. http://www.nytimes.com/2015/02/17/world/americas/drought-pushes-sao-paulo-brazil-toward-water-crisis.html.

Rosenthal, Elisabeth. "Water Is New Battleground in Drying Spain." *The New York Times*, June 3, 2008, A1, A12.

"Water Levels Still Dropping Near São Paulo." NASA Earth Observatory, March 7, 2015. http://earthobservatory.nasa.gov/IOTD/view.php?id=85460&src=eoa-iotd.

Wines, Michael. "Worst Drought in Half Century Shrivels the Wheat Belt of China." *The New York Times*, February 25, 2009. http://www.nytimes.com/2009/02/25/world/asia/25drought.html.

Wines, Michael. "Colorado River Drought Forces a Painful Reckoning for States." *The New York Times*, January 5, 2014. http://www.nytimes.com/2014/01/06/us/colorado-river-drought-forces-a-painful-reckoning-for-states.html.

See also: Carbon Cycle Feedbacks; Desertification; Drought, United States; Drought Worldwide; El Niño and La Niña; Extreme Weather; Heat Waves, Evidence of; Heat Waves, Forecasts of; Permafrost; Pliocene Paleoclimate; Sea Ice, Arctic; Temperatures, Global; Temperatures, Greenhouse Gas Levels and

CARBON CYCLE FEEDBACKS

Many climate scientists believe that the middle of the 21st century will witness a dramatic acceleration in global warming. In part, this acceleration will result from exhaustion of various natural sinks that have been absorbing greenhouse gases. At about the same time, various feedback loops are also expected to accelerate natural increases in atmospheric greenhouse gas levels and worldwide temperatures. These include several natural processes that add greenhouse gases to the atmosphere such as melting permafrost in the Arctic and, in the far future, possible gasification of solid methane deposits (clathrates) in the oceans, as well as an increasingly dark Arctic Ocean that will absorb more heat as its ice cap melts.

In each case, human-provoked warming caused by an overload of greenhouse gases in the atmosphere is expected to aggravate the natural feedback loops like a bank account drawing an environmentally dangerous form of compound interest. Evidence is accumulating that these processes have already begun. The danger, according to many people who are familiar with the paleoclimatic record, is that once this journey has begun in earnest, any return trip may become a matter of many centuries as well as copious human and natural pain and suffering.

In addition to possible increases in greenhouse gases from gasifying permafrost, other feedback mechanisms are expected to enhance the effects of greenhouse warming during the 21st century. Two of the most important are increasing amounts of water vapor (itself a greenhouse gas), as well as changes in the planet's *albedo*, or reflectivity, as land and sea surfaces once covered by ice are replaced by darker-colored bare land and open water.

One of global warming's most troubling aspects is its tendency to compound through feedback loops. Thus, its effects are not linear but accelerate as temperatures rise. Sea ice that melts, for example, exposes open water, which allows a darker surface to absorb more heat. Warming in the Arctic also melts permafrost and adds carbon dioxide and methane to the atmosphere. Extremes of weather such as heat waves, droughts, and floods can impede plant growth and decrease carbon absorption. The power of feedbacks is such that the worldwide rate of global warming may double, at least, during the half-century to come if greenhouse gases continue to increase in Earth's atmosphere at current rates, according to climate simulation models analyzed by Steven Smith and his colleagues at the Pacific Northwest National Laboratory in Richland, Washington ("Global Warming" 2015).

In the journal *Nature* (2013, 147), Quirin Schiermeier wrote:

"Land plants create a huge carbon 'sink' as they suck CO_2 out of the air to build leaves, wood and roots. The sink varies from year to year, but on average it soaks up one-quarter of the annual CO_2 emissions from the burning of fossil fuels. And events such as droughts, wildfires and storms are likely to 'cause a pronounced decline' in the sink," said Markus Reichstein, a carbon cycle scientist at the Max Planck Institute of Biogeochemistry in Jena, Germany.

Satellite observations have been combined with information from carbon dioxide measurement towers, providing a suggestion that extreme weather events reduce plant productivity in Europe as much 15 percent because of human emissions. Heat waves can turn forests that are usually carbon sinks into carbon sources. "In 2003 alone, a record-breaking heat wave in Europe led to the release of more CO_2 than is normally locked up over four years," Schiermeier wrote (2013, 1247). Drought also impedes plants' ability to combat insect and pathogen damage.

By 2007, methane was bubbling up through melting permafrost, and Inuit hunters sometimes lighted impromptu fires to warm themselves with the gas (Funk 2007, 54). "We are taking risks with a system we don't understand that is absolutely loaded with carbon," said Steven Kallick, a Seattle-based expert on the boreal forests for the Pew Charitable Trusts. "The impact could be enormous" (Struck 2007). "With permafrost, it may take longer for change to get moving. But it may keep moving, even if we get our emissions under control," said Antoni Lewkowicz, a professor of geography at the University of Ottawa. "It's like a big boulder. Once you get it moving, it won't stop" (Struck 2007).

Methane Leak Rates Vastly Underreported?

Just how much carbon dioxide does combustion of natural gas (which is principally methane) produce? One key factor is the proportion of it that leaks as it is being extracted and enters the atmosphere directly as a heat-retaining greenhouse gas. The debate over leakage rates has been raging for several years, but a 2015 study asserted that at least one device used to measure such leaks was underestimating them by amounts so large that the measurements were almost useless.

Touché Howard, who works at Indaco Air Quality Services in Durham, North Carolina, described a University of Texas study that reported measurements of methane emissions natural gas production sites as part of a national inventory. "Unfortunately," Howard wrote in *Energy Science and Engineering* (2015), "Their study appears to have systematically underestimated emissions. They used the Bacharach Hi-Flow® Sampler (BHFS) which in previous studies has been shown to exhibit sensor failures leading to underreporting of NG emissions." These errors were extremely large, "perhaps ten-fold to one hundred-fold for a particularly large leak," Howard told *The New York Times* (Schwartz 2015, August 4). "The presence of such an obvious problem in this high-profile landmark study highlights the need for increased quality assurance in all greenhouse-gas measurement programs," Howard wrote (Schwartz 2015, August 4).

As the methane supply system has been scrutinized for leaks, even the industry's leaders have been surprised at how inefficient it is. For example, facilities that gather gas from several wells were found by a study released in 2015 to be leaking 100 billion cubic feet a year in the United States, enough gas to power 37 coal-fired power plants and heat 3.2 million homes (Marchese et al. 2015). The study was partially sponsored by the industry, which possesses technology to greatly reduce the leakage. "This ultimately helps us perform better," said John Christiansen, a spokesman for Anadarko Petroleum. The research would help the company "get that methane back in the sales line," he added, "which is ultimately in our best interest—and everybody's best interest" (Schwartz 2015, August 18).

Further Reading

Howard, Touché. "University of Texas Study Underestimates National Methane Emissions at Natural Gas Production Sites Due to Instrument Sensor Failure." *Energy Science and Engineering*, August 4, 2015. http://onlinelibrary.wiley.com/doi/10.1002/ese3.81/abstract. doi: 10.1002/ese3.81.

Marchese, Anthony J., et al. "Methane Emissions from United States Natural Gas Gathering and Processing." *Environmental Science and Technology Letters*, August 18, 2015. http://pubs.acs.org/doi/abs/10.1021/acs.est.5b02275.

Schwartz, John. "Methane Leaks May Greatly Exceed Estimates, Report Says." *The New York Times*, August 4, 2015. http://www.nytimes.com/2015/08/05/science/methane-leaks-may-greatly-exceed-estimates-report-says.html.

Schwartz, John. "Methane Leaks in Natural-Gas Supply Chain Far Exceed Estimates, Study Says." *The New York Times*, August 18, 2015. http://www.nytimes.com/2015/08/19/science/methane-leaks-in-natural-gas-supply-chain-far-exceed-estimates-study-says.html.

The Endurance of Feedbacks

Even after carbon dioxide and methane levels in the atmosphere stop rising, feedbacks will produce adverse conditions for decades to centuries. The amount of time that temperatures and sea levels may continue to rise after carbon dioxide levels stabilize is not known with any degree of certainty, however. According to one scientific observer, "The rise in mean global SAT [surface average temperature] and the world ocean level (due to thermal expansion) may . . . continue for several centuries after the stabilization of the carbon dioxide concentration [in the atmosphere], due to the gigantic thermal inertia of the oceans." The same observer also believes that "The response of ice sheets to earlier climate changes may continue for several centuries after the climate stabilizes" (Kondratyev et al. 2004, 45).

Geophysical evidence suggests that Earth has suffered bouts of severe warming in the distant past from natural causes that were intensified by the release of stores of greenhouse gases. Considerable scientific inquiry is now aimed at determining just how much human-provoked warming might cause the process to reach a "runaway" status in which the feedbacks take control and force warming out of control.

Frozen bogs in Siberia contain an estimated 70 billion tons of methane. If the bogs become drier as they warm, according to one observer, "the methane will oxidize and the emissions will be primarily CO_2. But if the bogs stay wet, as they have been recently, the methane will escape directly into the atmosphere. . . . with 20 times the heat-trapping power of carbon dioxide" (Romm 2007, 69). Some 600 million tons of methane are emitted each year from human and natural sources, so if even a small fraction of the 70 billion tons of methane in the Siberian bogs is released, warming will accelerate dramatically.

Another example of a feedback loop that will reinforce warming temperatures (and probably drought) occurs in the tropics. A rain forest or jungle creates its own weather. In our time, however, humans seeking food and shelter have competed with nature so that now roughly 20 percent of the Amazon rain forest has been destroyed, and another 20 percent has been damaged by logging to such an extent that sunlight can reach the forest floor and cause significant drying. According to some studies, "Models suggest that when 50 percent of the forest is destroyed (The year 2050 would be a reasonable estimate for that, at present rates) we will reach a tipping point at which drought and heat will combine to at the rain forest and its canopy, reducing local rainfall and further accelerating the drought and local temperature rise, ultimately causing the release into the atmosphere of huge amounts of carbon currently locked in Amazon soils and vegetation, another fearsome feedback loop at work" (Romm 2007, 72).

Albedo: "The Dirty Snow Effect"

Albedo (Latin for "whiteness") measures a surface's ability to reflect light and heat. The concept is important in global warming because changes in albedo play an important role in changing temperatures in any given location and thus the amount of heat absorbed. In the Arctic, albedo influences the speed at which ice or permafrost melts. Changes in albedo are among the factors contributing to a rate of warming in the Arctic during the last 20 years that has been eight times the rate of warming during the previous 100 years ("Recent Warming" 2003). Recent increases in the number and extent of boreal forest fires also have been increasing the amount of soot in the atmosphere, which also changes albedo.

The melting of ocean-borne ice in polar regions can accelerate overall warming as it changes surface albedo. The darker a surface, the more solar energy it absorbs. Seawater absorbs 90 to 95 percent of incoming solar radiation, whereas snow-free sea ice absorbs only 60 to 70 percent of solar energy. If the sea ice is snow covered, the amount of absorbed solar energy decreases substantially to only 10 percent to 20 percent. Therefore, as the oceans warm and snow and ice melt, more solar energy is absorbed, leading to even more melting. "It is feeding on itself now, and this feedback mechanism is actually accelerating the decrease in sea ice," said Mark Serreze of the University of Colorado (Toner 2003).

Writing in *Philosophical Transactions of the Royal Society A* (Great Britain), James Hansen and colleagues said the Earth's climate is remarkably sensitive to albedo-driven influences on climate ("forcings"). A small forcing can produce a large effect. As they wrote,

> This allows the entire planet to be whipsawed between climate states. One feedback, the "albedo flip" . . . provides a powerful trigger mechanism. A climate forcing that "flips" the albedo of a sufficient portion of an ice sheet can spark a cataclysm. Ice sheet and ocean inertia provide only moderate delay to ice sheet disintegration and a burst of added global warming. Recent greenhouse gas . . . emissions place the Earth perilously close to dramatic climate change that could run out of our control, with great dangers for humans and other creatures. Carbon dioxide (CO_2) is the largest human-made climate forcing, but other trace constituents are important. (Hansen et al. 2007)

Further Reading

Hansen, James E., et al. "Climate Change and Trace Gases." *Philosophical Transactions of the Royal Society A*. July 15, 2007. http://rsta.royalsocietypublishing.org/content /365/1856/1925.
"Recent Warming of Arctic May Affect World-Wide Climate." National Aeronautics and Space Administration Press Release, October 23, 2003. http://www.gsfc.nasa .gov/topstory/2003/1023esuice.html (no longer available).
Toner, Mike. "Arctic Ice Thins Dramatically, NASA Satellite Images Show." *Atlanta Journal-Constitution*, October 24, 2003, 1A.

Surprises in the Carbon Cycle

In cold water, methane clathrates form crystal structures that are somewhat similar to water ice. Warming temperatures could destabilize the clathrates and release some of their stored methane. Roughly 10 trillion tons of methane is trapped under pressure in crystal structures in permafrost or on the edges of the oceans' continental shelves, "Earth's largest fossil fuel reservoir," according to Gerald Dickens, a geologist at James Cook University in Townsville, Australia (Pearce 1998). The greenhouse potential of all the methane stored in clathrates on the continental shelves and in permafrost worldwide is roughly equal to all of the world's coal reserves (Cline 1992, 34).

Atmospheric scientist Roger Revelle has estimated that, with a 3°C rise in global average temperature, methane emissions from clathrates would increase half a gigaton per year worldwide. Over a century, this rate could be enough to double the amount of methane in the atmosphere. Add to this another 12 gigatons of methane that could be released by clathrates liberated from ocean bottoms under the Arctic Ocean once the ice cap now covering them melts. "It is possible," writes Jonathan Weiner, "that [this] . . . feedback effect is already underway and the rise in Earth temperatures in the last hundred years has already sprung many giga-tons of methane from their molecular prisons at the bottom of the sea" (Weiner 1990, 118).

Warming temperatures also change the behavior of Earth's hydrological cycle. Warmer ocean water removes less carbon dioxide from the atmosphere than cooler water, so warming oceans may feed on themselves in coming years. Water vapor is also a potent absorber of heat in the atmosphere. A doubling of carbon dioxide in the atmosphere has been estimated as increasing its water content by 30 percent, raising temperatures an additional 1.4°C (Hansen 1981, 957). Many models proj-ect a rise in cloudiness and attendant atmospheric moisture in a warmer, more humid world. George M. Woodwell raises the possibility of a rapid surge in global warming beyond any possibility of human control:

> The possibility exists that the warming will proceed to the point where biotic releases from the warming will exceed in magnitude those controlled directly by human activity. If so, the warming will be beyond control by any steps now considered reasonable. We don't know how far we are from that point because we do not know sufficient detail about the circulation of carbon dioxide among the pools of the carbon cycle. We are not going to be able to resolve those questions definitely soon. Meanwhile, the concentration of heat-trapping gases in the atmosphere rises. . . . (Woodwell 1990, 130)

Given Woodwell's expectations, the peoples of Earth are approaching a point of no return with regard to climatic feedbacks. Deforestation is accelerating around the world because of growing populations and levels of material affluence. Use of fossil fuels, which has increased at an annual rate of roughly 5 percent during most of this century, shows no signs of stabilizing, much less falling by half in the next 30 years. China alone projects burning enough fossil fuel (mainly coal) by 2025 to account for around half the current consumption of fossil fuels by everyone on Earth (Leggett 1990, 27).

When Will Feedbacks Take Control?

How much time remains before critical feedback loops lock into place? In Janu-ary 2005, a world task force of senior politicians, business leaders, and academics warned that the point of no return—including widespread agricultural failure, water shortages and major droughts, increased disease, sea level rise and the death of forests—will occur as the average world temperature increases 2°C above the aver-age prevailing in 1750, shortly before the Industrial Revolution began (Byers and Snowe 2005, 1). By 2015, temperatures already had risen an average of 1.0°C. The report also asserted that the tipping point will occur as the atmospheric con-centration of carbon dioxide passes 400 parts per million (ppm). The level was 379 ppm in 2004, and the threshold of 400 ppm was crossed in 2015. Given half a century for thermal inertia to realize the effects of CO_2 at 400 ppm, serious dam-age is now factored into the system. The report was assembled by the Institute for

Public Policy Research in the United Kingdom, the Center for American Progress in the United States, and the Australia Institute. The group's chief scientific adviser was Rakendra Pachauri, chairman of the U.N. Intergovernmental Panel on Climate Change (McCarthy 2005).

The report concluded:

> Above the 2-degree level, the risks of abrupt, accelerated, or runaway climate change also increase. The possibilities include reaching climatic tipping points leading, for example, to the loss of the West Antarctic and Greenland ice sheets (which, between them, could raise sea level more than 10 meters over the space of a few centuries), the shutdown of the thermohaline ocean circulation (and, with it, the Gulf Stream), and the transformation of the planet's forests and soils from a net sink of carbon to a net source of carbon. (McCarthy 2005)

The science of warming feedbacks is not settled knowledge. In areas such as the role of organic decomposition in soils under warmer conditions, a robust debate continues. Knorr and colleagues have written in *Nature* that "the sensitivity of soil carbon to warming is a major uncertainty in projections of carbon dioxide concentration and climate" (Knorr et al. 2005, 298) because their findings indicate that "the long-term positive feedback of soil decomposition in a warming world may be even stronger than predicted by global models" (Knorr et al. 2005, 298). In the meantime, the very idea that organic soil decomposition is sensitive to temperature at all has been challenged by other scientists (Giardina and Ryan 2000). For climate models, the solution of this debate is no small matter because soils contain twice as much carbon as the atmosphere (Powlson 2005, 204), so the rate at which warming may accelerate exchange from one to the other is and will continue to be an important factor as other feedbacks add more greenhouse gases to the air that sustains us.

"The biggest lag is in the political system," said geoscientist Michael Oppenheimer of Princeton University. At least two decades have passed since scientists have pointed out the seriousness of the threat, he said, and another 20 years may pass before a worldwide program is established that is up to the task. In the meantime, the window of time before feedbacks take control narrows. "We can't really afford to do a 'wait and learn' policy," Oppenheimer said. "The most important question is: when do we commit to [contain global warming to] 2 [°C]. Really, there isn't a lot of headroom left. We better get cracking." The current pace, said Roger Pielke Jr. of the University of Colorado at Boulder, "isn't going to do it" (Kerr 2007, 1231).

Further Reading

Byers, Stephen, and Olympia Snowe. *Meeting the Climate Challenge: Recommendations of the International Climate Change Task Force.* London: Institute for Public Policy Research, January 2005.

Cline, William R. *The Economics of Global Warming.* Washington, DC: Institute for International Economics, 1992.

Funk, McKenzie. "Cold Rush: The Coming Fight for the Melting North." *Harper's*, September 2007, 54–55.

Giardina, C., and M. Ryan. "Evidence that Decomposition Rates of Organic Carbon in Mineral Soil Do Not Vary with Temperature." *Nature* 404 (2000): 858–861.

"Global Warming Could Speed Up." *Nature* 519 (March 12, 2015): 132.

Hansen, James E., et al. "Climate Impact of Increasing Atmospheric Carbon Dioxide," *Science* 213 (1981): 957–956.

Kerr, Richard. "How Urgent Is Climate Change?" *Science* 318 (November 23, 2007): 1230–1231.

Knorr, W., et al. "Long-Term Sensitivity of Soil Carbon Turnover to Warming." *Nature* 433 (January 20, 2005): 298–301.

Kondratyev, Kirill, Vladimir F. Krapivin, and Costas A. Varotsos. *Global Carbon Cycle and Climate Change*. Berlin, Germany: Springer/Praxis, 2004.

Leggett, Jeremy, ed. *Global Warming: The Greenpeace Report*. New York: Oxford University Press, 1990.

McCarthy, Michael. "Countdown to Global Catastrophe." *The Independent* (London), January 24, 2005, 1.

Pearce, Fred. "Nature Plants Doomsday Devices." *The Guardian* (United Kingdom), November 25, 1998. http://go2.guardian.co.uk/science/912000568-disast.html (no longer available).

Powlson, David. "Will Soil Amplify Climate Change?" *Nature* 433 (January 20, 2005): 204–205.

Romm, Joseph. *Hell and High Water: Global Warming—the Solution and the Politics—and What We Should Do*. New York: William Morrow, 2007.

Schiermeier, Quirin. "Wild Weather Can Send Greenhouse Gases Spiralling." *Nature* 496 (April 11, 2013): 147. http://www.nature.com/news/wild-weather-can-send-greenhouse -gases-spiralling-1.12764.

Struck, Doug. "Icy Island Warms to Climate Change." *Washington Post*, June 7, 2007, A1. http://www.washingtonpost.com/wp-dyn/content/article/2007/06/06/AR2007060602783 .html?referrer=email.

Weiner, Jonathan. *The Next One Hundred Years: Shaping the Fate of Our Living Earth*. New York: Bantam Books, 1990.

Woodwell, George M. "The Effects of Global Warming." Pp. 116–132 in Jeremy Leggett, ed., *Global Warming: The Greenpeace Report*. New York: Oxford University Press, 1990.

See also: Desertification; Drought, United States; Drought, Worldwide; El Niño and La Niña; Extreme Weather; Heat Waves, Evidence of; Heat Waves, Forecasts of; Permafrost; Pliocene Paleoclimate; Sea Ice, Arctic; Temperatures, Global; Temperatures, Greenhouse Gas Levels and; Thermal Inertia

CLIMATE CHANGE, ABRUPT NATURE OF

Climatologists have been sharing the disquieting idea that small shifts in global conditions may lead to sudden and abrupt climate changes. The National Academy of Sciences has warned that global warming could trigger "large, abrupt, and unwelcome" climatic changes that could severely affect ecosystems and human society (McFarling 2001). "We need to deal with this because we are likely to be surprised," said Richard Alley, a Pennsylvania State University climate expert. "It's as if climate change were a light switch instead of a dimmer dial," Alley said (McFarling 2001). The report, which was commissioned by the U.S. Global Change Research Program, includes a plea for more research on the links between the land, oceans, and ice that may trigger abrupt change. Alley also suggests that many of today's models of climate change are too simple because they do not include such changes (McFarling 2001).

In *Abrupt Climate Change* (2002), Alley, who has become an expert on the subject, wrote that climate may (and already has) changed rapidly "when gradual causes push the Earth system across a threshold" (Alley 2002, v). The temperature record for Greenland, according to Alley's research, more resembles a jagged row of sharp teeth than a gradual passage from one climatic epoch to another. According to Alley, "Model projections of global warming find increased global precipitation, increased variability in precipitation, and summertime drying in many continental interiors, including grain belt regions. Such changes might produce more floods and more droughts" (Alley 2002, 114).

Human Forcings Play an Important Role

"Although abrupt climate change can occur for many reasons, it is conceivable that human forcing of climate change is increasing the probability of large, abrupt events," wrote a team led by Alley (Alley et al. 2003, 2005). At times, they wrote, regional temperature changes one-third to one-half as large as those associated with 100,000-year ice-age cycles have actually taken place over a single decade (Alley et al. 2003; Alley 2000, 1331). An intense drought that played a major role in destroying classic Mayan civilization may be an example of such a change (Alley et al. 2003). The paleoclimatic record shows that regional temperature changes of 8 to 16°C have been known to have occurred within a few years. Such changes were most likely during the beginning or end of ice ages (Alley et al. 2003, 2006).

The Speed of Human-Provoked Warming

Human forcing of climate by accelerating emission of greenhouse gases has no precedent in Earth's natural history. Shaun A. Marcott and colleagues used fossils of tiny marine organisms to analyze surface temperatures as far back as 11,300 years ago and found recent warming to be unprecedented, "lurching," as one account phrased it, "from near-record cooling to a heat spike" ("Study Finds" 2013). This study assembled the longest continuous record of Earth's average temperatures. The earliest previous record of this type extended back about 2,000 years.

Early Holocene (10,000 to 5,000 years ago) warmth is followed by ~0.7°C cooling through the middle to late Holocene (less than 5,000 years ago), culminating in the coolest temperatures of the Holocene during the Little Ice Age, about 200 years ago. This cooling is largely associated with about a 2°C change in the North Atlantic. Current global temperatures of the past decade have not yet exceeded peak interglacial values but are warmer than during about 75 percent of the Holocene temperature history. Intergovernmental Panel on Climate Change model projections for 2100 exceed the full distribution of Holocene temperature under all plausible greenhouse gas emission scenarios. (Marcott et al. 2013, 1198)

This is, they said, "further evidence that modern-day global warming isn't natural, but the result of rising carbon dioxide emissions that have rapidly grown since the Industrial Revolution began roughly 250 years ago" ("Study Finds" 2013). The opening years of the 20th century (1900 to 1910) were among the coolest in the last 11,300 years. A century later (2000 to 2010), temperatures were among the warmest since the end of the last ice age. "In 100 years, we've gone from the cold end of the spectrum to the warm end of the spectrum," Marcott said. "We've never seen something this rapid. Even in the ice age the global temperature never changed this quickly."

Natural forcings tend to warm the atmosphere much more slowly than human provocations. "Marcott's data indicates that it took 4,000 years for the world to warm about 1.25 degrees from the end of the ice age to about 7,000 years ago. The same fossil-based data suggest a similar level of warming occurring in just one generation: from the 1920s to the 1940s" ("Study Finds" 2013).

Michael Mann, a climate scientist at Pennsylvania State University, wrote the 2,000-year study (which is well known for "the hockey stick" that shows the dramatic upward curve in temperature over the last century). "Scientists," he said, "may have to go back 125,000 years to find warmer temperatures potentially rivaling today's" ("Study Finds" 2013). Other scientists, one of whom is Jeff Severinghaus of the Scripps Institution of Oceanography, said that present-day warming is rivaled by another warm spike that was unaided by human use of fossil fuels about 12,000 years ago. However, said Katharine Hayhoe, an atmospheric scientist at Texas Tech University, "we have, through human emissions of carbon dioxide and other heat-trapping gases, indefinitely delayed the onset of the next ice age and are now heading into an unknown future where humans control the thermostat of the planet" ("Study Finds" 2013).

Further Reading

Marcott, Shaun A., et al. "A Reconstruction of Regional and Global Temperature for the Past 11,300 Years." *Science* 339 (March 8, 2013): 1198–1201.

"Study Finds Climate Change a New Phenomenon in Planet's 11,000-Year Cooling Trend." Associated Press in CBS News. March 7, 2013. http://www.cbsnews.com /8301-205_162-57573132/study-finds-climate-change-a-new-phenomenon-in -planets-11000-year-cooling-trend.

Alley has described "threshold transitions" as being comparable to leaning over the side of a canoe: "Leaning slightly over the side of a canoe will cause only a small tilt, but leaning slightly more may roll you and the craft into the lake" (Alley et al. 2003, 2005). At just the right point, a small "forcing" may set into motion an enormous climatic change. Thermohaline circulation and El Niño changes may be notable stress points in the global system, they assert.

James E. Hansen, long-time director of NASA's Goddard Institute for Space Studies (now retired), agreed with Alley and his colleagues:

The occurrence of abrupt climate changes this century is practically certain, if we continue with business-as-usual greenhouse gas emissions. This assertion is based on the magnitude and speed of the human-induced changes of atmospheric composition (which dwarf natural changes) and the rate of global warming that will result from such atmospheric changes (which exceeds any documented natural global warming event). The magnitude of the expected total climate change under business-as-usual is so large that it almost surely passes a number of thresholds. Although, it is impossible to say when these thresholds will be passed, we give examples here of abrupt changes that seem highly likely under business-as-usual and we discuss additional more speculative possibilities. (Hansen 2006, 30)

In May 1997, 21 nationally prominent ecologists warned President Bill Clinton that rapid climate change because of global warming could ruin ecosystems on which human societies depend. In the United States, the scientists said,

Rapid climate change could mean the widespread death of trees, followed by wildfires and . . . replacement of forests by grasslands. National parks and forests could become inhospitable to the rare plants and animals that are preserved there—and where the parks are close to developed or agricultural land, the species themselves may disappear for lack of another safe haven. Worldwide, fast-rising sea levels could inundate the marshes and mangrove forests that protect coastlines from erosion and serve as filters for pollutants and nurseries for ocean fisheries. "The more rapid the rate [of change] the more vulnerable to damage ecosystems will be," the scientists told the president. "We are performing a global experiment [with] little information to guide us." (Basu 1997)

Rapid Changes and Feedback Mechanisms

The abrupt nature of climate change can be enhanced by several feedback mechanisms that compound the effects of any single "forcing." Arctic snow and sea ice, for example, cover a large portion of Earth's surface with a light-reflective surface that reflects sunlight and heat back into space. If large parts of the Arctic ice cap melt in the summer, darker liquid sea surfaces will accelerate warming.

The possibility of such abrupt changes complicates the task of policy makers in two ways. It could mean that the amount of time available to adjust to climate change will be much shorter than many government officials have believed. It also increases the uncertainty of predictions, indicating that future climate cannot simply be projected forward in a straight line from the present. "We're a little spoiled by the last 30 years," said John M. Wallace, an atmospheric scientist at the University of Washington. "Many years, we're just barely breaking the previous record" (McFarling 2001).

"We know from ice-core records and deep-sea sediment records that the Earth's climate is capable of changing much more quickly than we had previously thought," said Jeff Severinghaus of the University of California (Webb 1998). "In some cases," said Severinghaus, "the climate warmed abruptly in less than 10 years . . . up to possibly 10 [°C]" (Webb 1998). Severinghaus's findings were presented at the 1998 climate-change conference in Buenos Aires. Severinghaus continued: "It is possible that by increasing greenhouse gases, we will induce such a change and that, instead of the smooth warming that's being anticipated over the next 50 years, that we'll instead go along for a while with very little warming and then all of a sudden in a matter of three or five or ten years we'll have a very large catastrophic warming" (Webb 1998).

A rapid rise in temperatures could change Earth's ecosystem fundamentally:

One must go back in time 5 [million] to 15 million years to the late Tertiary to find a time that was 3 or 4 [°C] warmer than now. During periods when there was no permanent pack ice in the Arctic, climatic and vegetational region and boundaries were displaced as much as 1,000 to 2,000 kilometers north of their present position (a displacement which we may replicate during the next 100 years). (Abrahamson 1989, 15)

During the period Abrahamson describes, intense aridity was the norm from what is now North and South Dakota to Missouri and Alabama, as well as throughout central and southern Africa. These changes may be similar to those that will be experienced by generations to come.

Warming More Rapid Than Expected

Scientists working for the National Atmospheric and Oceanic Administration (NOAA) in 2000 released compilations of global temperatures for the last half of the 20th century that revealed a speed of warming that most climatologists had not expected until late in the 21st century. The rate of warming (1°F over the entire century) increased to a rate of 4°F during the century's last quarter, according to calculations of Tom Karl and associates, published in the March 1, 2000 edition of *Geophysical Research Letters* (Karl et al. 2000). This is roughly the rate of increase that several climate models were then forecasting for the second half of the 21st century. "The next few years could be very interesting," Karl told the *Los Angeles Times*. "It could be the beginning of a new increase in temperatures" ("Analysis"

2000). Tom Wigley, a senior scientist at the National Center for Atmospheric Research in Boulder, Colorado, countered Karl's assessment by saying that warming was strengthened by frequent El Niño events, which he said are not human-induced. "Those months were unusual," he said, "but they weren't unusual due to human influences" ("Analysis" 2000).

A study using the MIT Integrated Global Systems Model, a computer simulation of global economic activity and climate processes developed by the Joint Program on the Science and Policy of Global Change, indicates that without rapid changes in humankind's behavior, "likelihood of how much hotter the Earth's climate will get in this century . . . will be about twice as severe as previously estimated six years ago—and could be even worse than that" ("MIT" 2009; Sokolov et al. 2009).

The Speed of Change in the Arctic

Evidence from ice cores taken in Greenland has altered the scientific view of the speed of climatic change at higher latitudes. The older assumption that climate changes exceedingly slowly in the polar regions is built into the English language when we say that something moves with "glacial speed." On the contrary, the ice-core record indicates that several rapid warmings and coolings have convulsed the Arctic during the last 100,000 years.

The last ice age may have ended far more quickly than previously thought. Jeffery P. Severinghaus of the Scripps Institute of Oceanography described a new method of analyzing gases trapped in Greenland's ice sheet indicating that temperatures in the area rose approximately 16°F within a decade or two at the end of the last great ice age. "The old idea was that the temperature would change over a thousand years, but we found it was much faster," said Severinghaus (Webb 1998). "We know that over the next one hundred years the Earth will probably warm because of the greenhouse effect," Severinghaus continued. "There is a remote possibility that we might trigger one of these abrupt climate changes. This certainly gives us pause" (Webb 1998, 15).

Past rapid changes in regional climates took place mainly without major human provocation. Factors contributing to these rapid changes may have included variations in the sun's radiance, changes in angle of Earth's axis, natural variations in atmospheric levels of carbon dioxide and methane (among other trace gases), changes in atmospheric water-vapor levels, and dust ejected from volcanoes and other sources. As Adams et al. note,

> Climate has a tendency to remain quite stable for most of the time and then suddenly "flip," at least sometimes over just a few decades, due to the influence of . . . various triggering and feedback mechanisms. . . . Such observations suggest that even without anthropogenic climate modification there is always an axe hanging over our head, in the form of random very large-scale changes in the natural climate system; a possibility that policy makers should perhaps bear in mind with contingency plans and international treaties designed to cope with sudden famines on a greater scale than any experienced

in written history. By starting to disturb the system, humans may simply be increasing the likelihood of sudden events which could always occur. . . . To paraphrase W.S. Broecker; "Climate is an ill-tempered beast, and we are poking it with sticks." (Adams et al. 1999, 1)

Further Reading

Abrahamson, Dean Edwin. "Global Warming: The Issue, Impacts, Responses." Pp. 3–34 in Dean Edwin Abrahamson, ed., *The Challenge of Global Warming*. Washington, DC: Island Press, 1989.

Adams, Jonathan, Mark Maslin, and Ellen Thomas. "Sudden Climate Transitions During the Quaternary." *Physical Geography* 23 (1999): 23:1–36.

Alley, Richard B. "Ice-Core Evidence of Abrupt Climate Changes." *Proceedings of the National Academy of Sciences of the United States of America* 97(4) (February 15, 2000): 1331–1334.

Alley, Richard B., ed. *Abrupt Climate Change: Inevitable Surprises*. Washington, DC: National Academy Press, 2002.

Alley, Richard B., et al. "Abrupt Climate Change." *Science* 299 (March 28, 2003): 2005–2010.

Alley, Richard B., et al. "Ice-Sheet and Sea-Level Changes." *Science* 310 (October 21, 2005): 456–460.

"Analysis: Climate Warming at Steep Rate." *Omaha World-Herald*, February 23, 2000, 12.

Basu, Janet. "Ecologists' Statement on the Consequences of Rapid Climatic Change—May 20, 1997." http://www.dieoff.com/page104.htm.

Hansen, James E. "Declaration of James E. Hansen." *Green Mountain Chrysler-Plymouth-Dodge-Jeep, et al., Plaintiffs v. Thomas W. Torti, Secretary of the Vermont Agency of Natural Resources, et al., Defendants*. Case Nos. 2:05-CV-302 and 2:05-CV-304, Consolidated. United States District Court for the District of Vermont. August 14, 2006. http://www.giss.nasa.gov/~dcain/recent_papers_proofs/vermont_14aug20061_textwfigs.pdf. (no longer available).

Karl, Thomas R., Richard W. Knight, and Bruce Baker. "The Record-Breaking Global Temperatures of 1997 and 1998: Evidence for an Increase in the Rate of Global Warming." *Geophysical Research Letters* 27 (March 1, 2000): 719–722.

McFarling, Usha Lee. "Scientists Now Fear 'Abrupt' Global Warming Changes." *Los Angeles Times*, December 12, 2001, A30.

"MIT: Climate Change Odds Much Worse Than Thought." NASA Earth Observatory, May 19, 2009. http://earthobservatory.nasa.gov/Newsroom/view.php?id=38818&src=eoa-manews.

Sokolov, A. P., et al. "Probabilistic Forecast for 21st Century Climate Based on Uncertainties in Emissions (Without Policy) and Climate Parameters." *Journal of Climate*, June 2009. doi: 10.1175/2009JCLI2863.1.

Webb, Jason. "World Temperatures Could Jump Suddenly." Reuters, November 4, 1998. http://bonanza.lter.uaf.edu/~davev/nrm304/glbxnews.htm.

See also: Atmospheric Circulation; Carbon Cycle Feedbacks; El Niño and La Niña; Emissions, Carbon Dioxide, and Peat; Extreme Weather; Heat Waves, Evidence of; Heat Waves, Forecasts of; Permafrost; Pliocene Paleoclimate; Sea Ice, Arctic; Temperatures, Global; Temperatures, Greenhouse Gas Levels and

CLIMATE CHANGE, GREENLAND

On close examination of paleoclimatic records, scientists who have examined ice cores from Greenland have determined that its climate can change quickly, giving cold comfort to anyone contemplating the stability of worldwide sea levels. The island's climate "flipped" to a different state within one to three years at the onset of the current interglacial period, for example. A northward shift of the Intertropical Convergence Zone may have triggered abrupt shifts in Northern Hemisphere atmospheric circulation, leading to changes of 2 to 4°C in Greenland temperatures from one year to the next (Steffensen et al. 2008, 654). Other scientists have found evidence that Greenland's surface has been covered with pine forests within a few hundred thousand years of the present; its ice mass was a fraction of today's size.

During the winters from 2003 to 2007, Greenland lost two to three times as much ice in summer melt as it regained during winter snows (Witze 2008, 798). Greenland is a relic of the last ice age "stranded out of time" (Witze 2008, 798). Even without human-induced global warming, its glaciers would not reform under current conditions. Ice loss is also accelerating irregularly year by year. During the 2007 melting season, with temperatures 4 to 6°C higher than the previous 30-years average, 500 billion tons of ice vanished, 30 percent more than the previous year, and 4 percent more than the previous record in 2005 (Witze 2008, 799). During years of record melting such as 2005 and 2007, high-pressure systems over Greenland kept storms away, clearing skies and allowing the sun to shine for extended hours.

By the summer of 2007, Greenland's ice cap was studded by more than 1,000 shallow meltwater lakes, some as wide as five kilometers, "like Minnesota, except white," wrote Alexandra Witze in *Nature* (2008, 800). Tens of millions of cubic meters of water swirl from these lakes to the base of the ice sheet in a matter of days, opening huge waterfalls where none previously existed. "It has been only in the last five years that we have realized that—hey—the ice sheet is falling apart," said Ian M. Howat of the University of California–Santa Cruz (Witze 2008, 801).

Ice melt in Greenland has been aggravated by occasional temperature spikes. During April 2016, for example, persistent high pressure pushed temperatures to as high as 20°C (36°F) above the averages for 2000 to 2010. "The most remarkable aspect here is the incredible departure from 2001–2010 average, especially deep in the ice sheet interior," said Santiago de la Peña, a research scientist at Ohio State University. "There have been occasional warming events in the past during spring over Greenland," he noted, "but they affected only local areas and were not as intense" ("Widespread Warmth" 2016).

Decay of Greenland's Zachariæ Isstrøm Glacier

By 2015, Greenland's ice was decaying at an accelerating rate. One massive glacier, Zachariæ Isstrøm, once eventually melted will raise world sea levels some 18 inches. Following three years of decay, it was already calving icebergs more quickly as it enters a less stable state that will eventually drain meltwater from a 91,780 square kilometer (35,440 square mile) area of northeast Greenland. The glacier's ice field, which accounts for 5 percent of Greenland's ice sheet, has been losing some 5 billion tons of ice per year.

A paper in *Science*, published late in 2015, said:

After eight years of decay of its ice shelf, Zachariæ Isstrøm, a major glacier of northeast Greenland that holds a 0.5-meter sea-level rise equivalent, entered a phase of accelerated retreat in fall 2012. The acceleration rate of its ice velocity tripled, melting of its residual ice shelf and thinning of its grounded portion doubled, and calving is now occurring at its grounding line. Warmer air and ocean temperatures have caused the glacier to detach from a stabilizing sill and retreat rapidly along a downward-sloping, marine-based bed. Its equal-ice-volume neighbor, Nioghalvfjerdsfjorden, is also melting rapidly but retreating slowly along an upward-sloping bed. The destabilization of this marine-based sector will increase sea level rise from the Greenland ice sheet for decades to come. (Mouginot et al. 2015)

"North Greenland glaciers are changing rapidly," said Jeremie Mouginot of the University of California–Irvine. "The shape and dynamics of Zachariæ Isstrøm have changed dramatically over the last few years. The glacier is now breaking up and calving high volumes of icebergs into the ocean, which will result in rising sea levels for decades to come" ("Zachariæ Isstrøm" 2015).

Mouginot and colleagues found that Zachariæ Isstrøm is losing ice mass mainly from below as relatively warmer ocean water mixes with an increasing volume of meltwater from the ice sheet. "Zachariæ Isstrøm is being hit from above and below," said study coauthor Eric Rignot. "The top of the glacier is melting away as a result of decades of steadily increasing air temperatures, while its underside is compromised by currents carrying warmer ocean water, and the glacier is now breaking away into bits and pieces and retreating into deeper ground" ("Zachariæ Isstrøm" 2015).

Further Reading

Mouginot, J., et al. "Fast Retreat of Zachariæ Isstrøm, Northeast Greenland." *Science*. Published online November 15, 2015. http://www.sciencemag.org/content/early/2015/11/11/science.aac7111.abstract?sid=47d39afa-a6a5-415e-918c-0735fab5c92f. doi: 10.1126/science.aac7111.

"Zachariæ Isstrøm Glacier, Greenland." NASA Earth Observatory, December 4, 2015. http://earthobservatory.nasa.gov/IOTD/view.php?id=87086.

After summer melt collects in lakes on the ice sheets, the water then finds fissures in the ice called *moulins* that conduct the water to the glacier's base, lubricating its movement toward the sea. The more ice melt, the faster the glacier moves. Sometimes surface lakes disappear down moulins nearly instantly (Appenzeller 2007, 68). Increases in the speed by which glaciers flow to the sea is itself evidence of a rapidly warming climate over a longer time period.

In early 2008, Andrew C. Revkin of *The New York Times* sketched this picture of Greenland's eroding ice cap:

For a lengthening string of warm years, a lacework of blue lakes and rivulets of meltwater have been spreading ever higher on the ice cap. The melting surface darkens, absorbing up to four times as much energy from the sun as unmelted snow, which reflects sunlight. Natural drainpipes called moulins carry water from the surface into the depths, in some places reaching bedrock. The process slightly, but measurably, lubricates and accelerates the grinding passage of ice toward the sea. Most important, many glaciologists say, is the breakup of huge semisubmerged clots of ice where some large Greenland glaciers, particularly along the west coast, squeeze through fjords as they meet the warming ocean. As these passages have cleared, this has sharply accelerated the flow of many of these creeping, corrugated, frozen rivers. (Revkin 2008)

Such rapid changes have many glaciologists "a little nervous these days—shell-shocked," said Ted Scambos, the lead scientist at the National Snow and Ice Data Center (NSIDC) in Boulder, Colorado, and a veteran of both Greenland and Antarctic studies (Revkin 2008).

Speed of Ice Melt Surprises Glaciologists

The speed at which Greenland's ice has been melting has surprised even veteran glaciologists. Robert Lee Hotz of the *Los Angeles Times* described the reaction of Jay Zwally, who works at NASA's Goddard Space Flight Center in Greenbelt, Maryland, who has been studying the polar regions since 1972:

Gripping a bottle of Jack Daniel's between his knees, Jay Zwally savored the warmth inside the tiny plane as it flew low across Greenland's biggest and fastest-moving outlet glacier. Mile upon mile of the steep fjord was choked with icy rubble from the glacier's disintegrated leading edge. More than six miles of the Jakobshavn had simply crumbled into open water. The Jakobshavn Isbrae glacier, on the west coast of Greenland at 69 degrees N. latitude, has retreated about 27 miles in 160 years and 6 miles since the year 2000. "My God!" Zwally shouted over the hornet whine of the engines. From satellite sensors and seasons in the field, Zwally, 67, knew the ice sheet below in a way that few could match. Even after a lifetime of study, the raffish NASA glaciologist with a silver dolphin in one pierced ear was dismayed by how quickly the breakup had occurred. (Hotz 2006)

At the same time, the Petermann glacier in northern Greenland continued to break up. An 11-square-mile (29-square-kilometer) piece of this floating ice mass broke away between July 10 and 24, 2008. The glacier already had lost 33 square miles of floating ice during 2000 and 2001. The entire Petermann glacier, all of which floats on the sea, covers roughly 500 square miles and measures 10 miles by 50 miles.

Working with the Byrd Polar Research Center, Jason Box, an associate professor of geography at Ohio State University, and his colleagues, graduate students Russell Benson and David Decker, have been studying cracks in the glacier that may portend future breakup. "If the Petermann glacier breaks up back to the upstream rift, the loss would be as much as 60 square miles (160 square kilometers)," Box said, representing a loss of one-third of the massive ice field ("Satellite Images" 2008).

Soot and Ice Melt

Albedo (brightness) plays an important role in the rate at which glacial ice melts. Soot transported by prevailing winds from an increasing number of wildfires in North America is accelerating the erosion of Greenland's ice sheet ("Soot" 2014).

The rate of ice melt on Greenland has been accelerated by cryoconite, a sheen of soot and other airborne debris that blows in from as far away as Central Asia's deserts. Each year's snowfall covers the cryoconite, but when melting outpaces snowfall, which it has in recent years, several layers may be exposed during one season, raising albedo because of its dark color, which speeds melting even more. This imported dark matter may come from forest fires, coal-fired power plants, and diesel engines. This layer of soot is not new to our time, but it has increased with population and industrialization. Although soot is less than 5 percent of the cryoconite, it turns it black and speeds melting. "What we have is a vicious, constantly cycle," said Carl Egede Boggild, geophysicist and native Greenlander. "It's like pulling a black curtain over the ice" (Jenkins 2010, 43).

This process helps meltwater lakes form on top of the ice sheet during the melting season. Moulins, vertical shafts under the meltwater lakes, suck the water downward. One two-square-mile lake, an estimated 11 billion gallons of water, was sucked dry in 84 minutes (Jenkins 2010, 43).

The surface of the Greenland's ice sheet has grown noticeably less reflective during the last decade because of increasing dust blowing in from its melting margins, according to NASA satellite observations. Even the center of the ice cap has become darker because, according to Jason Box of Ohio State University, "as temperatures rise, snow grains clump together and reflect less light than the many-faceted, smaller crystals. Additional heat rounds the sharp edges of the crystals, and round particles absorb more sunlight than jagged ones" ("Greenland's Ice" 2012).

By 2012, some parts of the ice sheet were reflecting close to 20 percent less light than a decade earlier. According to a NASA analysis, "Lower-elevation areas of the ice sheet have darkened more than the colder, higher-altitude interior. Each summer, winter snow retreats from the edge of the ice sheet. Dark pools of meltwater form on the surface of the ice, and windblown dust and other particles also collect near the surface, making it even less reflective" ("Greenland's Ice" 2012).

Ice Accumulates Slowly but Erodes Quickly

Paleoclimatic records indicate that Greenland's ice sheet tend to accumulate slowly, but can erode very quickly. Glaciologists describe one exceptionally intense interglacial 410,000 to 400,000 years ago, when global sea levels may have been 6 to 13 meters above today, "implying substantial mass loss from the Greenland ice sheet (GIS). the south GIS was drastically smaller . . . than it is now, with only a small residual ice dome over southernmost Greenland. . . . [suggesting] that the GIS lost about 4.5 to 6 meters of sea-level-equivalent volume. . . . This is evidence for late-Quaternary GIS collapse after it crossed a climate/ice-sheet stability threshold that may have been no more than several degrees above preindustrial temperatures" (Reyes et al. 2014).

James Hansen and Makiko Sato (2012) have raised the possibility that a business-as-usual rise in atmospheric greenhouse gases to the end of the 21st century may provoke a "climate forcing . . . so large that nonlinear ice sheet disintegration should be expected and multimeter sea level rise not only possible but likely. They assert that amplifying feedbacks that come into play as ice sheets begins to disintegrate may "lead to 1 meter sea level rise by 2067 and 5 meters by 2090."

"Very Close to the Threshold"

"We may be very close to the threshold where the Greenland ice cap will melt irreversibly," said Tavi Murray, professor of glaciology at the University of Wales. Professor Tulaczyk added, "The observations that we are seeing now point in that direction" (Lean 2005).

Over the last few years, Greenland's melt zone—where summer warmth turns snow on the edge of the ice cap into slush and ponds of water—has expanded inland, reaching elevations more than a mile high in some places, said Konrad Steffen, a glaciologist at the University of Colorado. Some tongues of floating ice, where glaciers protrude into the sea, are thinning rapidly. Measurements during 2004 by Steffen and others on the Petermann glacier in northern Greenland indicated that more than 150 feet of thickness had melted away under that tongue in one year. "If other ice streams start to react in a similar way," he said, "then we will actually produce much more freshwater," reported Andrew Revkin in *The New York Times* (2004).

Modeling by Jonathan Gregory of the Center for Global Atmospheric Modeling in Reading, England, "suggests that as ice is lost, portions of the surface of Greenland's interior will heat and up at lower elevations where the air is warmer. Less snowfall and more rain would cause the ice to disappear at a faster rate than it is being replaced, leading in turn to further drops in elevation" (Schiermeier 2004).

Models further suggest that the ice sheet would not re-form even if temperatures cooled in the future. The ice sheet creates its own climate, "depending on itself to exist" (Schiermeier 2004). According to these models, a warming of 3°C could initiate eventual melting of the entire ice sheet over 1,000 years or more, raising global sea levels by some seven meters. Gregory and colleagues have written that "concentrations of greenhouse gases probably will have reached levels before the year 2100 that are sufficient to raise the temperature past this warming threshold" (Gregory et al. 2004). Models are complicated by several factors. For example, warmer temperatures initially could increase snowfall and delay glacial melting. Ice melt also may depress the Gulf Stream, which would also cause cooling over Greenland.

Evidence published in *Geophysical Research Letters* late in 2005 (Howat et al. 2005) described a sudden thinning of Greenland's Helheim glacier on the island's east coast. Professor Slawek Tulaczyk of the Department of Earth Sciences at the University of California–Santa Cruz, said that the glacier retreated four and a half miles between 2000 and 2005. As this glacier has retreated and thinned, effects have spread inland "very fast indeed," said Tulaczyk. "If the 2005 speedup also produces strong thinning, then much of the glacier's main trunk may unground, leading to further retreat," the authors wrote (Howat et al. 2005). Because the center of the Greenland ice cap is only 150 miles away, researchers fear that it, too, may soon be affected (Lean 2005).

Further Reading

Appenzeller, Tim. "The Big Thaw." *National Geographic*, June 2007, 56–71.

"Greenland's Ice Is Growing Darker." NASA Earth Observatory. January 10, 2012. http:// earthobservatory.nasa.gov/IOTD/view.php?id=76916&src=eoa-iotd.

Gregory, Jonathan M., Philippe Huybrechts, and Sarah C. B. Raper. "Threatened Loss of the Greenland Ice Sheet." *Nature* 428 (April 8, 2004): 616.

Hansen, James, and Makiko Sato. *Update of Greenland Ice Sheet Mass Loss: Exponential?* December 26, 2012. http://www.columbia.edu/~jeh1/mailings/2012/20121226_Gree nlandIceSheetUpdate.pdf.

Hotz, Robert Lee. "Greenland's Ice Sheet Is Slip, Sliding Away." *Los Angeles Times*, June 24, 2006, n.p.

Howat, I. M., et al. "Rapid Retreat and Acceleration of Helheim Glacier, East Greenland." *Geophysical Research Letters* 32 (2005): L22502. doi: 10.1029/2005GL024737.

Jenkins, Mark. "True Colors: The Changing face of Greenland." *National Geographic*, June 2010, 36–43.

Lean, Geoffrey. "The Big Thaw: Global Disaster Will Follow If the Ice Cap on Greenland Melts." *The Independent* (London), November 20, 2005. http://arizonaenergy.org/News_06 /News_Feb06/The%20Big%20Thaw—%20Global%20Disaster%20Will%20Follow%20 If%20the%20Ice%20Cap%20on%20Greenland%20Melts.htm.

Revkin, Andrew. "An Icy Riddle as Big as Greenland." *The New York Times*, June 8, 2004. www.nytimes.com/learning/teachers/featured_articles/20040609wednesday.html.

Revkin, Andrew C. "In Greenland, Ice and Instability." *The New York Times*, January 8, 2008. http://www.nytimes.com/2008/01/08/science/earth/08gree.html.

Reyes, Alberto V., et al. "South Greenland Ice-Sheet Collapse during Marine Isotope Stage 11." *Nature* 510 (June 26, 2014): 525–528. doi: 10.1038/nature13456.

"Satellite Images Show Continued Breakup of Greenland's Largest Glaciers, Predict Disintegration in Near Future." NASA Earth Observatory, August 20, 2008. http://earthobservatory.nasa.gov/Newsroom/MediaAlerts/2008/2008082027361.html (no longer available).

Schiermeier, Quirin. "A Rising Tide: The Ice Covering Greenland Holds Enough Water to Raise the Oceans Six Meters—and It's Starting to Melt." *Nature* 428 (March 11, 2004): 114–115.

"Soot Drives Greenland Melting." *Nature* 509 (May 29, 2014): 537.

Steffensen, Jørgen Peder, et al. "High-Resolution Greenland Ice Core Data Show Abrupt Climate Change Happens in Few Years." *Science* 321 (August 1, 2008): 680–684.

"Widespread Warmth Envelops Greenland." NASA Earth Observatory, May 18, 2016. http://earthobservatory.nasa.gov/IOTD/view.php?id=88048&src=eoa-iotd.

Witze, Alexandra. "Climate Change: Losing Greenland." *Nature* 452 (April 17, 2008): 798–802.

See also: Animal Life, Arctic; Arctic Hunters; Fisheries; Glacial Erosion; Global Warming, Greenland; Ice Melt, Greenland; Ice Science, Greenland; Inuit; Sea Level Rise

COLD SPELLS

Global warming is not linear. As greenhouse gases retain heat and raise surface temperatures generally and over time, many other climatic influences come into play. Weather remains variable over the short range. Some areas may cool as most of the world warms. During the winters of 2013–2014 and 2014–2015, for example, much of the central and eastern United States experienced brutal cold, snow, and ice—even as the rest of the planet posted some of its highest temperatures on record. Boston recorded more than 100 inches of snow, and New York City had its coldest winter in 80 years; on April 2, 2015, the fashion section in *The New York Times* headlined a story "Goodbye Winter Coat, and Good Riddance." At the same time, California roasted in record heat and drought, which provoked mandatory restrictions on water usage. The year 2014, in fact, was then the warmest on the instrumental record and notable for setting this high mark in the absence of a Pacific Ocean El Niño, which usually boosts world temperatures.

Why was Eastern North America so cold during the winter? Arctic air was being displaced southward by erratic movements in the Jet Stream that brought it northward, often as far as Alaska, then plunging southward over the Plains, bringing Arctic air. In a double punch, the upper-air current that steers storms then flowed northeastward off the Atlantic Coast, entwining Gulf Stream warmth and moisture with the Arctic air on the continent, spawning frequent "nor'easters," with strong winds, cold rain, snow, and ice. This atmospheric arrangement has acquired the names *Arctic Oscillation* and *polar vortex*. It has become rather persistent, and has a role in the prolonged hot, dry weather that has plagued the United States Southwest, including California.

Lake-Effect Snow Blasts

Lake-effect snow blitzes have become more common as warm lake water fuels their formation much later into the fall. Historically, the formation of ice on the lakes shut down the lake effect. In November 2001, having had no snow and only 1.5 inches until the third week of December, residents of Buffalo, New York, greeted Christmas Eve with nearly two feet. During the ensuing week, Buffalo had another nearly three-foot storm; with a few other smaller storms, Buffalo ended the year with its snowiest month on record (83 inches), nearly all of which fell in *one* week. How is seven feet of snow in one week evidence of global warming? Lake Erie's water was much warmer than usual; when cold air moved in suddenly, it set up the most ferocious period of lake-effect snowfall in Buffalo's recorded history to that time.

During October 2006, a weird combination of summer and winter weather swept off the lakes. Douglas M. George, a resident on Oneida Castle 30 miles east of Syracuse, recalled the event:

A nice clear day here in Oneida—but 100 miles west the folks in Rochester-Buffalo have found out what global warming really means. A truly unique storm struck that region last night with a foot of "thunder snow" coating the ground, trees, roads and power lines. This was very strange as the water temperature of Lake Erie is a relatively warm 60 degrees but a frigid upper air mass from central Canada flowed through over the waters and dumped its load of moisture as soon as it reached land. Most of the trees in the northeast have their leaves so the weight of the snow-one of the wet, heavy kind-has resulted in many cracking and falling down. . . . Parts of the New York lakeshore near Buffalo got two feet of snow, which quickly turned to mountains of slush. (George 2006)

Another account described the same storm:

Heavy wet snow started falling about 2 p.m. Thursday. Snow built up on the leaves and branches of trees, bending and breaking even the healthiest of trunks. Heightening the clash of seasons, lightning flashed through the night and loud claps of thunder mingled with the cracks and pops of trees giving way under the weight of the snow. "It sounded like gunshots," said Percy Jackson, 18, who spent the day off from school shoveling sidewalks. The falling trees severed power lines, damaged cars and houses, and blocked countless streets. (Staba 2006)

The storm reached east to the Rochester suburbs, snapping power lines and cutting power to almost 400,000 homes and businesses, 70 percent of the city. Three people died during the storm, two in automobile accidents. The third was struck by a falling tree laden with heavy, wet snow. Don Paul, a meteorologist at WIVB-TV4 in Buffalo, who had worked in the area for more than 20 years, said, "Of all the events I've seen here, this storm involves the most widespread devastation in the most populated area. . . . It's absolutely an historic storm" (Staba 2006).

Further Reading

George, Douglas M. Personal communication, October 13, 2006.
Staba, David. "Snowstorm Blankets Buffalo, Killing at Least 3." *The New York Times*, October 14, 2006. http://www.nytimes.com/2006/10/14/nyregion/14storm.html.

According to NASA,

The Arctic Oscillation is a climate pattern that influences winter weather in the Northern Hemisphere. It describes the relationship between high pressure in the mid-latitudes and low pressure over the Arctic. When the pressure systems are weak, the difference between them is small, and air from the Arctic flows south, while warmer air seeps north. ("Arctic Oscillation Chills U.S." 2010)

At least one scientist believes the *polar vortex*—a persistent, massive, low-pressure zone rotating counterclockwise around the North Pole—may be related to shrinking sea ice and warmer waters in the Arctic Ocean. One theory advanced by Rutgers University climate scientist Jennifer Francis holds that melting Arctic sea ice's changes in Arctic weather has consistently distorted the jet stream and pushed Arctic air southward. However, Ryan Maue, a meteorologist who works with Weather Bell Analytics, dismisses Francis's theory, based on what he regards as flaws in her database.

A slower jet stream with more north–south amplitude (producing what are called *Rossby waves*) tends to lock extreme weather features into place and entrench both warm, hot regimes as well as cold, stormy ones. Thus, record heat and drought in the U.S. Southwest, including California, is part of the same pattern that provoked record cold and snowfall in the U.S. Northeast. James Screen of the U.K. University of Exeter and Ian Simmonds of Australia's University of Melbourne in 2014 found "a strong statistical correlation between amplified, slow Rossby waves and months from 1979 to 2012" with extreme weather events (Gramling 2015, 819). In 2013, Vladimir Petoukhov of the Potsdam Institute for Climate Impact Research in Germany correlated summer heat waves in Europe with slow-moving, high-amplitude Rossby waves (Gramling 2015, 819).

In January 2011, another year with a pronounced Arctic Oscillation across much of North America, snow fell in the American Deep South as severe snow and ice swept up the East Coast. In the midst of Arctic high pressure, International Falls, Minnesota, set a low temperature record of −46°F (or −43°C) on January 21. At the same time, NASA's Earth Observatory reported that "areas north of the United States and southern Canada . . . were above normal. In fact, unusual warmth forced residents of Iqaluit, capital of the Canadian territory of Nunavut, to cancel their New Year's snowmobile parade" ("Arctic Oscillation Chills North America" 2011). NASA maps showed a swath of temperatures that were far below average across the eastern and central regions of the United States and temperatures that were much above usual across most of the Canadian Arctic.

The same static "blocking highs" that pull Arctic air southward over the northeastern United States can also convey unusually large amounts of warm, moist air northward over Greenland, where it accelerates melting of the ice sheet. Francis theorizes that the melting of Artic sea ice thus sometimes also contributes to melting in Greenland. "When [blocking highs] do happen, and they kind of set up in just the right spot, they bring a lot of warm, moist air from the North Atlantic up over Greenland, and that helps contribute to increased cloudiness and warming of the surface," Francis said. "When that happens, especially in the summer, we tend to see these melt events occur" (Harvey 2016). One scientific team (Hanna et al. 2016) found that "[since] 1981 there are significant GBI [Greenland blocking index] increases in all seasons and annually, with the strongest monthly increases in July and August. A recent clustering of high GBI values is evident in summer, when seven out of the top 11 values in the last 165 years, including the two latest years 2014 and 2015, occurred since 2007."

Cooler Areas in a Warming World

The winter of 2014–2015 was the warmest on the instrumental record to that time, but on world composites of satellite photographs displaying mainly red hues two areas of dark blue—record low temperatures—stood out in stark contrast. One area over much of northeastern North America had been caught (one of several such winters) in the jet stream bow of the Arctic Oscillation, dragging cold air from northerly latitudes. Another area south and east of Greenland is even more pronounced and indicates accelerating ice melt and freshening of seawater there. This area came to the attention of several scientists, who were building a case for slowdown in the Atlantic meridonal overturning circulation (AMOC), which theory had said would provoke such cooling. As Stefan Rahmstorf and colleagues wrote in *Nature Climate Change*,

> Possible changes in [AMOC] provide a key source of uncertainty regarding future climate change. Maps of temperature trends over the 20th century show a conspicuous region of cooling in the northern Atlantic. Here we present multiple lines of evidence suggesting that this cooling

may be due to a reduction in the AMOC over the 20th century and particularly after 1970. Since 1990 the AMOC seems to have partly recovered. This time evolution is consistently suggested by an AMOC index based on sea surface temperatures, by the hemispheric temperature difference, by coral-based proxies and by oceanic measurements. We discuss a possible contribution of the melting of the Greenland ice sheet to the slowdown. Using a multiproxy temperature reconstruction for the AMOC index suggests that the AMOC weakness after 1975 is an unprecedented event in the past millennium. Further melting of Greenland in the coming decades could contribute to further weakening of the AMOC. (Rahmstorf et al. 2015)

Further Reading

Rahmstorf, Stefan, et al. "Exceptional Twentieth-Century Slowdown in Atlantic Ocean over Turning Circulation." *Nature Climate Change.* March 23, 2015. http://www.nature.com/articles/nclimate2554.epdf.

In some years, cold and snow have been shaped by atmospheric collusion of El Niño in the Pacific Ocean (usually a warming influence) and the Arctic Oscillation, a high-pressure system over Greenland that opens the gates of the Arctic in middle and eastern North America. Whereas El Niño fuels storms by providing more warmth and moisture, the Arctic Oscillation provides cold air. The result is back-to-back blizzards and near-record cold across parts of the U.S. Midwest and East.

While people in the eastern half of the United States were shivering, Greenland's temperatures were above average. As snow and ice piled up in Chicago, at the Winter Olympics near Vancouver, B.C., snow had to be imported to skiing venues with helicopters and trucks because the mountains were too warm for natural snow. Climate-change deniers were not watching events in Greenland, however. In the meantime, global greenhouse gas levels and temperatures kept rising. In Australia, where it was summer, the temperature reached 106°F in Sydney and 108°F in Melbourne. The hot wind was blowing record highs out of the interior of an increasingly dry, fire-ravaged continent.

NASA reported in November 2009 that the period of July through October worldwide was the warmest on the instrumental record (to that time). During the decade 2000–2009 in the continental United States, record high temperatures for individual dates occurred twice as often as record lows. From January 1, 2000, to September 30, 2009, 291,237 daily record highs and 142,420 daily record lows were recorded, according to the National Center for Atmospheric Research.

The 2000–2009 decade was also the warmest on the instrumental record worldwide—a third of a degree warmer than the 1990s. That is a large amount in decadal record keeping, "providing fresh evidence that the planet may be warming at a potentially disastrous rate," according to the National Climatic Data Center. The year 2005 was the warmest on record, at 1.11°C above average (Johansen

2010). The decade of 2000–2009, in fact, included the two warmest years on record to that time, according to NASA. The year 2009 was the second warmest on the instrumental record (since 1880) after 2005. "When we average temperature over five or 10 years to minimize that variability," said James E. Hansen, then the director of NASA's Goddard Institute for Space Studies, one of the world's leading climatologists, "we find global warming is continuing unabated" (Johansen 2010). Hansen's point later was made emphatically in 2014–2016 as world temperatures raced upward.

The Arctic Oscillation also sometimes chills Europe. The same conditions that brought snow and cold to Chicago and Omaha in December 2010 also buried Britain and France in snow. NASA's Earth Observatory commented, "The unusual cold brought heavy snow to Northern Europe, stopping flights and trains early in December. Cold temperatures and snow also closed roads and schools in the eastern United States and Canada during the first week of December" ("Arctic Oscillation Chills U.S." 2010).

The same pattern occurred again during the winters of 2012 through 2015. Once again, as Justin Gillis wrote in *The New York Times* (2014), "For people throughout the Eastern United States who spent January slipping, sliding and shivering, here is a counterintuitive fact: For the Earth as a whole, it was the fourth-warmest January on record. It was, in fact, the 347th consecutive month with temperatures above the 20th-century average." (The United States occupies only 2 percent of the world's surface, so its temperatures have little impact on the world average.) South Africa, Brazil, large parts of Europe, and China were much above average in January 2014.

As New Yorkers and Chicagoans shoveled and shivered, temperatures were 10 to 15°F above average in drought-stricken California and more than 15°F higher than usual in Alaska. With NASA's temperature maps riven with red (indicating above-average temperatures), the eastern United States stood out as an odd pocket of below-average blue. No one there set a January record for cold, however, although Alabama, having been afflicted by two major ice storms, had its fourth-coldest January on the instrumental record.

Even with a boost from the Arctic Oscillation, severe cold spells in the United States were becoming less frequent, according to Greg Carbin, a meteorologist for NOAA. The cold spell of January 2014 was the worst in 17 years (and was largely repeated a year later) but set few records. The two decades from 1970 to 1989 contained 17 cold periods as intense as January 2014, one of two since 2000. According to Jeff Masters, meteorology director of the private firm Weather Underground, "It's become a lot harder to get these extreme (cold) outbreaks in a planet that's warming." "These types of events have actually become more infrequent than they were in the past," said Carbin, who works at the Storm Prediction Center in Norman, Oklahoma. "This is why there was such a big buzz because people have such short memories" (Borenstein 2014).

According to the NASA Earth Observatory, in February 2015 the Arctic Oscillation was so firmly locked in that large swaths of Arizona, Colorado, Idaho, Nevada, Oregon, Utah, and Wyoming experienced temperatures that were more than 10°C (18°F) above average. Other states in the Midwest, mid-Atlantic region, and New

England were 10°C below normal. Meanwhile, many cities saw air temperature records fall. Worcester, Massachusetts, with an average air temperature of −10°C (14°F), had its coldest month on record. Records also fell in Bangor, Maine; Marquette, Michigan; and Syracuse, Buffalo, and Rochester, New York. In many cases, records were not simply broken—they were obliterated. Syracuse broke the old record by 3°F (1.6°C) and Bangor did so by 2.3°F even though such records are usually broken by just a fraction of a degree. Meanwhile, San Francisco, Seattle, Portland, Las Vegas, and Salt Lake City all experienced their warmest winter months on record. This pattern was being held in place by a persistent ridge of high pressure in the northeastern Pacific Ocean, which became so firmly anchored that observers nicknamed it the "ridiculously resilient ridge." This pattern has persisted over several winters ("Going Hot" 2015).

Further Reading

"Arctic Oscillation Chills North America, Warms Arctic." NASA Earth Observatory, January 26, 2011. http://earthobservatory.nasa.gov/IOTD/view.php?id=48882&src=eoa-iotd.

"Arctic Oscillation Chills U.S. and Europe." NASA Earth Observatory, December 17, 2010. http://earthobservatory.nasa.gov/IOTD/view.php?id=47880&src=eoa-iotd.

Borenstein, Seth. "Scientists: Americans Are Becoming Weather Wimps." *Omaha World-Herald*, January 10, 2014, 3A. http://www.omaha.com/article/20140109/AP01/301099 812 (no longer available).

Gillis, Justin. "Freezing January for Easterners Was Not Felt Round the World." *The New York Times*, February 20, 2014. http://www.nytimes.com/2014/02/21/science/earth/more -bite-left-to-winter-but-it-hasnt-been-as-bad-as-you-think.html.

"Going Hot and Cold in February." NASA Earth Observatory, March 10, 2015. http:// earthobservatory.nasa.gov/IOTD/view.php?id=85460&src=eoa-iotd.

Gramling, Carolyn. "Arctic Impact: Is the Melting Arctic Really Bringing Frigid Winters to North America and Eurasia?" *Science* 347 (February 20, 2015): 818–821.

Hanna, E., et al. "Greenland Blocking Index 1851–2015: A Regional Climate Change Signal." *International Journal of Climatology* (May 2016), JOC-15-0742.R1.

Harvey, Chelsea. "Dominoes Fall: Vanishing Arctic Ice Shifts Jet Stream, Which Melts Greenland Glaciers." *Washington Post*, May 2, 2016. https://www.washingtonpost.com/news /energy-environment/wp/2016/05/02/dominoes-fall-vanishing-arctic-ice-shifts-jet -stream-melts-greenland-glaciers/?utm_term=.a8421ba10c66.

Johansen, Bruce E. "Has Our Nasty Winter Ended Global Warming? We Should Be So Lucky." *The Nebraska Report*, April 2010. http://nebraskansforpeace.org/nasty-winter -global-warming (no longer available).

See also: Atmospheric Circulation; Climate Change, Abrupt Nature of; Drought, Worldwide; El Niño and La Niña; Thermal Inertia

EL NIÑO AND LA NIÑA

The El Niño and La Niña (El Niño Southern Oscillation, or ENSO) climate cycle is Earth's most prominent source of short-term climate change. It alters upper-air storm tracks, brings droughts to areas that usually receive abundant rainfall, and delivers deluges to some of the driest deserts in the world. El Niño, which warms the Pacific

Ocean near the equator west of South America and then extends westward, generally raises worldwide temperatures. La Niña, which occurs in the same area, usually has the opposite effect. These conditions often set in very precisely within days of Christmas, thus their name—a reference in Spanish to the Christ child.

Ocean Warming, El Niño, and Marine Diseases

Beginning in the 1980s, as oceans began to warm rapidly, scientists began to assess the increasing toll of diseases affecting various forms of marine life and often the people who eat them. Such waves of disease peaked during El Niño periods when ocean waters became exceptionally warm.

In November 1987, a wave of illnesses afflicted more than 150 people who ate seafood harvested along the shores of Canada's Prince Edward Island. Those who suffered endured headaches, vomiting, diarrhea, nausea, abdominal cramps, short-term memory, problems breathing, and seizures. The malady quickly acquired a name—*amnesic shellfish poisoning*. At least four people died. All of those afflicted with this condition had eaten cultivated mussels from the same inlet, Cardigan Bay, which had experienced an algal bloom that produced domoic acid. Later, this was associated with a worldwide El Niño event and warmer than usual temperatures in the Gulf Stream (Epstein and Ferber 2011, 127–128).

During the mid-1990s, the same conditions affected North America's western coast. Birds were found dead or dying, sea lions were disoriented, and Native peoples who ate mussels and razor clams became ill.

Further Reading

Epstein, Paul R., and Dan Ferber. *Changing Planet, Changing Health: How the Climate Crisis Threatens Our Health and What We Can Do about It*. Berkeley: University of California Press, 2011.

Climate Change and El Niño

El Niño, a warming of the Pacific Ocean, raises temperatures worldwide and can play a role in shifting patterns of drought and deluge that have intensified as global temperatures warm. Floods afflict the usually dry west coast of South America, while the Amazon River valley dries. Drought provokes wildfires in the usually wet jungles of Indonesia, and Ethiopia suffers famine when seasonal rains fail. Australia braces for wildfires, as California finds itself occasionally flooded after having suffered its worst drought in at least 1,000 years.

First described in the late 1880s by a Peruvian navy captain who reported a warm "corriente del Niño" (ocean current of the Christ child) off Peru's coast,

El Niño may occur more often as temperatures warm. Some scientists assert that a warming atmosphere does not cause El Niños, but does contribute to an irregular rise in temperatures provoked by rising greenhouse gas emissions. "We have no reason at this point to think that El Niño itself is responding to the forcing from greenhouse gases," said Richard Seager, a climate scientist at the Lamont–Doherty Earth Observatory of Columbia University. "You can think of them as independent and adding to each other" (Gillis 2015).

Other studies have asserted a relationship. A study by Wenju Cai, a physical oceanographer with the Commonwealth Scientific and Industrial Research Organization in Aspendale, Australia, suggests that by the end of the century extreme El Niños such as the 1997–1998 and 2015–2016 events may occur twice as often as in the early 21st century (Cai et al. 2014).

Further Reading

Cai, Wenju, et al. "Increasing Frequency of Extreme El Niño Events Due to Greenhouse Warming." *Nature Climate Change* 4 (February 2014): 111–116.
Gillis, Justin. "2015 Likely to Be Hottest Year Ever Recorded." *The New York Times,* October 22, 2015. http://www.nytimes.com/2015/10/22/science/2015-likely-to-be -hottest-year-ever-recorded.html.

By 2016, scientists were able to detect human evidence of global warming in detailed, long-term ocean temperature records derived from corals on Christmas Island in Kiribati and other islands in the tropical Pacific Ocean. These records indicate that late 20th and early 21st century El Niño events reflect not just the natural ocean-atmosphere cycle but a new factor: global warming caused by human activity.

Christopher Pala wrote in *Science* (2016, 1210), "Over the last 7,000 years, El Niños, which warm the eastern Pacific, waxed and waned. Then, during the 20th century, their intensity began to climb. The trend is likely to continue, boding ever-more-destructive El Niños in the future. The finding helps settle a long-standing debate about the role of global warming in these events, which had been hard to resolve because records are short and spotty in the remote parts of the Pacific where El Niño hits hardest."

In 2015 and 2016 global temperatures surged, with a powerful assist from a strong El Niño pattern in the equatorial Pacific Ocean, which has raised ocean temperatures in that area as much as 3°C (5.4°F) above 20th-century averages, an all-time record. Global air temperatures set unprecedented highs on the instrumental record (since 1880) in 2015, far above previous El Niño years such as 1997 and 1998, indicating an underlying warming trend from rising greenhouse gas levels in the atmosphere. "The bottom line is that the world is warming," said Jessica Blunden, a climate scientist with NOAA in Asheville, North Carolina. "We're seeing it all across the Indian Ocean, in huge parts of the Atlantic Ocean, in parts of the Arctic oceans,"

Dr. Blunden said in an interview. "It's just incredible to me. I've never seen anything like this before" (Gillis 2015).

During the first three months of 2016, worldwide temperatures continued to set records as heat accumulated in the Pacific Ocean during the El Niño (which ended in November 2015) continued to diffuse into the atmosphere. Temperatures by that time had set worldwide records for 11 consecutive months at levels nearly 0.8°F above the previous El Niño–aided peak in 1997–1998.

The so-called Godzilla El Niño of 2015–2016 ended with some improvement in Sierra Nevada snowpacks—but less than an average year (87 percent). The previous year, snowpack had been only 5 percent of average (Fountain 2016). The Sierra Nevada snowpack provides approximately 30 percent of California's water. Southern California received only a handful of damaging deluges and remained in drought. Warming temperatures may have affected atmospheric circulation, moving Hadley cells (and the jet stream) farther north than in previous El Niños. The Pacific Northwest, which is usually relatively dry during such periods, received more rain and snow than during other El Niños.

During December 2015, January 2016, and February 2016 (meteorological winter), temperatures not only set world records but also did so by the largest margins (anomalies) since record keeping began around 1880. February's global temperature was 1.35°C above the 1951–1980 average, exceeding the previous record anomaly set in January, 1.13°C, according to NASA's Goddard Institute for Space Studies. December 2015 was 1.11°C above the same set of averages. Higher latitudes were the warmest: "Much of Alaska into western and central Canada, as well as eastern Europe, Scandinavia and much of Russia were at least 4 [°C] (roughly 7 degrees Fahrenheit) above February averages, according to NASA/GISS," Weather .com reported (2016). Alaska had its second warmest winter on record, almost 11°F above average. At the end of February, Anchorage had no snow on the ground for the first time on record. By April 2016, with El Niño waning, record-high global temperatures had been set for seven consecutive months by substantial margins, according to NASA.

Justin Gillis wrote in *The New York Times* (2015, October 22) that the combined effects of El Niño and greenhouse warming "probably [are] contributing to dry weather and forest fires in Indonesia, to an incipient drought in Australia and to a developing food emergency across parts of Africa, including a severe drought in Ethiopia. Those effects are likely to intensify in coming months as the El Niño reaches its peak and then gradually subsides." In Zambia, the worst drought in several decades severely limited industrial production because the country relies on hydroelectricity for 95 percent of its power.

El Niño and Tropical Cyclones

As temperatures rose rapidly in 2015, Hurricane Patricia became the strongest tropical cyclone in the recorded history of the Western Hemisphere, with sustained winds of 200 miles per hour, when it struck Mexico's western coast during the third week of October. The storm exploded from a tropical storm to a category 5 hurricane in one day, the most rapid intensification of any such storm on record. The hurricane's development was stoked by 86°F water temperatures produced by El Niño conditions. "Patricia is estimated to have intensified [by 100 miles per hour] in the past 24 hours," the National Hurricane Center said in one update. "This is a remarkable feat, with only Linda of 1997 intensifying at this rate in the satellite era" (Samenow 2015). The storm came ashore near Puerto Vallarta on October 23.

Two weeks or so after Hurricane Patricia revved up overnight to category 5 off Mexico's west coast, a cyclone named Chapala made an extremely rare landfall in Yemen on the southern Arabian peninsula on November 4. Like Patricia, this storm earlier had exploded in one day from tropical storm status to become a category 4 cyclone over a bathtub-warm Indian Ocean. At its peak, NASA satellite photos showed the enormous storm's swirling clouds reaching from west of Mumbai to the Arabian peninsula. With extremely heavy rain falling on steep terrain (three to four years' the area's annual average of about 2 inches in a desert), flash floods inundated the war-ravaged port of Al Mukalla, population 300,000 (Weather.com 2015). It was the first cyclone-intensity storm in the area since 1945, and the most intense on record. The storm rapidly weakened as it neared land because dry air was sucked into its circulation pattern. Following its first cyclone since records began in 1945, Yemen was struck by *another* one within the same week—Cyclone Megh.   .

Warming and El Niño Cycles

With a warming climate, the number of El Niño and La Niño events seems to increase, and the two events tend to switch back and forth more emphatically, according to a study in *Nature Climate Change* (Kim et al. 2014). "Most of the severe La Ninas will follow severe El Niños, resulting in wide, annual swings between opposite extreme weather events," according to an analysis of the paper in *Nature* ("Big Swings" 2015).

Seon Tae Kim and colleagues wrote:

The destructive environmental and socioeconomic impacts of the El Niño [and] Southern Oscillation (ENSO) demand an improved understanding of how ENSO will change under future greenhouse warming. Robust projected changes in certain aspects of ENSO have been recently established. However, there is as yet no consensus on the change in the magnitude of the associated sea surface temperature (SST) variability, commonly used to represent ENSO amplitude, despite its strong effects on marine ecosystems and rainfall worldwide. Here we show that the response of ENSO SST amplitude is time-varying, with an increasing trend in ENSO amplitude before 2040, followed by a decreasing trend thereafter. . . . The nine most realistic models identified show a strong consensus on the time-varying response and reveal that the non-unidirectional behaviour is linked to a longitudinal difference in the surface warming rate across the Indo-Pacific basin. Our results carry important implications for climate projections and climate adaptation pathways. (Kim et al. 2014)

Sea surface temperature changes also can move opposite the usual, as "anomalies propagate eastwards during extreme El Niño events, prominently in the post-1976 period. . . . Despite decades of research, its behavior continues to challenge scientists." Some science indicates that eastward-moving El Niños increase as climate warms (Santoso et al. 2013, 126).

> Here we trace the cause of the asymmetry to the variations in upper ocean currents in the equatorial Pacific, whereby the westward-flowing currents are enhanced during La Niña events but reversed during extreme El Niño events. Our results highlight that propagation asymmetry is favored when the westward mean equatorial currents weaken, as is projected to be the case under global warming. By analyzing past and future climate simulations of an ensemble of models with more realistic propagation, we find a doubling in the occurrences of El Niño events that feature prominent eastward propagation characteristics in a warmer world. Our analysis thus suggests that more frequent emergence of propagation asymmetry will be an indication of the Earth's warming climate. (Santoso et al. 2013, 126)

The Pliocene Epoch's climatic conditions were similar to that of Earth with warming as projected by the Intergovernmental Panel on Climate Change (IPCC) for 2100 with thermal inertia worked in (in the case of the oceans, perhaps 100 to 200 years later). These include reduced temperature differences between the tropics and polar regions (especially the Arctic), a nearly permanent El Niño–like condition in the Pacific Ocean, oceans roughly 25 meters higher than today, and little or no year-round ice in the Arctic. "[T]entative evidence that the tropical belt has been expanding poleward over the pasty few decades make our finings especially relevant to current discussions about global warming," Chris M. Brierley and colleagues wrote in *Science* (2009, 1717).

Scientists have shown that an El Niño "discharges heat into the eastern North Pacific basin two to three seasons after its wintertime peak," sometimes leading to intensified tropical cyclones called *typhoons* (Jin et al. 2014, 82). Because they enhance upper-atmospheric wind shear by warming parts of the Pacific Ocean, El Niño conditions impede the development of tropical cyclones (hurricanes) in the North Atlantic Ocean.

As Jin et al. wrote, "As a result of the time involved in ocean transport, El Niño's equatorial subsurface 'heat reservoir', built up in boreal winter, appears in the eastern North Pacific several months later during peak tropical cyclone season (boreal summer and autumn). By means of this delayed ocean transport mechanism, ENSO provides an additional heat supply favorable for the formation of strong [tropical storms]" (Jin et al. 2014, 82).

El Niño, Surface Temperatures, and Politics

The year 2014 was the first on the instrumental record to set a record global high temperature without the aid of a strong El Niño. In fact, several scientists said that fact alone distinguished the year as a temperature landmark. *The New York Times* quoted Gavin A. Schmidt, director of NASA's Goddard Institute for Space Studies in New York City, as saying that "the next strong El Niño would probably rout all temperature records." "Why do we keep getting so many record-warm years?" Schmidt asked in an interview. "It's because the planet is warming. The basic issue is the long-term trend, and it is not going away" (Gillis 2014). In 2015 and 2016, Schmidt's forecast was realized.

El Niños have already played a role in the politics surrounding climate science. The 1990s included several El Niño episodes, the strongest of which peaked in 1998, along with record high global temperatures by a wide margin. Temperatures cooled somewhat in subsequent years, before slowly rising back to the record high of 1998 between 2010 and 2014—in the general absence of El Niño conditions. Climate-change deniers called his pause a "hiatus" that brought into question the entire "theory" (as they termed it) that temperatures were rising in concordance with greenhouse gas levels. However, as 1998's record was approached 15 years later under neutral or La Niña (cooler ocean) conditions, the trajectory of the average remained upward.

"Most climate scientists argue that the climate models never predicted steady, uninterrupted warming," wrote Nate Cohn in *The New York Times* (2014).

> Annual global temperatures always rise and fall on either side of the longer-term average, in much the same way that the stocks rise and fall from day to day, even during a strong market. They believe, based on computer simulations of hiatus periods and measurements from new floating sensors, they can account for the "missing" heat, much of which they believe is deep in the ocean, more than 700 meters below the water's surface. (Cohn 2014)

A strong El Niño might provoke record highs by a substantial margin later in the decade.

"But El Niño has the potential to do more than offer a one-time jolt to climate activists," wrote Cohn, who is primarily a political analyst (2014).

> It could unleash a new wave of warming that could shape the debate for a decade, or longer. In this chain of events, a strong El Niño causes a shift in a longer cycle known as the Pacific Decadal Oscillation, which favors more frequent and intense El Niños during its "warm" or "positive" phase. The oscillation has been "negative" or "cool" since the historic El Niño of 1998.

Cohen speculated that a resurgent El Niño could influence 2016's presidential campaign in the United States. Temperatures did rise to record levels that year, but global warming was largely ignored as Donald J. Trump won, waging a campaign of personal denigration.

El Niño and Carbon-Creating Peat Fires

Humans harvest peat bogs, especially in Southeast Asia, for use as fuel. Widespread use, however, has made peat an important source of carbon dioxide that can exceed (in some regions) that of transportation, industry, and energy generation.

An important case study has been provided by Indonesian fires that polluted air over Southeast Asia during the El Niño years of 1997 and 1998, which provoked drought in Southeast Asia. Roughly 60,000 kilometers of peat swamps, an area twice that of Belgium, dried and burned in Indonesia in 1997. Susan Page, with Britain's University of Leicester, together with colleagues in England, Germany, and Indonesia, analyzed satellite photos and data gathered on the ground to estimate how much of the fire area's living vegetation and peat deposits burned (Cowen 2002; Page et al. 2002).

Robert Cowen of the *Christian Science Monitor* sketched the situation:

In Indonesia, nature and human activity had prepared a massive subsurface fuel reservoir. Tropical forests built up thick peat deposits as vegetation died and decayed over many centuries. Forest clearance and drainage for logging and farming have tended to dry the peat. Drought due to the 1997 El Niño was all that was needed to make the circumstances right for a sustained conflagration when forest-clearing fires were lit that year. (Cowen 2002)

This situation was repeated during the strong El Niño of 2015–2016.

El Niños also often reverse the upper-air flow over South America from east–west to west–east, provoking drought in the Amazon River valley and heavy rain on the continent's west coast, which is usually extremely dry. The change profoundly disrupts ecosystems in both areas. Hot, dry weather turns the Amazon into a net producer of carbon dioxide. In a usual year, the Amazon absorbs 700 million tons of carbon dioxide, but the Amazon forest *added* 200 million tons of carbon dioxide to the atmosphere in the El Niño year of 1995.

Further Reading

"Big Swings in Weather to Come." *Nature* 517 (January 29, 2015): 530–531.

Brierley, Chris M., et al. "Greatly Expanded Tropical Warm Pool and Weakened Hadley Circulation in the Early Pliocene." *Science* 323 (March 27, 2009): 1714–1718.

Cohn, Nate. "How El Niño Might Alter the Political Climate." *The New York Times*, May 20, 2014. http://www.nytimes.com/2014/05/20/upshot/how-el-nino-might-alter-the-political-climate.html.

Cowen, Robert C. "One Large, Overlooked Factor in Global Warming: Tropical Forest Fires." *Christian Science Monitor*, November 7, 2002, 14.

Fountain, Henry. "Sierra Nevada Snow Won't End California's Thirst." *The New York Times*, April 12, 2016. http://www.nytimes.com/2016/04/12/science/california-snow-drought-sierra-nevada-water.html.

Gillis, Justin. "U.N. Says Lag in Confronting Climate Woes Will Be Costly." *The New York Times*, January 16, 2014. http://www.nytimes.com/2014/01/17/science/earth/un-says-lag-in-confronting-climate-woes-will-be-costly.html.

Gillis, Justin. "2015 Likely to Be Hottest Year Ever Recorded." *The New York Times,* October 22, 2015. http://www.nytimes.com/2015/10/22/science/2015-likely-to-be-hottest-year-ever-recorded.html.

Jin, F.-F., J. Boucharel, and I.-I. Lin. "Eastern Pacific Tropical Cyclones Intensified by El Niño Delivery of Subsurface Ocean Heat." *Nature* 516 (December 4, 2014): 82–85.

Kim, Seon Tae, et al. "Response of El Niño Sea Surface Temperature Variability to Greenhouse Warming." *Nature Climate Change,* August 2014. http://www.researchgate.net/publication/264424604_Response_of_El_Nio_sea_surface_temperature_variability_to_greenhouse_warming.

Page, Susan E., et al. "The Amount of Carbon Released from Peat and Forest Fires in Indonesia During 1997." *Nature* 420 (November 7, 2002): 61–65.

Pala, Christopher. "Corals Tie Stronger El Niños to Climate Change." *Science* 354 (December 9, 2016): 1210.

Samenow, Jason. "'Potentially Catastrophic' Category 5 Hurricane Patricia to Slam West Coast of Mexico Friday." *Washington Post*, October 22, 2015. https://www.washingtonpost.com/news/capital-weather-gang/wp/2015/10/22/extremely-dangerous-category-4-hurricane-patricia-to-slam-west-coast-of-mexico-friday/?wpmm=1&wpisrc=nl_headlines.

Santoso, Agus, et al. "Late-Twentieth-Century Emergence of the El Niño Propagation Asymmetry and Future Projections." *Nature* 504 (2013): 126–130.

Weather.com. "Cyclone Chapala in the Arabian Sea Likely to Be Rare, Destructive Landfall in Yemen." November 3, 2015. http://www.weather.com/storms/hurricane/news/cyclone-chapala-yemen-oman-arabian-peninsula.

Weather.com. "February 2016 Was the Most Abnormally Warm Month Ever Recorded, Topping January 2016, NOAA and NASA Say." March 14, 2016. https://weather.com/news/climate/news/record-warmest-february-global-2016.

See also: Desertification; Drought, United States; Drought, Worldwide; Extreme Weather; Heat Waves, Evidence of; Heat Waves, Forecasts of; Ocean Circulation; Temperatures, Global; Temperatures, Greenhouse Gas Levels and

EMISSIONS, CARBON DIOXIDE, AND PEAT

Human burning of peat bogs, especially in Southeast Asia, has become an important source of carbon dioxide that can exceed (in some regions) that of transportation, industry, and energy generation. An important case study has been provided by Indonesian fires that polluted air over Southeast Asia during the El Niño years of 1997 and 1998 (and again in 2015–2016), which provoked drought in Southeast Asia. Roughly 60,000 kilometers of peat swamps, an area twice that of Belgium, dried up and burned in Indonesia in 1997 (Richardson 1997, 5). The University of Leicester's Susan Page, together with colleagues in England, Germany, and Indonesia, analyzed satellite photos and data gathered on the ground to estimate how much of the fire area's living vegetation and peat deposits burned (Cowen 2002; Page et al. 2002).

Robert Cowen of the *Christian Science Monitor* sketched the situation:

In Indonesia, nature and human activity had prepared a massive subsurface fuel reservoir. Tropical forests built up thick peat deposits as vegetation died and decayed over many centuries. Forest clearance and drainage for logging and farming have tended to dry the peat. Drought due to the 1997 El Niño was all that was needed to make the circumstances right for a sustained conflagration when forest-clearing fires were lit that year. (Cowen, 2002)

Research indicates that human disturbance of peat swamp forests in Southeast Asia is responsible for the increased release of carbon dioxide previously stored for several thousand years. Writing in *Nature* (2013, 660), Sam Moore and colleagues explained:

Tropical peat lands contain one of the largest pools of terrestrial organic carbon, amounting to about 89,000 teragrams (1 teragram is a billion kilograms). Approximately 65 percent of this carbon store is in Indonesia, where extensive anthropogenic degradation in the form of deforestation, drainage and fire are converting it into a globally significant source of atmospheric carbon dioxide. Here we quantify the annual export of fluvial organic carbon from both intact peat swamp forest and peat swamp forest subject to past anthropogenic disturbance. We find that the total fluvial organic carbon flux from disturbed peat swamp forest is about 50 percent larger than that from intact peat swamp forest.

By carbon-14 dating of dissolved organic carbon (which makes up over 91 percent of total organic carbon), we find that leaching of dissolved organic carbon from intact peat swamp forest is derived mainly from recent primary production (plant growth). In contrast, dissolved organic carbon from disturbed peat swamp forest consists mostly of much older (centuries to millennia) carbon from deep within the peat column. When we include the fluvial carbon loss term, which is often ignored, in the peat land carbon budget, we find that it increases the estimate of total carbon lost from the disturbed peat lands in our study by 22 percent. We further estimate that since 1990 peat land disturbance has resulted in a 32 percent increase in fluvial organic carbon flux from southeast Asia—an increase that is more than half of the entire annual fluvial organic carbon flux from all European peat lands.

The emitted carbon dioxide from these fires was massive, especially when added to the many other fires that have burned around the globe, notably during 2002 in North America during an intense drought. The work of Page and colleagues has major implications for climate-change modeling:

Tropical peat lands are one of the largest near-surface reserves of terrestrial organic carbon, and hence their stability has important implications for climate change. In their natural state, lowland tropical peat lands support a luxuriant growth of peat swamp forest overlying peat deposits up to 20 meters thick. Persistent environmental change—in particular, drainage and forest clearing—threatens their stability, and makes them susceptible to fire. This was demonstrated by the occurrence of widespread fires throughout the forested peat lands of Indonesia during the 1997 El Niño event. (Page et al. 2002, 61)

In Indonesia, layers of peat as thick as 20 meters (66 feet) cover an area of some 180,000 square kilometers (112,000 square miles) in Kalimantan (Borneo), Sumatra, and Papua New Guinea (formerly Irian Jaya) (Richardson 2002, 5). Page and colleagues used satellite images of a 2.5 million hectare study area in Central Kalimantan from before and after the 1997 fires. According to their estimates, 32 percent of the area had burned, of which peat land accounted for 91.5 percent. An estimated 0.19 to 0.23 gigatons of carbon were released to the atmosphere through peat combustion, and another 0.05 gigaton was released when the overlying vegetation was burned. Extrapolating these estimates to Indonesia as a whole, the researchers estimated that between 0.81 and 2.57 gigatons of carbon were released to the atmosphere in 1997 as a result of burning peat and vegetation in Indonesia (Page et al. 2002, 61). According to the researchers, "This is equivalent to [between] 13 [and] 40 percent of the mean annual global carbon emissions from fossil fuels," which contributed measurably to the largest annual increase in atmospheric CO_2 concentration detected since records began in 1957 (Page et al. 2002, 61).

Fires and Global Greenhouse Gas Emissions

Jack Rieley of the University of Nottingham, United Kingdom, believes that burning peat in Borneo is a major factor in rapidly rising global atmospheric carbon dioxide levels. As farmers continue to clear the forests by burning, the bogs catch fire and release carbon for months afterward. A biologist from Borneo told *New Scientist* late in 2004 that the fires had returned after an earlier peak during an El Niño–provoked drought 1998. "During October [2004], the atmosphere around Palangka Raya has been covered in thick smoke, with visibility down to 100 meters. The schools have been shut and flights cancelled," said Suwido Limin from the University of Palangkaraya Raya in the Indonesian province of Central Kalimantan (Johansen 2005).

The fires in Indonesia had other environmental effects as well. Iron fertilization of Indian Ocean waters resulting from the massive wildfires may have played a crucial role in producing a red tide of historic proportions that severely damaged coral reefs, according to Nerilie J. Abram and associates, writing in *Science*. Their findings "highlight tropical wildfires as an escalating threat to coastal marine ecosystems" (Abram et al. 2003, 952).

Indonesia was only the worst of several wildfire sites during the summer of 1998. A "ring of fire circled the globe as rain forests burned: 2,150 square miles in Central America; 1,500 square miles in Mexico in the worst drought in 70 years; 20,000 square miles in Brazil, an area about the size of Slovakia. Fires also engulfed huge tracts of Siberian Russia, Canada, Kenya, Rwanda, Tanzania, Senegal, the Congo, and Florida in the United States" (Brandenburg and Paxson 1999, 241).

Largely because of peak burning and decomposition, which make up 90 percent of its carbon dioxide emissions, Indonesia by 1998 was the third-largest national source of anthropomorphic CO_2 in the world after China (6.16 billion metric tons in 2006) and the United States (5.81 billion). Indonesia's 2.23 billion metric tons of carbon dioxide are mainly produced when peat bogs are drained, dried, and then cleared and burned for agriculture, including booming palm oil plantations. Diffusion of peat into the atmosphere in Indonesia puts as much carbon dioxide into the air as the total emissions of Britain, Canada, and Germany combined, Or as much as car and airplane travel in the United States ("Indonesia" 2009).

The spread of human populations is aggravating fire dangers around the world. The fires that ravaged much of Indonesia during 1997 and 1998 were caused in part by drought-provoked El Niño conditions. They were intensified, however, by fires set by peasants hired by local speculators as they opened forestland for farming, grazing, and other forms of development. The fires were illegal under Indonesian law when they were set in protected areas—but not if they could be blamed on El Niño, a natural condition. At least 29 companies later were indicted for setting illegal fires in Indonesia's rain forests (Glantz 2003, 196–197).

Indonesia Peat Fires Continue

Haze from forest fires stoked by smoldering peat as well as slash-and-burn land clearance for agriculture have become a frequent feature of life in much of Southeast Asia. As Joe Cochrane reported for *The New York Times* (2015, October 8), as a new El Niño set in during 2015: "Savir Singh's taxi rolled into downtown Singapore, taking an overpass that provides a stunning view of the popular hotels and tourist attractions around Marina Bay. The only problem was that he could barely see them. Thick haze from forest fires set in neighboring Indonesia to clear land for agriculture has blanketed this island state for weeks, and has spread to Malaysia and southern Thailand." In 2015 and 2016, drought stoked by El Niño conditions made the haze more pervasive than people living in this area had ever seen.

Life was being disrupted across the region as flights were grounded and schools closed. People in the tens of thousands crowded hospitals with respiratory ailments, eczema, and allergies. A world-class swimming meet, the FINA Swimming World Cup, was canceled in October 2015 because of the choking, acrid haze that came and went, sometimes for weeks at a time. The haze will continue, wrote Cochrane (2015, October 8), "until rain forest deforestation is severely curtailed in Indonesia, where it remains rampant, and the Indonesian government bans the draining and clearing of peat land for agricultural use. Currently, Indonesian government policy allows peat land of less than about nine feet deep to be cleared."

After several weeks of intense haze with short breaks offered by occasional heavy rains, Luhut B. Pandjaitan, Indonesia's coordinating minister for political, legal, and security affairs, admitted that the government had made a mistake when it approved palm oil concessions on 14.8 million acres of peatlands during the preceding decade. The land was drained and burned to strip forest cover, causing fires that released a stifling haze during dry weather, and added immense amounts of carbon dioxide to the atmosphere (Cochrane 2015, October 29).

Peat Fires and Climate Models

Commenting on Page et al.'s study in *Nature*, David Schimel and David Baker of the National Center for Atmospheric Research in Boulder, Colorado, noted that two other independent studies of atmospheric carbon dioxide concentrations during that time period support their conclusion that the fires were a major contributor to atmospheric carbon dioxide levels. Schimel and Baker explained that computer climate simulations assume that processes that emit carbon dioxide and remove it from the atmosphere operate smoothly and continuously (Cowen 2002; Schimel and Baker 2002). Episodic events such as wildfires play havoc with such simulations.

Currently, no climate modeler knows exactly how to forecast catastrophic events in small areas that release carbon dioxide that was locked away in peat or other carbon and methane reservoirs on a world scale. Such events "can evidently have a huge impact on the global carbon balance" Schimel and Baker wrote (Cowen 2002). During 1997, the growth rate of carbon dioxide in the atmosphere was double the usual rate, reaching its highest level on record to that time, largely because of these peat fires. Most of the carbon injected into the atmosphere during the Indonesian fires resulted from peat burning rather than combustion of trees (Schimel and Baker 2002). Indonesia's peat is unusually dense and high in carbon content. Today's satellites can detect land-use patterns on a 50-kilometer resolution; an estimated 30-meter resolution will be required to factor effects such as Indonesia's peat fires into global climate models. This is important because relatively small-scale events "can have an appreciable effect on the carbon cycle" (Schimel and Baker 2002).

"Kick-Starting" Greenhouse Gases from Peat

In a series of experiments described in *Nature*, researchers found that the increase in atmospheric carbon dioxide observed in recent decades has had a direct impact on kick-starting the release of carbon from peat (Freeman et al. 2004, 195). "Under elevated carbon dioxide levels, the proportion of dissolved organic carbon derived from recently assimilated carbon dioxide was ten times higher than that of controlled cases," they wrote. "Concentrations of dissolved organic carbon appear far more sensitive to environmental drivers that affect net primary productivity than those affecting decomposition alone" (Freeman et al. 2004, 195). This research is the first to measure and forecast the release of carbon dioxide from peat bogs.

This research was among the first direct physical evidence of a positive feedback between carbon dioxide in the atmosphere and the huge stores of carbon locked up on land, with increases in one causing a corresponding rise in the other. Chris Freeman of the University of Wales in Bangor, leader of the research team, said, "We've got an enormous carbon store locked up in peat bogs which is equivalent to the entire store of carbon in the atmosphere and yet this store on land appears to have sprung a leak," Freeman said (Connor 2004). The amount of carbon dioxide being released from peat lands is accelerating roughly at a rate of 6 percent per year, according to the research of Freeman and colleagues. "By 2060 we could see more carbon dioxide being released into the atmosphere than is being released by burning fossil fuel," he said (Connor 2004).

Tests on peat samples taken from three different sites in Britain show that when the amount of carbon dioxide in the air around the samples increases, the peat itself emits up to 10 times the amount of carbon dioxide that it would have under usual conditions. Freeman said that peat bogs release carbon in a dissolved organic form. Emissions from peat lands into surrounding rivers and streams has increased by between 65 and 90 percent in past six years (Connor 2004).

"The rate of acceleration suggests that we have disturbed something critical that controls the stability of the carbon cycle on our planet," Freeman said. Dissolved organic carbon in rivers and watercourses also may react with the chlorine in water-treatment processes to produce potentially carcinogenic chemicals, he said (Connor 2004). "We've known for some time that CO_2 levels have been rising and that these could cause global warming. But this new research has enormous implications because it shows that even without global warming, rising CO_2 can damage our environment," he added (Connor 2004).

Carbon Feedback Loop Engaged

C. Freeman and colleagues summarized the situation regarding peat's contribution to the carbon cycle in *Nature* (2001): "We have observed a 65 percent increase in the dissolved organic carbon (DOC) concentration in freshwater draining from upland catchments in the United Kingdom over the past 12 years. Here we show that rising temperatures may drive this process by stimulating the export of DOC from peat lands. . . . [This process] may increase substantially as a result of global warming." In simple English, a dangerous carbon feedback loop has been engaged.

Methane emissions have been increasing from bogs in Sweden, according to an international research team led by the GeoBiosphere Science Center at Sweden's Lund University. The Abisko region in sub-Arctic Sweden, which they studied, has long-term records of climate, permafrost, and other environmental variables that they said make comparisons possible (Christensen et al. 2004). "In the present study, researchers wrote, "airborne infrared images were used to compare the distribution of vegetation in 1970 with that of 2000. Dramatic changes were observed, and the scientists relate them to the climate warming and decreasing extent of permafrost that was observed over the same period" ("Swedish Bogs" 2004). At one site, Stordalen, these researchers estimated an increase in methane emissions between 22 percent and 66 percent between 1970 and 2000, according to lead researcher Torben R. Christensen of Lund University's GeoBiosphere Science Center (Christensen et al. 2004; "Swedish Bogs" 2004).

Further Reading

Abram, N. J., et al. "Coral Reef Death during the 1997 Indian Ocean Dipole Linked to Indonesian Wildfires." *Science* 301 (August 15, 2003): 952–955.

Brandenburg, John E., and Monica Rix Paxson. *Dead Mars, Dying Earth.* Freedom, CA: The Crossing Press, 1999.

Christensen, Torben R., et al. "Thawing Sub-Arctic Permafrost: Effects on Vegetation and Methane Emissions." *Geophysical Research Letters* 31(4) (February 20, 2004): L04501. doi: 10.1029/2003GL018680.

Cochrane, Joe. "Southeast Asia, Choking on Haze, Struggles for a Solution." *The New York Times,* October 8, 2015. http://www.nytimes.com/2015/10/09/world/asia/indonesia -forest-fires-haze-singapore-malaysia.html.

Cochrane, Joe. "Rain In Indonesia Dampens Fires That Spread Toxic Haze." *The New York Times,* October 29, 2015, A4.

Connor, Steve. "Peat Bog Gases Accelerate Global Warming." *The Independent* (London), July 8, 2004, 9.

Cowen, Robert C. "One Large, Overlooked Factor in Global Warming: Tropical Forest Fires." *Christian Science Monitor,* November 7, 2002, 14.

Freeman, C., C. D. Evans, and D. T. Monteith. "Export of Organic Carbon from Peat Soils." *Nature* 412 (August 23, 2001): 785.

Freeman, C., et al. "Export of Dissolved Carbon from Peatlands under Elevated Carbon Dioxide Levels." *Nature* 430 (July 8, 2004): 195–198.

Glantz, Michael H. *Climate Affairs: A Primer.* Washington, DC: Island Press, 2003.

Johansen, Bruce. "Global Warming Approaching." Z Magazine, April 1, 2005. https://zcomm .org/zmagazine/global-warming-approaching-by-bruce-johansen/.

Moore, Sam, et al. "Deep Instability of Deforested Tropical Peatlands Revealed by Fluvial Organic Carbon Fluxes." *Nature* 493 (January 31, 2013): 660–663.

Page, Susan E., et al. "The Amount of Carbon Released from Peat and Forest Fires in Indonesia during 1997." *Nature* 420 (November 7, 2002): 61–65.

Schimel, David, and David Baker. "Carbon Cycle: The Wildlife Factor." *Nature* 420 (November 7, 2002): 29–30.

"Swedish Bogs Flooding Atmosphere with Methane; Thawing Sub-Arctic Permafrost Increases Greenhouse Gas Emissions." American Geophysical Union. February 10, 2004. http://www.scienceblog.com/community/article2366.html (no longer available).

See also: Deforestation; Methane Burp; Methane Hydrates; Permafrost; Wildfires; Wildfires, United States; Wildfires, Worldwide

GLACIAL EROSION

Rising temperatures have been melting ancient glaciers on the high Alps, causing devastating summer rockslides that have endangered the lives of many climbers, including 70 on July 14, 2003, which required one of the largest mass rescues in the area's history. Most climbers were plucked from the Matterhorn, which was racked by two major landslides that day. According to one observer, "Those climbing its slopes could have been forgiven for thinking the crown jewel of the Alps had started falling apart under their feet" (McKie 2003).

According to Robin McKie, writing in *The Observer* in London, "The great mountain range's icy crust of permafrost, which holds its stone pillars and rock faces together, and into which its cable car stations and pylons are rooted, is disappearing" (McKie 2003). Several recent Alpine disasters, including the avalanches that killed more than 50 people at the Austrian resort of Galtur during 1999, have been blamed on melting permafrost. During August 2003, the freezing level in the Alps rose to 13,860 feet (4,200 meters), almost 4,000 feet above its usual summer maximum of 3,000 meters (9,900 feet) (Capella 2003).

Roughly half of the recent Alpine ice erosion can be attributed to natural climate change, according to a study. "This doesn't question the actuality, and the seriousness, of man-made climate change in any way," said Matthias Huss, a University of Fribourg (Switzerland) glaciologist who led the study. "But what we do see is that current glacier retreat might be due to natural climate variations as it is to anthropogenic greenhouse warming" (Schiermeier 2010). Among the natural changes may be a 60-year cycle called the Atlantic Multidecadal Oscillation (AMO) that affects ocean circulation and temperatures. In 100-plus years (1910 to the present), 30 glaciers in the Swiss Alps monitored in this study lost half their mass on average. Changes in the AMO in the 1910s and 1970s produced small mass gains. The rest of the time, the AMO cycle has worked together with human greenhouse gas emissions to erode the glaciers' mass. The human role has accelerated over time, Huss said.

More Danger Ahead

Scientists attending the 2003 International Permafrost Association conference in Zurich, Switzerland, said that conditions in the high Alps could get more dangerous in coming years, assuming continued warming. "I am quite sure what happened on the Matterhorn . . . was the result of the Alps losing its permafrost. We have found that the ground temperature in the Alps around the Matterhorn has risen considerably over the past decade. The ice that holds mountain slopes and rock faces together is simply disappearing. At this rate, it will vanish completely—with profound consequences," said civil engineer Professor Michael Davies of Dundee University and a conference organizer (McKie 2003).

Air-temperature increases in the Alps are being amplified fivefold underground. A borehole dug at Murtel in southern Switzerland has revealed that frozen subsurface soils have warmed by more than 1°C since 1990. In addition to general air-temperature rises that are heating the ground, increased evaporation caused by warmer summers also has caused heavier snows that insulate the soil and keep it warmer in winter. Ice also becomes more unstable as it warms, raising the danger of devastating landslides.

Melting permafrost in the Alps and other European mountain ranges does much more than spoil mountain climbers' treks. Landslides sometimes threaten alpine villages and ski resorts. Fear has been expressed that some villages may have to be abandoned. Rivers also may be blocked by debris, causing flash floods when these unstable mounds of earth subsequently collapse. According to a report in the The Guardian (U.K.), Charles Harris of the earth sciences department at Cardiff University, who coordinates research for the European Union, said that the main areas at risk are the Alps in Switzerland, Austria, France, Germany, and Italy, where the mountains are densely populated and the slopes are especially steep. According to this account, among the places being monitored is the Murtel-Corvatsch mountain above fashionable St. Moritz, and the Schilthorn, which towers above the Muran and Gandeg resorts near Zermatt (Brown 2001). Harris said that the Swiss Alps had warmed by 0.5°C to 1.0°C since the mid-1980s (Clover 2001).

Melting Glaciers Yield Baskets, Arrows, and Mummies

As temperatures warm and glaciers melt, some of the world's highest mountains are yielding mummies, skeletons, pieces of old aircraft, and many other things—so much so and in such wild variety that a new academic field has organized with its own periodical, *The Journal of Glacial Archaeology*. Publication editor and University of New Mexico anthropologist E. James Dixon professes a sense of urgency to catalog findings before they decompose in warming air. "For every discovery that is made, there are thousands coming out of the ice and are decomposing very rapidly," Dixon said. "In the ice, some of the most delicate artifacts are preserved. We've found baskets, arrow shafts with the feathers intact and arrowheads and lashings perfectly preserved" (Johnson 2015).

Anthropologists and archaeologists are finding more than old baskets and arrow shafts. Tim Johnson of McClatchy Newspapers reported that Mexico's tallest volcano, Pico de Orizaba, North America's third highest peak at 18,491 feet, is "performing an all-natural striptease," yielding the corpses of climbers who died in attempts to ascend its slopes.

"Late in February [2015]," reported Johnson, "a climbing party circled the jagged crater atop Orizaba. One of them slipped, and they later said he skidded down and came to a stop. When he got up, he saw a head poking out of the snow," said Hilario Aguilar Aguilar, a veteran climber. It was a mummified climber, a member of a Mexican expedition hit by an avalanche in November 1959. Some climbers fell near the Chimicheco Ridge, their bodies frozen in an icy time machine, to be discovered more than half a century later (Johnson 2015).

Elsewhere, now-bare Andean slopes have yielded Incan mummies, and mountains in the Canadian Yukon have bared piles of vile-smelling caribou dung that had been frozen for several thousand years. Two German tourists in 1991 nearly stumbled on a 5,300-year-old dummy in the Italian Alps that came to be known as Otzi the Iceman, who may have been a shepherd.

Further Reading

Johnson, Tim. "As Globe Warms, Melting Glaciers Revealing More Than Bare Earth." McClatchy Newspapers, June 19, 2015. http://www.mcclatchydc.com/news/nation -world/world/article25186561.html.

An organization called Permafrost and Climate in Europe (PACE) monitors the effects of climate change on the stability of mountains. As PACE literature contends, "The combination of ground temperatures only slightly below the freezing point, [along with] high ice contents and steep slopes, makes mountain permafrost particularly vulnerable to even small climate changes" (Brown 2001).

Flood Threats below the Alps

Melting glacial ice has been increasing the volume of the Rhine, Rhone, and Po rivers. Since 1850, the volume of Europe's glaciers has shrunk by 50 percent (Toner 2002). The great rivers of Europe that rise in the Alps may decline from swollen summer torrents to trickles as global warming melts ancient mountain ice fields during the 21st century. David Collins of Salford University presented findings to a Royal Geographical Society conference in Belfast that indicate that the ice now melting in the Alps accumulated during the Little Ice Age between the 15th and 18th centuries.

"The [present] combination of warmer summers and drier winters, meaning less snow to feed the glaciers, has meant that the vast bank of ice on the mountain tops is disappearing," said Collins. "The ice is like money in the bank, if you keep drawing more than you put in, eventually it runs out" (Brown 2002). Before the rivers dry up, warming will supply a final torrent as the glaciers melt. The summer flows of the rivers fed from the Alps, including the Rhine, Rhone, Po, and Inn (which feeds into the Danube) recently have been higher than they have been for centuries. The excessive water flow has been good news for those living on riverbanks in southern and Eastern Europe who drew off the excess water for irrigation and domestic use. In France, the river water was used to cool nuclear power stations.

Forecasts by Collins and his team showed that these boom times for water supply may soon end. When all the ice goes, the summer flow of the rivers will be almost entirely dependent on rainfall. Under some climate models, rainfall in southern Europe probably will decline under warmer conditions. "We can see serious potential problems but it is very hard to be precise because weather patterns could change again," he said (Brown 2002).

"Some of the glaciers—for example, there are a number of small ones at Gornergrat, near Zermatt—are now below the snow line in summer. This means they are doomed. The ice they are made of was laid down in snowfall two or three centuries ago and is melting away faster each year," Collins reported (Brown 2002). Collins said the reduction in the glaciers in the Alps had been matched by an increase in glacier size in the Jotunheimen range in Norway because of increased precipitation in northern Europe—in this case, falling as snow—had blanketed those glaciers and protected them from any temperature increase. This had led to a net increase in the size of glaciers over the same period as those in the Alps were retreating.

The people of Macugnaga (pronounced maa-COON-yaga), an Italian Alpine resort village, long ago learned to cope with the floods that sometimes accompany the melting snow in the spring. "But nothing," according to on account, "prepared them for the catastrophic flood threat they now face—a glacier rapidly melting from unusually warm temperatures" (Konviser 2002).

During July 2002, as many as 300 officials and volunteers struggled under a state of emergency "to prevent a gigantic glacier-fed lake from breaking through the giant ice wall that confines it" (Konviser 2002). If they failed, a devastating wall of water carrying chunks of glacier and mountainside could surge through this verdant valley. Known technically as a *glacier lake outburst flood*, "it's an event previously seen only in the Himalayas where the slopes of the mountains are steeper. But scientists say the threat is both real, and a warning of things to come if the global-warming trend continues" (Konviser 2002).

"It's a dangerous situation because the border of the lake is ice, which isn't stable," said Claudia Smiraglia, a professor of physical geography at Milan University. "The glacier is always in motion" (Konviser 2002). Reporting for the *Boston Globe*, Bruce I. Konviser described the potential scope of the threat:

> If the water escapes, the 650 residents of Macugnaga and as many as 7,000 vacationers, depending on the time of year—would have approximately 40 minutes to gather their belongings and get to higher ground before the wave of water and mountain wipes out [many], if not all, of the manmade structures, according to Luka Spoletini, a spokesman for the Italian government's Department of Civil Protection. (Konviser 2002)

Further Reading

Brown, Paul. "Melting Permafrost Threatens Alps; Communities Face Devastating Landslides from Unstable Mountain Ranges." *The Guardian* (U.K.), January 4, 2001, 3.

Brown, Paul. "Geographers' Conference: Ice Field Loss Puts Alpine Rivers at Risk: Global Warming Warning to Europe." *The Guardian* (U.K.), January 5, 2002, 9.

Capella, Peter. "Europe's Alps Crumbling; Glaciers Melting in Heat Wave." Agence France Presse, August 7, 2003 (LEXIS).

Clover, Charles. "Geographers' Conference: Alps May Crumble as Permafrost Melts." *Telegraph* (London), January 4, 2001, 12.

Konviser, Bruce I. "Glacier Lake Puts Global Warming on the Map." *Boston Globe*, July 16, 2002, C1.

McKie, Robin. "Decades of Devastation Ahead as Global Warming Melts the Alps: A Mountain of Trouble as Matterhorn Is Rocked by Avalanches." *The Observer* (London), July 20, 2003, 18.

Schiermeier, Quirin. "Glaciers' Wane Not All Down to Humans." *Nature* 465 (June 10, 2010): 677.

Toner, Mike. "Meltdown in Montana; Scientists Fear Park's Glaciers May Disappear within 30 Years." *Atlanta Journal-Constitution*, June 30, 2002, 4A.

See also: Glaciers, The Andes; Glaciers, Central Asia; Glaciers, The Himalayas; Glaciers, North America; Ice Melt, Antarctica; Sea Ice, Arctic; Sea Level Rise; Temperatures, Winter

GLACIERS, THE ANDES

Hundreds of Andean glaciers are retreating, and scientists say their erosion is a direct result of rising temperatures. During three decades (1970–2000), Peru's glaciers lost almost a quarter of their 1,225-square-mile surface (Wilson 2001). An analysis of 268 mountain stations between 1° north and 23° south latitude along the tropical Andes indicates a temperature rise of 0.11°C per decade, compared to a global average of 0.6°C between 1939 and 1998 (Bradley et. al. 2006, 1755). Glacial retreat, much of it rapid, has been observed all along the spine of the Andes and on adjacent high plateaus.

Benjamin Morales, the dean of Peru's glaciologists, calls Peru's glaciers—80 percent of the world's tropical ice pack) "the world's most sensitive thermometers," because they react to the smallest changes in temperatures (Wilson 2001). Many of the Peruvian glaciers that are now melting formed more than a million years ago. Today, some of these glaciers are melting so quickly that "residents have watched a usually painstakingly slow geological process with their own eyes. Since 1967, when Peru began monitoring its glaciers, scientists estimate, the ice caps have lost 22 percent of their volume, enough to fill more than 5.6 million Olympic-size swimming pools" (Wilson 2001). Peru's ice caps continued to melt through at least 2015, to a point where fears were being expressed for the water supplies of Lima and other cities.

Glaciologist Lonnie Thompson has estimated that many of Peru's glaciers could disappear by 2020 (Wilson 2001). In Bolivia, the mass of glaciers and mountain snowcaps has shrunk 60 percent since 1978, raising a specter of water shortage for La Paz, home of 1.5 million people, as well as nearby El Alto (Forero 2002).

Peru's Glaciers Erode

The 18,700-foot-high Quelccaya ice cap in the Andes of southeastern Peru has been steadily shrinking at an accelerating rate, losing 10 to 12 feet a year between 1978 and 1990, as much as 90 feet a year between 1990 and 1995, and 150 feet a year between 1995 and 1998. The glacier retreated between 100 and 500 feet, depending on location, between 1999 and 2004. The Peruvian National Commission on Climate Change forecast in 2005 that Peru would lose all its glaciers below 18,000 feet by 2015. Within 40 years, the commission said that all of Peru's glaciers will be gone (Regaldo 2005, A1).

Within 25 years ending in 2012, 1,600 years' worth of ice had melted in the Peruvian Andes, according to scientists' measurements of the Quelccaya ice cap at 5,670 meters (roughly 18,000 feet) above sea level, the largest surviving ice cap in tropical regions of Earth. The scientists have been unearthing remains of plants that were buried under advancing ice thousands of years in the past. By dating the decay rate of the plants, the scientists can gauge the ice sheet's movements. Using these calculations (Thompson et al. 2013, 945), the scientists estimated that the ice cap is now smaller than it has been for at least 6,300 years. Dating those plants using a radioactive form of carbon in the plant tissues that decays at a known rate has given scientists an unusually precise way to determine the history of the ice sheet's margins.

Thompson describes the Andes' glaciers on which he has worked for more than a quarter century in familial terms: "When I go back to Quelccaya in Peru, where I've been 26 times, it's like visiting a patient dying of cancer. You know there's no hope; you can only watch it shrink away. So my work has become a salvage operation—to capture history before it disappears forever." With some sadness, Thompson says that data do not do much to change human behavior. "It's human nature to deal only with what's on our plate today. When people lose their houses or crops to fires, droughts, tornadoes—when they lose everything they've worked for—they'll say 'whoa! What's going on here.' And that's already starting to happen. At some point, the discussion will change very rapidly. It'll seem like it happened overnight" (Walters 2013, 62).

The Andes' Water Supplies

Glaciers in the Andes sustain human life by supplying millions of people with water, a fact that is always on the minds of scientists who study them. Douglas R. Hardy, a University of Massachusetts researcher who works in the region, said, "How much time do we have before 50 percent of Lima's or La Paz's water resources are gone?" (Gillis 2013). Drinking water is already drying up in parts of Bolivia:

> When the tap across from her mud-walled home dried up in September, Celia Cruz [of El Alto, Bolivia] stopped making soups and scaled back washing for her family of five. She began daily pilgrimages to better-off neighborhoods, hoping to find water there. Though she has lived here for a decade and her husband, a construction worker, makes a decent wage, money cannot buy water. "I'm thinking of moving back to the countryside; what else can I do?" said Ms. Cruz, 33, wearing traditional braids and a long tiered skirt as she surveyed a courtyard dotted with piglets, bags of potatoes and an ancient red Datsun. "Two years ago this was never a problem. But if there's not water, you can't live." (Rosenthal 2009)

Warmer temperatures and changing rainfall patterns are causing glaciers to be less dependable sources of water for El Alto and other Andean cities and towns. In some cases, constructing reservoirs could prolong the usefulness of limited water, but even they will dry up eventually with continued glacier melting in what is otherwise a dry land.

"The effects are appearing much more rapidly than we can respond to them, and a reservoir takes five to seven years to build. I'm not sure we have that long," said Edson Ramirez, a Bolivian glaciologist who has documented and projected the glaciers' retreat for two decades. The Chacaltaya glacier at 17,500 feet, until recently expected to last until 2020, disappeared in 2009. The area had a ski area until 2005, when it closed as the ice and snow retreated. The ski lodge still stands, mute and empty, stocked with rental gear (Rosenthal 2009).

In the cities, water supplies dwindle but urban populations continue to grow. As Elisabeth Rosenthal reported for *The New York Times*,

> "A lot of us think about not having kids anymore," said Margarita Limachi Álvarez, 46, a blue Andean cap with earflaps pulled over her head. "Without water or food, how would we survive?" A hundred miles away, in a middle-class neighborhood of El Alto, water has also become a gnawing concern. From September through November, the taps gave forth at best eight hours a day, often with little pressure. (Rosenthal 2009)

For 25 years, Thompson has been tracking a particular Peruvian glacier, Qori Kalis, where the pace of shrinkage accelerated enormously since the late 1990s. From 1998 to 2000, the glacier receded an average of 508 feet a year, according to Thompson. "That's thirty-three times faster than the rate in the first measurement period," he said, referring to a previous study of the glacier that covered the years 1963 to 1978 (Revkin 2001). The Qori Kalis outlet glacier in Peru has been retreating about 660 feet (200 meters) a year (as of 2004–2005), 40 times the rate when measurement began in the 1970s (Braasch 2007, 37).

Water from hundreds of glaciers in a section of the Andes known as the Cordillera Blanca ("White Range") drives Peru's rural economy. The water runoff moistens wheat and potatoes along the mountain slopes. It also lights residents' houses with electricity generated by a hydroelectric plant on the river (Wilson 2001).

As hydropower generated from glacier runoff becomes undependable, these areas may come to rely more on fossil fuels for power. Raymond S. Bradley and colleagues (2006, 1756) point out that similar conditions apply in areas of eastern Africa and New Guinea with some reliance on glaciers that (as in the Andes) may disappear during the next 20 years.

Lima, a city of about 8 million people situated in the Atacama, one of the driest deserts on Earth, receives nearly all of its water during a six-month dry season from glacial ice melt. Within a few decades, at current melting rates, Lima's people will experience severe water shortages. The same water is used to generate much of Lima's electricity. Even as its wells go dry and mountain glaciers shrink, Lima has been adding 200,000 residents a year. Bolivia's capital La Paz and Ecuador's capital of Quito face similar problems (Lynas 2004, 236–237). Quito receives part of its drinking water from a rapidly retreating glacier on Volcano Antizana. Within a few years, the life-giving waters supplying all three cities may diminish to a trickle for the last time if freezing levels continue to rise.

Many Peruvians who face drought in the long term are being lulled into a sense of plenty in the short term by increasing glacial runoff. According to Scott Wilson, writing in the *Washington Post*, the short-term glacial run-off "has made possible plans to electrify remote mountain villages, turn deserts into orchards and deliver potable water to poor communities. In some mud-brick villages scattered across the valley, new schools will open and factories will crank up as the glacier-fed river increases electricity production" (Wilson 2001).

"In the long run . . . these long-frozen sources of water will run dry," said Cesar Portocarrero, a Peruvian engineer who worked for Electroperu, the government-owned power company, who has monitored Peru's water supply for 25 years (Revkin 2002). In the meantime, evidence of changing climate has appeared in Portocarrero's hometown, Huaraz, a small city at 10,000 feet in the Andes. "I was doing work in my house the other day [in 2002] and saw mosquitoes," Portocarrero said. "Mosquitoes at more than 3,000 meters [almost 10,000 feet]. I never saw that before. It means really we have here the evidence and consequences of global warming" (Revkin 2002).

Ninety-nine percent of the Chacaltaya glacier in Bolivia has disappeared since 1940, said World Bank engineer Walter Vergara, in his 2007 report, "The Impacts of Climate Change in Latin America." One of the highest glaciers in South America, Chacaltaya is one of the first glaciers to melt because of climate change. Since 1970, glaciers in the Andes have lost 20 percent of their volume, according to a report by Peru's National Meteorology and Hydrology Service. Eventual melting could curtail water supplies to 30 million people, the report said ("Melting" 2008).

El Niño conditions in the equatorial Pacific, which may become more frequent as temperatures warm, cause drought in the Andes, eroding glaciers and providing little new snow to replace melting. The albedo of the glaciers also diminishes as they become grayer from dust.

Ice melt in Peru also may make some areas more vulnerable to the frequent earthquakes that afflict the area. "Glaciers usually melt into the rock, filling in fissures with water that expands and freezes when the temperatures drop. What scientists fear is that, with increased melting, more water and larger ice masses are pulling apart the rock and making the ice cap above more susceptible to the frequent seismic tremors that rock the area" (Wilson 2001).

Further Reading

Braasch, Gary. *Earth under Fire: How Global Warming Is Changing the World*. Berkeley: University of California Press, 2007.

Bradley, Raymond S., et al. "Threats to Water Supplies in the Tropical Andes." *Science* 312 (June 23, 2006): 1755–1756.

Forero, Juan. "As Andean Glaciers Shrink, Water Worries Grow." *The New York Times*, November 24, 2002, A3.

Gillis, Justin. "In Sign of Warming, 1,600 Years of Ice in Andes Melted in 25 Years." *The New York Times*, April 4, 2013. http://www.nytimes.com/2013/04/05/world/americas/1600-years-of-ice-in-perus-andes-melted-in-25-years-scientists-say.html.

Lynas, Mark. *High Tide: The Truth about Our Climate Crisis*. New York: Picador/St. Martins, 2004.

"Melting Andean Glaciers Could Leave 30 Million High and Dry." Environment News Service, April 28, 2008. http://www.ens-newswire.com.

Regaldo, Antonio. "The Ukukus Wonder Why a Sacred Glacier Melts in Peru's Andes." *Wall Street Journal*, June 17, 2005, A1, A10.

Revkin, Andrew C. "A Message in Eroding Glacial Ice: Humans Are Turning Up the Heat," *The New York Times*, February 19, 2001, A1.

Revkin, Andrew C. "Forecast for a Warmer World: Deluge and Drought." *The New York Times*, August 28, 2002, A10.

Rosenthal, Elisabeth. "In Bolivia, Water and Ice Tell of Climate Change." *The New York Times*, December 14, 2009. http://www.nytimes.com/2009/12/14/science/earth/14bolivia.html.

Thompson, L. G., et al. "Annually Resolved Ice Core Records of Tropical Climate Variability over the Past ~1800 Years." *Science* 340 (May 24, 2013): 945–950.

Vergara, Walter. "The Impacts of Climate Change in Latin America." 2007. https://www.thefreelibrary.com/The+impacts+of+climate+change+in+Latin+America.-a0221761359.

Walters, Pat. "Risk Takers: Ice Investigator." *National Geographic*, January 2013, 58–67.

Wilson, Scott. "Warming Shrinks Peruvian Glaciers; Retreat of Andean Snow Caps Threatens Future for Valleys." *Washington Post*, July 9, 2001, A1.

GLACIERS, CENTRAL ASIA

Roughly 15,000 glaciers border India, China, and Nepal, many on the Tibetan plateau, which supplies the Ganges, Indus, and Brahmaputra rivers, among others. The "roof of the world" have been melting more quickly than at any time in recorded history, according to research surveying conditions through 2006 published late in November 2008 by Lonnie Thompson and colleagues in *Geophysical Research Letters*. Studying ice melt on the Naimona'nyi glacier, Thompson's team was stunned to learn that all snow built up since 1944 had melted. "We were very surprised not to find the 1962–1963 horizon, and even more surprised not to find the 1951–1952 signal," Thompson said. In more than 20 years of sampling glaciers all over the world, this was the first time both markers were missing. As more heat is trapped in the atmosphere, said Thompson, it holds more water vapor. "This humidity condenses as it rises, releasing heat in mountains. At the highest elevations, we're seeing something like an average of 0.3 degrees Centigrade warming per decade," Thompson said ("Tibetan Glaciers" 2008).

"Seasonal meltwater serves as the main source of power for an increasing number of hydroelectric dams on the rivers served by the glaciers," wrote Javaid R. Laghari in *Nature* (2013). But hydropower faces a difficult future in south Asia because of climatic, environmental and politico-economic factors. In 2010, floods caused by excessive monsoon rains as well as melting snow and ice submerged as much as one-fifth of Pakistan's land area, killing 2,000 people. "The frequency of tropical cyclones is predicted to increase as a result of global warming. Because rain, rather than snow, falls on mountains in spring, river flows will peak before the main growing season. Summers will increasingly see dry streams, withered and abandoned crops, dead fish and low groundwater levels" (Laghari 2013, 617).

Glaciers in Central Asia's largest mountain range, the Tien Shan, contribute to the water supplies of Kazakhstan, Kyrgyzstan, Uzbekistan, and parts of China. Since the 1960s, they have lost 18 percent of their area and 27 percent of their mass. Data collected and interpreted by Daniel Farinotti of the GFZ German Research Center for Geosciences indicate that 3,000 square kilometers had melted ("Substantial" 2015). When glaciers melt, future aridity is almost inevitable. In 2013, India had plans to build as many as 60 dams on the Chenab River in Jammu and Kashmir to capture melt water for power and irrigation. The first dam, the Baglihar, went online in 2008.

"Asian Brown Cloud" Sullies Glaciers

The "Asian brown cloud" from low-level pollution in India and China is changing the color of glaciers in the Hindu Kush, Himalayas, and other mountain ranges, speckling them with soot, changing their albedo, and speeding melting. Jane Qiu, writing in *Nature* (2010), described the work of Xu Baiqing, an environmental scientist, who measured black carbon deposited on the glaciers during a half century "and found increased emissions since the 1990s, coinciding with rapid industrial growth in the region." Working with Angela Marinoni and colleagues at the Institute of Atmospheric Sciences in Bologna, Italy, "high concentrations of aerosols, including black carbon [were found] above 5,000 meters in the Nepalese Himalayas, which caused significant atmospheric warming. They calculate that deposition of black carbon could increase snow and ice melting of a typical Himalayan glacier by 12 to 34 percent by reducing its ability to reflect light" (Qiu 2010, 141).

Qiu reported:

As a consequence, glacial lakes are getting larger and more numerous, causing more floods. A study led by Yongwei Sheng, an ecologist at the University of California, Los Angeles, shows that the area of such lakes on the plateau has increased by 26 percent since the 1970s, with a devastating effect on surrounding pastures. Outbursts of glacial lakes have caused more than 40 floods in the Himalayas since the 1950s, and more are likely in the coming decades, says Pradeep Mool, a remote-sensing specialist at the International Center for Integrated Mountain Development (ICIMOD) in Kathmandu. (Qiu 2010, 142)

According to Xu et al. (2009), "We find evidence that black soot aerosols deposited on Tibetan glaciers have been a significant contributing factor to observed rapid glacier retreat. Reduced black soot emissions, in addition to reduced greenhouse gases, may be required to avoid demise of Himalayan glaciers and retain the benefits of glaciers for seasonal freshwater supplies." Xu and colleagues observed that concentrations of black soot (mainly from cooking fires as well as from coal-fired power plants) have increased rapidly since the 1990s, along with increasing industrial activity and accelerating glacier retreat around the Himalayas. They describe how seasonal factors intensify its effects:

[The] impact of albedo [reflectivity] change is magnified in the spring, at the start of the melt season, because it allows melt to begin earlier. Then, as melting snow tends to retain some aerosols, the surface Asian haze, which peaks during November–March, spreads northeastward along the south side of the Himalayas. Thus, highest black soot concentration in un-melted snow occurs at the time of maximum snow extent, accelerating spring melt and lengthening the melt season. (Xu et al. 2009)

This study concludes that black soot may be responsible for as much regional warming during the past century as carbon dioxide. In other words, in the greenhouse equation, coal-fired power facilities count twice.

Water for Billions Imperiled

Some 90 percent of the Tibetan plateau's glaciers have retreated during the past century; their rate of decay has increased substantially since the mid-1990s, according to "Global Outlook For Ice and Snow," a report released during 2007 by the United Nations Environment Program. The snowpack in this region could shrink another 43 percent to 81 percent by 2100, according to projections from the Intergovernmental Panel on Climate Change (IPCC) (Gartner 2007). Thousands of Himalayan glaciers feed several major rivers, partially sustaining one-sixth of Earth's population, about 1.4 billion people, downstream. Their retreat threatens the region's drinking-water supply, agricultural production, and vulnerability to disease and floods. Glaciers on the Himalayas contain 100 times as much ice as do the Alps and provide more than half the drinking water for 40 percent of the world's population in seven major Asian rivers, including the Ganges and the Indus.

Walter W. Immerzeel and colleagues described the relationship of melting glaciers to food and water supplies in the center of Asia (2010, 1382):

More than 1.4 billion people depend on water from the Indus, Ganges, Brahmaputra, Yangtze, and Yellow rivers. Upstream snow and ice reserves of these basins, important in sustaining seasonal water availability, are likely to be affected substantially by climate change, but to what extent is yet unclear. Here, we show that meltwater is extremely important in the Indus basin and important for the Brahmaputra basin, but plays only a modest role for the Ganges, Yangtze, and Yellow rivers. A huge difference also exists between basins in the extent to which climate change is predicted to affect water availability and food security. The Brahmaputra and Indus basins are most susceptible to reductions of flow, threatening the food security of an estimated 60 million people.

Approximately half of the drinking-water wells that supply Afghanistan's Kabul basin may run dry within 50 years because of increasing population and rising temperatures, according to research by scientists from that country and the U.S. Geological Survey and the Afghanistan Geological Survey ("Afghanistan's Kabul Basin" 2010). Temperatures in the region have risen by 1°C since the 1950s, causing thousands of glaciers to retreat by an average of 30 meters a year. The worst recorded collapse of one of these dams occurred in 1954 when 300,000 cubic meters of water and rock poured without warning into China in a 40 meter-high flood surge from the Sangwang dam on the Tibet–Nepal border. The city of Gyangze, 120 kilometers away, was destroyed. The dead totaled many thousands (Brown 2002).

The Indian Space Research Organization has used satellite imaging to measure changes in 466 glaciers and has found a more than 20-percent reduction in their size between 1962 and 2001. Another study found that the Parbati glacier, among the largest, was retreating by 170 feet a year during the 1990s. Another glacier, Dokriani, lost an average of 55 feet a year. Temperatures in the northwestern Himalayas have risen by 2.2°C since the late 1980s (Sengupta 2007).

Ice-Melt Complexities

Not all glaciers have been melting, and not all have been melting at the same rate. Even as many Himalayan glaciers retreat, some in the Karakoram Range (near the intersection of Pakistan, China, India, and Afghanistan) have experienced growth spurts, a matter useful to global warming contrarians who have argued that

Himalayan glaciers as a whole are not retreating. In 2015, however, Jane Qiu wrote in *Science*, "New research based on satellite images and on-the-ground records shows that what looks like a steady advance is in many cases just a temporary surge. And even though a surging glacier appears to be expanding, it may be losing mass, its upper reaches collapsing even as its snout thickens temporarily. Such surging glaciers may say little about climate change."

Anecdotal reports in the media have created an impression that the roughly 10,000 glaciers in the Himalayas have been retreating rapidly. A summary of measurements from the ground and by satellite released late in 2009 by India's Ministry of Environment and Forests by senior glaciologist Vijay Kumar Raina, formerly of India's Geological Survey, presented a more complex picture, one that asserted that "many" (precise numbers were not offered) glaciers had stabilized or grown in recent years. The report also takes issue with a report by the IPCC in 2007 that melting glaciers in this area portend severe water shortages for India by 2035. In turn, Syed Iqbal Hasnain of the Energy and Resources Institute in New Delhi said that the government's report illustrated "an ostrich-like attitude in the face of impending apocalypse" (Bagla 2009, 924).

Further Reading

"Afghanistan's Kabul Basin Faces Dry and Thirsty Future." Environment News Service, June 18, 2010. http://www.ens-newswire.com/ens/jun2010/2010-06-18-02.html.

Bagla, Pallava. "No Sign Yet of Himalayan Meltdown, Indian Report Finds." *Science* 326 (November 13, 2009): 924–925.

Brown, Paul. "Scientists Warn of Himalayan Floods: Global Warming Melts Glaciers and Produces Many Unstable Lakes." *The Guardian* (London), April 17, 2002, 13.

Gartner, John. "Climbers Bring Climate Change from Mountaintop to Laptops." Environment News Service, July 31, 2007. http://www.ens-newswire.com/ens/jul2007/2007-07-31-04.asp (no longer available).

Immerzeel, Walter W., Ludovicus P. H. van Beek, and Marc F. P. Bierkens. "Climate Change Will Affect the Asian Water Towers." *Science* 328 (June 11, 2010): 1382–1385.

Laghari, Javaid R. "Melting Glaciers Bring Energy Uncertainty." *Nature* 502 (October 30, 2013): 617–618.

Qiu, Jane. "Measuring the Meltdown." *Nature* 468 (November 11, 2010): 141–142.

Sengupta, Somini. "Glaciers in Retreat." *The New York Times*, July 17, 2007. http://www.nytimes.com/2007/07/17/science/earth/17glacier.html.

"Tibetan Glaciers Melting at Stunning Rate." November 24, 2008. Discovery.com. http://dsc.discovery.com/news/2008/11/24/tibet-glaciers-warming-02.html (no longer available).

Xu, Baiqing, et al. "Black Soot and the Survival of Tibetan Glaciers." *Proceedings of the National Academy of Sciences*. Published online before print December 8, 2009. doi: 10.1073/pnas.0910444106.

GLACIERS, THE HIMALAYAS

A study published in May 2015 in *The Cryosphere* (an open-access journal published by the European Geosciences Union) projected that if greenhouse gas emissions continue to rise at what have become usual rates, "glaciers in the Everest region of the Himalayas could melt away in this century." Researchers based in Nepal, France, and the Netherlands joined efforts to project that "[glacier] volume could be reduced between 70 percent and 99 percent by 2100" ("Glacier Ice" 2015). The research included projections for glaciers in the Dudh Kosi basin of Nepal, including Mount Everest, that were based on historical models from 1961 to 2007, with estimates of ice-mass change from 1992 to 2008. "The modeled glacier sensitivity to future climate change is high," the scientists concluded (Shea et al. 2015, 1105). "Our results indicate that these glaciers may be highly sensitive to changes in temperature, and that increases in precipitation are not enough to offset the increased melt," said Shea ("Glacier Ice" 2015).

Rivers fed by glaciers on mountains running from the Hindu Kush through the Karakoram and Himalayan regions supply water and some energy to one-quarter of Earth's people, as well as their farms and industries. Between 2003 and 2009, Himalayan glaciers lost an estimated 174 gigatons of water each year, which contributed to catastrophic floods along the Ganges, Indus, and Brahmaputra Rivers. Warming temperatures have accelerated melting. Melting glaciers provide flush water supplies (even floods) for a time, but when they disappear other problems such as drought may ensue. This is as true for the Missouri River valley of North America as it is for the Ganges, Indus, and Yellow Rivers (among others) in eastern and southern Asia. Similarly, residents of Lima, Peru, will eventually face water shortages as glaciers in the Andes erode, serving up a hot, often dry, and inhospitable future to many densely populated areas.

In India, where glacial water supplies 700 million people, research regarding the melting of 9,000 Himalayan glaciers has lagged because of a lack of funding. Anecdotal evidence, including increasing runoff, indicates accelerated melting. Snowpacks have suffered as monsoon rains (which fall as snow at high elevations) have become more sporadic "for reasons that many scientists ascribe to the world's changing climate" (Filkins 2016, 63).

Temperatures have risen as much as 4 to 5°F during the last 100 years in the Himalayas, about triple the global average of 1.4°F (0.75°C). "You can be sure that if the climate is changing—and it is—then glaciers are changing and the danger is shifting," said U.S. hydrologist Jeff Kargel of the University of Arizona. Kargel is leading a global project to measure and map the tens of thousands of Himalayan glaciers through satellite data. "It doesn't necessarily mean it's getting worse, it just means you don't know."

Grazing lands in Tibet are already drying in an area that gives rise to the Yellow, Yangtze, and Mekong Rivers. Madoi County was once known as *qianhu xian* ("county of a thousand lakes"), one of the richest counties on the plateau with abundant fish and livestock. Now, reported Jane Qiu in *Nature* (2016, 142): "The wetlands are drying up and sand dunes are replacing the prairies, which means that less water flows into the Yellow River. Such changes on the plateau have contributed to recurring water shortages downstream: the Yellow River often dries up well before it reaches the sea. . . ." Several factors contribute to these changes, including China's grazing policies, a steadily warming climate, road building, and air pollution.

Temperatures in Tibet have been rising by 0.3–0.4°C per decade since 1960, about twice the worldwide average.

Headwaters of the Ganges

By 2002, the snout of the Himalayan glacier that feeds the mighty Ganga (the Ganges River) had developed giant fractures and crevices along a 10-kilometer stretch, which indicated massive ice melting. During 15 years of researching such phenomena, Syed Iqbal Hasnain, who heads the Glacier Research Group at Delhi's Jawaharlal Nehru University, had never seen such a rapid deterioration of the frozen massif. "If the rate continues, we could see much of the Gangotri glacier and others in the Himalayas vanish in the next couple of decades," he said (Chengappa 2002, 40). Hundreds of millions of people who live within the watershed of the Ganges depend on watershed feed by Himalayan glaciers to some degree. Half of India's hydroelectric power is generated from glacial runoff (Lynas 2004, 238). The Indus River supplies 90 percent of the water used in desert areas of Pakistan from glaciers that have been rapidly losing mass for most of the 20th century (Lynas 2004, 238).

More than 40 lakes high in the Himalayas that have formed from rapidly melting glaciers have been over-flowing during the monsoon, sending millions of gallons of water and rocks cascading onto towns and villages (Brown 2002). These lakes are growing larger and more unstable as temperatures rise. As many as 44 such lakes have been identified, 20 in Nepal and 24 in Bhutan. There are thought to be hundreds more such "liquid time bombs" in India, Pakistan, Afghanistan, Tibet, and China (Brown 2002).

Surendra Shrestha, Asian regional coordinator for the United Nations Environment Program's early warning division, said, "These 44 could burst their banks with potentially catastrophic results for people and property hundreds of kilometers downstream" (Brown 2002). Nepal's Tsho Rolpa lake has grown sixfold in size since the 1950s and is now 2.6 kilometers long, 500 meters wide, and as deep as 107 meters. A flood from this lake could cause serious damage as far as 108 kilometers downstream in the village of Tribeni, threatening 10,000 lives (Brown 2002).

"The signal of future glacier change in the region is clear—continued and possibly accelerated mass loss from glaciers is likely given the projected increase in temperatures," said Joseph Shea, a glacier hydrologist at the International Center for Integrated Mountain Development in Kathmandu, Nepal, and leader of the study ("Glacier Ice" 2015). "We find that glaciers in the basin [are] more sensitive to temperature than anyone expected before," said Shea (Mooney 2015). Because of thermal inertia, glacial mass in the Himalayas will decline by 2100 even if major steps are taken to reduce greenhouse-gas emissions.

The western Himalayas are so steep that in places strikingly different biological regimes dominate within short distances: "alluvial grasslands, subtropical forests, conifers and alpine meadows lie stacked almost on top of each other, producing a spectacular range of vegetation" (Padma 2014). As temperatures warm, specific plant species have been migrating uphill in a range that extends from 300 to more than 6,000 meters as they "seek" familiar thermal habitats. Species that cannot move quickly risk extinction.

Vaneet Jishtu, a botanist at the Himalayan Forest Research Institute in Shimla in the north Indian state of Himachal Pradesh, leads a team that is cataloging this upward migration in an area that hosts one-tenth of Earth's known high-altitude species of plants and animals, as well as half those known in India. In *Nature*, T. V. Padma described changes in specific plant species (2014). For example, the flowering and fruiting cycles of pear and apple trees in the area have shifted as temperatures have risen, changing fruit sizes, taste, and color.

> Himalayan blue pines (*Pinus wallichiana*) are on the move too. They are now seen at heights of 4,000 meters, whereas two or three decades ago they grew at altitudes no higher than 3,000 meters, says a team led by Sher Singh Samant, head of the biodiversity and conservation team at the Himachal Pradesh unit of the G. B. Pant Institute of Himalayan Environment and Development. "In the past, heavy snowfall in higher regions prevented the upward shift of species. Now there is less snow and it has started melting faster, bringing about changes in vegetation and alpine meadows," he says. Even species among the glacier-deposited rocks at 4,500–5,500 meters are moving upwards to cooler climes, he notes.

Glaciers Retreat on and Near Mount Everest

In the Himalayas, the Rongbuk glacier on the north face of Mount Everest retreated between 170 and 270 meters from 1966 to 1997. The glacier from which Sir Edmund Hillary and Tenzing Norgay set out to climb Mount Everest in 1953 has retreated three miles up slope on its host mountain. Much of that glacier has turned to melt water, according to U.N. observers who visited the site (Williams 2002). Roger Payne, one of the observation team's leaders, said, "Back in 1953 when Hillary and Tenzing set off to climb Everest they stepped out of their base camp and straight on to the ice. You would now have to walk for over two hours [from the same site] to get on to the ice" (Williams 2002, 2). "Glaciers in the Himalayas are wasting at alarming and accelerating rates, as indicated by comparisons of satellite and historic data, and as shown by the widespread, rapid growth of lakes on the glacier surfaces," said Jeff Kargel of the U.S. Geological Survey, international coordinator for Global Land Ice Measurements from Space ("Glacial Retreat" 2002).

Melt water from the Himalayas' Imja glacier, 10 kilometers (6 miles) to the east–southeast of Mount Everest, has created a vast lake held back only by an unstable natural dam made of boulder debris that once marked the edge of the glacier. A collapse of this dam could send a wall of water as high as 100 meters surging down the valley that is inhabited by the Sherpas who assist climbing expeditions and ascents of Mount Everest. The valley is also the main approach route to the Everest base camp. "We know it's going to go shooting down the flood plain, and in a mountainous area, that's where the people live," said Lisa Graumlich, who directs the Big Sky Institute at Montana State University (McFarling 2002).

The Dig Tsho glacial outburst in Nepal in 1985 destroyed a hydroelectric plant, wiped out 14 bridges, and drowned dozens of villagers. The danger is so obvious, Graumlich said, that some Himalayan villages have installed primitive warning systems—basically a system of horns—in an attempt to save lives during the next flood (McFarling 2002). "We're just watching [glacial lakes] form in the Himalayas and Peru," said Alton C. Byers, director of research and education for the West Virginia–based Mountain Institute. "All you have to do is release that dam and you'll lose vast amounts of water in seconds" (McFarling 2002).

Climbing Becomes Riskier

Climate change is making climbing the world's highest mountains more risky because the behavior of ice and snow is becoming more unpredictable, a fact hammered home on April 18, 2014, when a huge, sudden avalanche of large ice chunks killed 16 Sherpas in the deadliest single avalanche on Mount Everest. Warming temperatures were listed as a culprit because ice was melting into a treacherous slurry. "While it is impossible to link any single event to long-term changes in the global climate, scientists say the future will likely hold more such dangers in high-altitude regions," the Associated Press reported ("Climate Change Likely" 2014). Avalanches are increasing in number and severity, as climbing terrain becomes less stable. "The Himalayas in particular could see more snow as warming oceans send more moisture into the air for the annual Indian monsoon that showers the 2,400-kilometer (1,500-mile) mountain range," the AP reported. The deadly avalanche in 2014 occurred at the Khumbu icefall, among Mount Everest's most dangerous ice, "as the edge of the slow-moving glacier is known to crack, cave and send huge chunks of ice tumbling without warning" ("Climate Change Likely" 2014).

"It's Mother Nature who calls the shots," Tim Rippel, an expedition leader, said in a blog post from Everest base camp as many of the 400 Sherpa guides were leaving, demanding better government compensation for the high risks they take in helping climbing companies ferry tourists up the peak. "The mountain has been deteriorating rapidly in the past three years due to global warming, and the breakdown in the Khumbu icefall is dramatic," he said. "We need to learn more about what is going on up there" ("Climate Change Likely" 2014). Many observations are anecdotal, with little on which to base scientific conclusions. Systemic study of conditions in the area is only a few years old, and no one is studying snow patterns on a large scale, said Nepalese glaciologist Rijan Bhakta Kayastha at Kathmandu University ("Climate Change Likely" 2014).

Apa Sherpa has climbed Everest more than any other person (21 times) and is a weather watcher who has described how weather has changed. Trails that used to be covered with thick layers of packed snow are now exposed rocky surface, often freezing and thawing under a sheath of slippery ice. Ice and rock-strewn rubble make for deadly avalanches as slopes freeze, thaw, and freeze again. "The danger level has significantly risen for climbers," said Apa, 53, who now lives in Draper, Utah, having climbed Mount Everest last in 2011 ("Climate Change Likely" 2014). When snow does fall, it tends to be heavier, raising avalanche risks.

By 2012, an increasing number of climbers and disintegration of the mountain provoked by drying and thawing was causing Mount Everest to become dangerously unstable in places. The resulting rockfalls, avalanches, and serac collapses were causing some Sherpas to refuse offers of work. Russell Brice, owner of Himalayan Experience, one of the largest climbing companies, canceled most of its season, saying, "We can no longer take the responsibility of sending you, the guides, and Sherpas through the dangerous icefall and up the rockfall-ridden Lhotse Face" (Wilkinson 2012).

Further Reading

Brown, Paul. "Scientists Warn of Himalayan Floods: Global Warming Melts Glaciers and Produces Many Unstable Lakes." *The Guardian* (London), April 17, 2002, 13.

Chengappa, Raj. "The Monsoon: What's Wrong with the Weather?" *India Today*, August 12, 2002, 40.

"Climate Change Likely to Make Everest Even Riskier." *Newsday*, April 23, 2014. http://www.newsday.com/travel/climate-change-likely-to-make-everest-even-riskier-1.780 0984 (no longer available).

Filkins, Dexter. "The End of Ice: Exploring a Himalayan Glacier." *The New Yorker*, April 4, 2016, 59–65.

"Glacier Ice in Everest Region Could Vanish by 2100." Environment News Service, June 3, 2015. http://ens-newswire.com/2015/06/03/glacier-ice-in-everest-region-could-vanish-by -2100/.

"Glacial Retreat Seen Worldwide." Environment News Service, May 30, 2002. http://ens -news.com/ens/may2002/2002-05-30-09.asp#anchor2 (no longer available).

Lynas, Mark. *High Tide: The Truth about Our Climate Crisis.* New York: Picador/St. Martins, 2004.

McFarling, Usha Lee. "Glacial Melting Takes Human Toll." *Los Angeles Times*, September 25, 2002, A4.

Mooney, Chris. "Climate Change Could Shrink Mount Everest's Glaciers by 70 Percent, Study Finds." *Washington Post,* May 27, 2015. http://www.washingtonpost.com/news/energy -environment/wp/2015/05/27/climate-change-could-shrink-glaciers-in-the-mount -everest-region-by-70-percent-study-finds/.

Padma, T. V. "Himalayan Plants Seek Cooler Climes." *Nature* 512 (August 28, 2014): 359. http://www.nature.com/news/himalayan-plants-seek-cooler-climes-1.15771.

Qiu, Jane. "Trouble in Tibet: Rapid Changes in Tibetan Grasslands Are Threatening Asia's Main Water Supply and the Livelihood of Nomads." *Nature* 529 (January 14, 2016): 142–145. http://www.nature.com/news/trouble-in-tibet-1.19139.

Shea, J. M., et al. "Modelling Glacier Change in the Everest Region, Nepal Himalaya." *The Cryosphere,* 9 (May 27, 2015): 1105–1128. http://www.the-cryosphere.net/9/1105/2015 /tc-9-1105-2015.html.

Wilkinson, Freddie. "Don't Climb Every Mountain." *The New York Times*, May 18, 2012, A27.

Williams, Frances. "Everest Hit by Effects of Global Warming." *Financial Times* (London), June 6, 2002, 2.

GLACIERS, NORTH AMERICA

Most glaciers in Alaska and northwestern Canada have been melting and contributing to global sea level rise, which is now estimated as being approximately 30 percent of Greenland's annual contributions ("Alaska's Biggest" 2015). Inland glaciers are losing mass more rapidly than those that calve into the ocean, however. C. F. Larsen, a researcher at University of Alaska at Fairbanks, and colleagues explained in *Geophysical Research Letters*, summarizing a study of 116 glaciers in the region (2015):

> Mountain glaciers comprise a small and widely distributed fraction of the world's terrestrial ice, yet their rapid losses currently drive a large percentage of the cryosphere's contribution to sea level rise. Regional mass balance assessments are challenging over large glacier populations due to remote and rugged geography, variable response of individual glaciers to climate change, and episodic calving losses from tidewater glaciers. In Alaska, we use airborne altimetry from 116 glaciers to estimate a regional mass balance of -75 ± 11 Gt yr^{-d} (1994–2013). Our glacier sample is spatially well distributed, yet pervasive variability in mass balances obscures geospatial and climatic relationships. However, for the first time, these data allow the partitioning of regional mass balance by glacier type. We find that tidewater glaciers are losing mass at substantially slower rates than other glaciers in Alaska and collectively contribute to only [6 percent] of the regional mass loss.

Although these glaciers do not calve directly into the ocean, their melt water eventually does reach it and affects sea levels.

Glaciers were melting so quickly in Alaska that their anticipated demise was prompting tourists to visit. The Travel Section of the New York *Times* headlined: "The Race to Alaska Before It Melts" (Egan 2005). Cities and towns across the entire state (including Anchorage, Fairbanks, Juneau, and Nome) reported record-high temperatures during the summer of 2004. At Portage Lake 50 miles south of Anchorage, "people came by the thousands to see Portage glacier, one of the most accessible of Alaska's frozen attractions. Except, you can no longer see Portage glacier from the visitor center. It has disappeared" (Egan 2005).

The retreat of Alaskan glaciers is illustrated most graphically by comparing photographs taken recently with those taken a century or more ago in the same locations. Bruce Molnia, a geologist with the U.S. Geological Survey, has gathered more than 200 glacier photos taken from the 1890s to the late 1970s.

The surfaces in Molnia's images now stand bare where "masses of ice were once surging down wide mountain passes into the sea, or were hanging from high and perilously steep faces," wrote David Perlman of the *San Francisco Chronicle*. "What remains from many of the retreating glaciers are stretches of open water or broad, snow-free layers of sediment" (Perlman 2004). "And as the glaciers disappear," Molnia said, "you get the amazing appearance of vegetation." On the tundra north of Alaska's Brooks Range, according to Molnia's observations (as cited by Perlman), "Explosive bursts of vegetation—willows, alders, birch, and many shrubs—are thriving where permafrost once kept the tundra surface frozen in winter. The growth of shrubs across the tundra has increased by 40 percent in less than 60 years, Tape said, and that perturbation is certainly due to the changing climate" (Perlman 2004).

While most glaciers in Alaska have been receding (a result of a general rise in temperatures), a few in wet maritime areas are gaining mass, an attribute they share with ice in some higher elevations of Norway and Sweden, where melting has been offset by increased snowfall, also a facet of climate change. Alaska's Hubbard glacier "is advancing so swiftly that it threatens to seal off the entrance to Russell fiord near Yukatat and turn the fiord into an ice-locked lake. Like a handful of other Alaska glaciers, the Hubbard is fed by a high-altitude snowfield that has not yet been affected by warmer temperatures" (Toner 2002). "The Hubbard is definitely an exception," said the U.S. Geological Survey's Bruce Molnia, who has been tracking 1,500 Alaska glaciers. "Every mountain group and island we have investigated is seeing significant glacier retreat, thinning or stagnation, especially at lower elevations. Ninety-nine percent of the named glaciers in Alaska are retreating" (Toner 2002).

The Mendenhall and Other Glaciers Retreat

One of the swiftest melting glaciers is the Mendenhall, 14 miles from Juneau, where torrents of water pouring off its shrinking mass have become a summer tourist attraction. In 2011, some 10 billion gallons of water poured off the glacier in three days, raising the levels of the river of the same name that flows through Juneau. The huge burst resulted from water breaching Suicide basin. Glaciologists call this process *jokulhlaup*, an Icelandic word that means "glacier leap." The torrents have become the talk of the town and a favorite site for many of roughly 400,000 tourists who visit annually, most on cruise ships that stop in Juneau's port. "We're a drive-up glacier," said Nikki Hinds, the assistant director at the Mendenhall Glacier Visitor Center, which is operated by the U.S. Forest Service. "In how many places can you have that?" (Johnson 2013).

Fewer than 20 of Alaska's several thousand valley glaciers were advancing after 2000. Glacial retreat, thinning, stagnation, and a combination of these changes characterize all 11 mountain ranges and three island areas that support glaciers in the state, according to U.S. Geological Survey scientist Bruce Molnia ("Alaskan Glaciers" 2001).

"The Earth recently emerged from a global climate event, called the 'Little Ice Age' during which Alaskan glaciers expanded significantly," explained Molnia. The Little Ice Age began to wane in the late 19th century. In some areas of Alaska, glacier retreat started during the early 18th century shortly before the Industrial Revolution began ("Alaskan Glaciers" 2001). "During the 20th century, most Alaskan glaciers receded and, in some areas, disappeared. But it is important to note that our data do not address whether or not any of these changes are human-induced," said Molnia, who warned against blaming the receding glaciers on any single cause, including human emissions of greenhouse gases ("Alaskan Glaciers" 2001).

Measurements by aircraft using global positioning satellites and laser altimeters show that Alaska's 5,000-square-kilometer Malaspina glacier is losing nearly a meter of thickness per year, the equivalent of three cubic kilometers of water. That glacier alone has 10 times the water of all the glaciers in the Alps, which have lost more than half of their ice volume since the 1850s. Climatologists have focused most of their attention on the great ice sheets of Greenland and Antarctica for signs of melting that could raise sea levels, but Meier believes the impact of smaller-scale glacial melting has been underestimated (Russell 2002).

From 1993 until his death in 2010, Keith Echelmeyer surveyed the size of 90 glaciers from northern Alaska to the Cascades of Washington. The pilot, mountaineer, and glaciologist used data from satellites and his own observations, often taken from a single-engine airplane. For reference, Echelmeyer used U.S. Geological Survey maps from the middle of the 20th century. He found that 90 percent of the glaciers were losing mass balance, with summer melting surpassing winter ice accumulation. "The glaciers in Alaska are giving us a clear picture that indeed something is happening to cause them to thin . . . that is climate-related," said Echelmeyer (Monastersky 2001, 31–32). Most of the glaciers are shrinking about a meter a year, but some—such as the Lemon glacier near Juneau—are losing two to three meters annually. Echelmeyer believed that several of the glaciers he surveyed would be gone in 50 to 100 years.

A study by Anthony Arendt and colleagues at the University of Alaska–Fairbanks used airborne laser altimetry to estimate volume changes of 67 glaciers in Alaska between the mid-1950s and mid-1990s. The profiles they developed were compared with contours on U.S. Geological Survey and Canadian topographic maps made from aerial photographs taken in the 1950s to early 1970s (Pianin 2002). According to a report by Arendt et al. in *Science* (2002) these glaciers, representing 20 percent of the glacial area in Alaska and neighboring Canada, have been melting at an average of six feet a year, and some have retreated as much as a few hundred feet annually, a rate that is likely to accelerate in coming years.

One study of Western Canadian glaciers by Garry Clarke of the University of British Columbia in Vancouver projects that they will shrink, on average, by 70 percent relative to 2005 by the end of the 21st century ("Few Canadian" 2015). By 2007, Alaska's Columbia glacier was discharging about two cubic miles of ice into Prince William Sound per year, according to Robert Anderson, a University of Colorado–Boulder geology professor who is affiliated with the Institute of Arctic and Alpine Research. The Columbia glacier has thinned as much as 1,300 feet in places and shrunk by nine miles since 1980, and it may recede another nine miles by 2025 ("Glaciers" 2007).

Glacier National Park Going Iceless?

Glacier National Park in northern Montana lost two-thirds of its ice over a century or so, raising the possibility that it could be free of its namesake glaciers within a half-century. The naturalist George Bird Grinnell campaigned for the creation of Glacier National Park in the late 19th century, and a 500-acre glacier there was named for him. Today, it has lost two-thirds of its mass. During the summer of 2015, several wildfires and temperatures near 100°F accelerated melting in and near the park.

When the 1.4 million-acre Glacier National Park was created early in the 20th century, it included more than 150 glaciers in rugged crags and valleys along the Continental Divide in northern Montana. A hundred years later, only 37 of them remained. The Grinnell glacier, for example, has been retreating more than 15 feet a year on average. By the middle of the 21st century, given current trends, the park with "glacier" in its name will have no permanent ice (Toner 2002).

The pace of melting has accelerated with the advent of the new century. "Even those of us who work on these glaciers are surprised at how quickly they are melting," said Dan Fagre of the U.S. Geological Survey. "At the current rate, we may see the complete disappearance of functioning glaciers in the park within 30 years" (Toner 2002). "Glaciers don't know anything about global warming or its causes, but they are excellent barometers of climate change," Fagre said. "They integrate temperature, solar radiation, and snowfall and express it as a big lump of ice. They don't have a political agenda. They just reflect what's happening" (Toner 2002).

Spring has been arriving earlier in the park every year, and summers are warmer and longer than ever before. Most years, more ice melts than can be replaced by winter snows. During the winter, precipitation now falls more often as rain than as snow, even at relatively high altitudes.

Weather can still be extremely variable. For example, after eight years of record warmth and drought, Glacier National Park was smothered in snow during the La Niña year of 2007–2008. By early July 2008, the park's Going-to-the-Sun Road in the park was still snowed in, a record-late date, and tourism was suffering from too much snow (Robbins 2008).

The winter of 2009–2010 was famously cold and snowy across much of the Midwest and the eastern United States, but it was mild and dry in the park, which lost two more of two dozen remaining glaciers in an area that 160 years earlier had 150 bodies of moving ice. A glacier is defined as at least 25 acres of moving ice.

Further Reading

Robbins, Jim. "Snow in July Is a Mixed Blessing for the Northern Rockies." *The New York Times*, July 2, 2008, A9, A10.

Toner, Mike. "Meltdown in Montana; Scientists Fear Park's Glaciers May Disappear within 30 Years." *Atlanta Journal-Constitution*, June 30, 2002, 4A.

In Glacier Bay, Alaska, 95 percent of the ice observed when the area was first mapped in the 1790s had melted by 2000. During the ensuing two centuries, the ice that once nearly covered the bay has receded more than 60 miles; by 2002, it was retreating nearly half a mile a year (Toner 2002). The first European explorers to visit the area named it after ice that covered nearly the entire harbor. By the 1880s, steady glacial retreat opened liquid water in a 40-mile-long bay. Today, Glacier Bay extends more than 60 miles ("Alaskan Glaciers" 2001).

Glacier Contributions to Sea Level Rise

Some scientists believe that rapid glacial melting in Alaska could be a harbinger of worldwide sea level rise. University of Colorado professor Mark Meier anticipates that the level of the world's oceans will rise between 7 and 11 inches by the end of the 21st century, more than twice the level anticipated in 2000 by the Intergovernmental Panel on Climate Change (IPCC) (Russell 2002). Meier said that the IPCC's previous prediction of a sea level increase from 2 inches to 4 inches by century's end was too low for several reasons, the most important of which is an underestimation of water that he believes will be contributed by the melting of glaciers in the Alps, southern Alaska, and the Patagonian mountains of South America (Russell 2002).

Anthony A. Arendt and colleagues calculated that Alaskan glaciers are generating nearly twice the annual volume of melting water as the Greenland ice sheet, the largest ice mass in the Northern Hemisphere. According to this study, the Alaskan melt is adding some two-tenths of a millimeter a year to worldwide sea levels. Alaskan ice melt accounts for around 9 percent of the sea level rise during the last century (Arendt et al. 2002, 382).

"The change we are seeing is more rapid than any climate change that has happened in the last 10 to 20 centuries," said Echelmeyer (Pianin 2002). The scientists did not speculate whether accelerating melting results from human-induced global warming, natural factors, or a combination. Long-time global-warming skeptic Sallie L. Baliunas of the Harvard–Smithsonian Center for Astrophysics in Cambridge, Massachusetts, contended that Alaskan glacial melting is the result of a dramatic but temporary shift in warm Pacific Ocean waters and wind patterns that began in 1976. "It doesn't have the fingerprints of enhanced greenhouse gas concentrations," she said (Pianin 2002). Whatever the cause, "[most] glaciers have thinned several hundred feet at low elevation in the last 40 years and about 60 feet at higher elevations," said the late Keith Echelmayer. The ice cover in the Arctic ocean itself is shrinking by an area the size of the Netherlands each year (Radford 2002).

Further Reading

"Alaskan Glaciers Retreating." Environment News Service, December 11, 2001. http://ens-news.com/ens/dec2001/2001L-12-11-09.html.

"Alaska's Biggest (Ice) Losers Are Inland." NASA Earth Observatory, July 7, 2015. http://earthobservatory.nasa.gov/IOTD/view.php?id=86168&src=eoa-iotd.

Arendt, Anthony A., et al. "Rapid Wastage of Alaska Glaciers and Their Contribution to Rising Sea Level." *Science* 297 (July 19, 2002): 382–386.

Egan, Timothy. "The Race to Alaska before It Melts." *The New York Times*, June 26, 2005. http://www.nytimes.com/2005/06/26/travel/the-race-to-alaska-before-it-melts.html.

"Few Canadian Glaciers Left by 2100." *Nature* 520 (April 9, 2015): 134.

"Glaciers and Ice Caps Quickly Melting into the Seas." Environment News Service, July 20, 2007. http://www.ens-newswire.com/ens/jul2007/2007-07-20-03.asp.

Johnson, Kirk. "Alaska Looks for Answers in Glacier's Summer Flood Surges." *The New York Times*, July 23, 2013. http://www.nytimes.com/2013/07/23/us/alaska-looks-for-answers -in-glaciers-summer-flood-surges.html.

Larsen, C. F., et al. "Surface Melt Dominates Alaska Glacier Mass Balance." *Geophysical Research Letters*, June 4, 2015. http://onlinelibrary.wiley.com/doi/10.1002/2015GL064349 /abstract.

Monastersky, Richard. "The Long Goodbye: Alaska's Glaciers Appear to Be Disappearing Before Our Eyes. Are They a Sign of Things to Come?" *New Scientist*, April 14, 2001, 30–32.

Perlman, David. "Shrinking Glaciers Evidence of Global Warming; Differences Seen by Looking at Photos from 100 Years Ago." *San Francisco Chronicle*, December 17, 2004, A18.

Pianin, Eric. "Study Fuels Worry over Glacial Melting; Research Shows Alaskan Ice Mass Vanishing at Twice Rate Previously Estimated." *Washington Post*, July 19, 2002, A14.

Radford, Tim. "85 Per Cent of Alaskan Glaciers Melting at 'Incredible Rate.'" *The Guardian* (U.K.), July 19, 2002, 9.

Russell, Sabin. "Glaciers on Thin Ice; Expert Says Melting to Be Faster Than Expected." *San Francisco Chronicle*, February 17, 2002, A4.

Toner, Mike. "Meltdown in Montana; Scientists Fear Park's Glaciers May Disappear within 30 Years." *Atlanta Journal-Constitution*, June 30, 2002, 4A.

See also: Glacial Erosion; Glaciers, The Andes; Glaciers, Central Asia; Glaciers, The Himalayas; Ice Melt, Antarctica; Sea Ice, Arctic; Sea Level Rise; Temperatures, Winter

GLOBAL WARMING, ANTARCTICA

Above Antarctica's Amundsen Sea, two enormous glaciers, Pine Island and Thwaites, are poised like corks in a bottle, disintegrating as relatively warm water laps at their bases and draining ice from the West Antarctic ice sheet. Both glaciers are sitting on a ridge of land that slopes below sea level, making them an unstable gateway to the entire West Antarctic ice sheet, which becomes more vulnerable as water temperatures rise. Jon Gertner wrote in the *New York Times Sunday Magazine* (2015) that glaciologist Eric Rignot foresees a time, in 30 to 40 years, when "people will be accustomed to watching Thwaites and Pine Island disintegrate constantly, iceberg by iceberg, into the ocean. And by then, he adds, their collapse 'will be a part of everyday life.'"

A "disaster scenario," as described by Richard Alley, a glaciologist at Pennsylvania State University, has the Pine Island glacier retreating enough to "make a hole in the side of the ice sheet. . . . The remaining ice would drain through that hole" (Melting 1998). Once enough ice had drained through the hole, the West Antarctic ice sheet might eventually collapse, raising average sea levels around the world 15 to 20 feet in a few years. Such an increase in mean sea level would flood roughly 30 percent of Florida and Louisiana; 15 percent to 20 percent of the District of Columbia, Maryland, and Delaware; and 8 percent to 10 percent of the Carolinas and New Jersey. Inundation of coastal areas would have a similar impact around the world (Schneider and Chen 1980). Among the flooded areas would be the centers of some of the world's great urban and commercial centers—from New York City to Mumbai (Bombay), Calcutta, and Manila.

During 1998, Eric Rignot wrote in *Science* that West Antarctica's Pine Island glacier was retreating at 1.2 kilometers a year (plus or minus 0.3 kilometers) and that its ice was thinning 3.5 meters per year (plus or minus 0.9 meter). "The fast recession of the Pine Island glacier, predicted to be a possible trigger for the disintegration of the West Antarctic ice sheet, is attributed to enhanced basal melting of the glacier's floating tongue by warm ocean waters," Rignot wrote (Rignot 1998, 549).

This glacier is widely believed to be "the ice sheet's weak point" (Kerr 1998, 499). Although the accelerated melting of this glacier does not portend an immediate disintegration of the West Antarctic ice sheet, Alley wrote that "most models indicate [that the retreat] would speed up if it kept going" (Kerr 1998, 499). One observer was quoted in this context as stating that a quick collapse of the West Antarctic ice sheet would "back up every sewer in New York City" (Kerr 1998, 500). Rignot speculated that warmer ocean waters were causing the bottom of the Pine Island glacier to rapidly melt. "This is one of the most sensitive ice sheets to climatic change. For many, many years we have neglected the importance of bottom melting," Rignot said (Melting 1998).

"The sudden appearance of thousands of small icebergs suggests that the shelves have essentially broken up in place and then flushed out by storms or currents afterward," said Scambos (Britt 1999). The Larsen and Wilkins ice shelves have been melting since the 1950s, and scientists had expected them to fall apart. The disintegration, however, occurred more quickly than anticipated. "We have evidence that the shelves in this area have been in retreat for 50 years, but those losses amounted to only about 7,000 square kilometers," said David Vaughan, a researcher with the Ice and Climate Division of the British Antarctic Survey. "To have retreats totaling 3,000 square kilometers in a single year is clearly an escalation. Within a few years, much of the Wilkins ice shelf will likely be gone" (Britt 1999).

Hemorrhaging Ice at "An Alarming Rate"

By 2012, the Pine Island glacier (PIG) was "hemorrhaging" ice at "an alarming rate" as relatively warm water eroded its mass from 3,000 feet below. AutoSub3, a robot submarine, explored the bottom side of the ice shelf in 2009, mapping the sea floor with sonar. The research vessel *Nathaniel B. Palmer* gathered data from the sub that revealed how the underside of the glacier had lost 19 cubic miles of ice during 2009 alone. The melting glacier was flowing into the ocean more quickly. From 1974 to 2009, the PIG thinned by 230 feet and accelerated more than 70 percent (Fox 2012).

Radar images taken from satellite observations of the Pine Island glacier during the 1990s indicate that the glacier has been shrinking rapidly. Its shrinking is important "because it could lead to a collapse of the West Antarctic ice sheet," said Eric Rignot, who led the study. "We are seeing a . . . glacier melt in the heart of Antarctica" (Melting 1998). "The continuing retreat of the Pine Island glacier could be a symptom of the WAIS [West Antarctic Ice Sheet] disintegration," said Craig Lingle, a glaciologist at the University of Alaska–Fairbanks, who is familiar with the study (Melting 1998).

By 2008, two-thirds of ice-mass loss from the West Antarctic ice sheet was stemming from the Pine Island glacier and its environs, where release of ice had doubled during the decade ending in 2008 (Gillet-Chaulet and Durand 2010, 794). By 2009, the Pine Island glacier was losing ice four times as quickly as a decade previously, according to satellite imagery. The disclosure alarmed scientists and led them to cut their estimates for the demise of this glacier to just one century rather than 600 years.

During October 2009, scientists flying over Antarctica found a deep-water channel beneath the Pine Island glacier that is probably a path for relatively warm water to melt ice from below, A major reason why it was losing more than 19 cubic miles of ice per year. Using satellite images, scientists spotted a series of large surface undulations on the ice shelf. Next they matched the undulations with the timing of warm water pulses in the waters adjacent to the ice shelf. When surface winds are strong, they stir the Southern Ocean and lift the warm water onto the continental shelf where the additional heat contributes to melt ("Unstable" 2010). A channel of relatively warm water running below the glacier also may be accelerating the melting. The channel conducts ocean water to the grounding line, melting the ice shelf from below.

In the meantime, late in 2009 an unmanned submarine found that an undersea ridge to which the Pine Island glacier may have been frozen in the past was now well below the underside of the ice, as relatively warm seawater continued to eat away at its base (Kerr 2010).

A modeling study published in 2010 argued that the Pine Island glacier had passed its tipping point, which could bring on collapse of parts of the West Antarctic ice sheet. The study by Richard Katz and colleagues at the University of Oxford projected that the glacier could lose half its mass in less than a century. Katz and M. Grae Worster wrote in Britain's *Proceedings of the Royal Society* (2010), "Our results indicate that unstable retreat of the grounding line over retrograde beds is a robust feature of models that evolve based on force balance at the grounding line. We conclude, based on our simplified model, that unstable grounding-line recession may already be occurring at the Pine Island glacier" (Katz and Worster 2010).

Warming Water Erodes Ice

Warming water in the Amundsen Sea continues to erode the glacier from below, "pushing the grounding line higher up the continental shelf" (Barley 2010). This model may understate the speed at which glacier's grounding line was retreating. "Ours is a simple model of an ice sheet that neglects some important physics," said Richard Katz. "The take-home message is that we should be concerned about tipping points in West Antarctica and we should do a lot more work to investigate" (Barley 2010).

NASA's Earth Observatory reported on October 19, 2012, that five days earlier "scientists flying over Antarctica's Pine Island glacier ice shelf . . . made a startling discovery: a massive rift running about 29 kilometers (18 miles) across a part of the glacier's floating tongue. The rift was 80 meters (260 feet) wide on average, and 50 to 60 meters (170 to 200 feet) deep." "When the crack reaches the other side of the ice shelf," said the report, "it will send a huge new iceberg drifting into Pine Island Bay" ("A Growing Rift" 2012).

Rifts in the Pine Island glacier form roughly every five years. "What makes this one remarkable is that it will lead to calving of a significantly larger iceberg than PIG has produced in the last few decades," said Joseph MacGregor, a research scientist at the University of Texas–Austin. "It is likely that the front of PIG will be farther back than any time in the recent past after the iceberg calves," he said ("A Growing Rift" 2012).

Another massive iceberg calved in July 2013 adjacent to open water in Pine Island Bay and the Amundsen Sea ("New Ice Island" 2013). Yet another one calved between November 9 and 11 the same year, developing from rifts that scientists had been watching for two years. Although the term *iceberg* may evoke an image of a chunk of ice, this one was really the size of a small island—35 kilometers by 20 kilometers (21 by 12 miles), or 700 square kilometers—roughly the size of Singapore ("Major Iceberg" 2013). As a whole, the PIG had been moving into the sea at roughly 4 kilometers a year, so the new "iceberg" was not a surprise, but it was unusually large, providing more evidence that "warmer seawater below the shelf will cause the ice grounding line to retreat and the glacier to thin and speed up" ("Major Iceberg" 2013).

Pine Island Glacier Melt Rates Vary

Although the Pine Island glacier has thinned and accelerated rapidly over the last several decades, melting can vary immensely from year to year. The PIG has been melting at an average rate of 100 meters (330 feet) a year (contributing by itself 7 percent of Earth's recent sea level rise). Why is the Pine Island glacier melting so quickly? The NASA Earth Island Observatory explained that it

> flows out of Antarctica's Hudson Mountains and floats over the ocean. Scientists think that the glacier is shrinking because the ocean water that flows under the glacier is warming, increasing melt at the base. . . . Sea ice abuts the floating glacier tongue except in three places along the front of the glacier. These ice-free areas are called *polynyas* [and are] present in these three locations when sea ice is present. The polynyas most likely form where warm ocean currents rise toward the ocean surface. ("Polynyas" 2011)

Within the overall pattern, however, melting decreased 50 percent between January 2010 and January 2012, with "large fluctuations in the ocean heat available in the adjacent bay and enhanced sensitivity of ice-shelf melting to water temperatures at intermediate depth, as a seabed ridge blocks the deepest and warmest waters from reaching the thickest ice" (Dutrieux et al. 2014, 174).

Is West Antarctican Ice Melt Irreversible?

By 2014, several scientists regarded the dissolution of the West Antarctic ice sheet as highly likely, probably beginning in the 22nd century, triggering a world sea level rise of 16 to 18 feet. This collapse was forecast in 1978 by glaciologist John H. Mercer (who died in 1987), who was an outlier at the time. In the decades since, his ideas have gathered credibility. By 2014, scientists were placing a date on the ice sheet's dissolution—200 to 900 years (Rignot et al. 2014).

A year after the studies projected the inevitability of the collapse of the West Antarctic ice sheet, another study published in the *Proceedings of the National Academy of Sciences* late in 2015 mapped it out. The paper concluded, "If the Amundsen Sea Sector of the ice sheet is destabilized—something that is now well underway—then

Some Antarctic Sea Ice Expands

Paradoxically, as Arctic sea ice has repeatedly set new record lows during the early 21st century, some areas of sea ice around East Antarctica have been expanding to a record surface area, giving joy to many climate-change contrarians. In 2012, for example, the National Snow and Ice Data Center reported that Antarctic sea ice covered 19.44 million square kilometers (7.51 million square miles). The previous record of 19.39 million kilometers (7.49 million square miles) was set in 2006.

The NASA Earth Observatory described a study by sea ice scientists Claire Parkinson and Donald Cavalieri of NASA's Goddard Space Flight Center, who surveyed an Antarctic sea ice increase averaging 17,100 square kilometers per year from 1979 to 2010. "Much of the increase occurred in the Ross Sea, with smaller increases in Weddell Sea and Indian Ocean. At the same time, the Bellinghausen and Amundsen Seas have lost ice," the Earth Observatory said. "The strong pattern of decreasing ice coverage in the Bellingshausen [and] Amundsen Seas region and increasing ice coverage in the Ross Sea region is suggestive of changes in atmospheric circulation," they noted ("Antarctic Sea Ice" 2012).

According to Parkinson, "Both hemispheres have considerable inter-annual variability, so that in either hemisphere, next year could have either more or less sea ice than this year. Still, the long-term trends are clear, but not equal: the magnitude of the ice losses in the Arctic considerably exceed the magnitude of the ice gains in the Antarctic" ("Antarctic Sea Ice" 2012).

Further Reading

"Antarctic Sea Ice Reaches New Maximum Extent." NASA Earth Observatory, October 11, 2012. http://earthobservatory.nasa.gov/IOTD/view.php?id=79369&src=eoa -iotd.

the entire marine part of West Antarctica will be discharged into the ocean." The projected timetable for this discharge is "centuries to millennia" (Feldmann and Levermann 2015). They left refinement of the timing to future studies, warning that it could happen more quickly. Their models project that at current rates of ice melt, the ice sheet's fate could be sealed within 60 years. By that time, melting may be self-sustaining, with relatively warm water eroding any ice reaching the ocean. They wrote:

> The Antarctic ice sheet is losing mass at an accelerating rate, and playing a more important role in terms of global sea-level rise. The Amundsen Sea sector of West Antarctica has most likely been destabilized. Although previous numerical modeling studies examined the short-term future evolution of this region, here we take the next step and simulate the long-term evolution of the whole West Antarctic ice sheet. Our results show that if the Amundsen Sea sector is destabilized, then the entire marine ice sheet will discharge into the ocean, causing a global sea-level rise of about three meters. We thus might be witnessing the beginning of a period of self-sustained ice discharge from West Antarctica that requires long-term global adaptation of coastal protection. (Feldmann and Levermann 2015)

As Ian Joughin and colleagues reported in *Science* (2014, 735),

> Resting atop a deep marine basin, the West Antarctic ice sheet has long been considered prone to instability. Using a numerical model, we investigated the sensitivity of Thwaites glacier to ocean melt and whether its unstable retreat is already under way. Our model reproduces observed losses when forced with ocean melt comparable to estimates. Simulated losses are moderate (<0.25 mm per year at sea level) over the 21st century but generally increase thereafter. Except possibly for the lowest-melt scenario, the simulations indicate that early-stage collapse has begun. Less certain is the time scale, with the onset of rapid (>1 mm per year of sea-level rise) collapse in the different simulations within the range of 200 to 900 years

Like a Cork in a Bottle

Some scientists compare the "tongue" of a glacier (where the body of ice meets the sea) to a cork in a bottle. "The tongue of the glacier or the cork in the bottle [does] not represent that much," said Claudio Teitelboim, director of the Center for Scientific Studies, a private Chilean institution that cooperates with NASA to survey the ice fields of Antarctica and Patagonia. "But once the cork is dislodged, the contents of the bottle flow out, and that can generate tremendous instability" (Rohter 2005).

Glaciers flowing into Antarctica's Amundsen Sea, which help drain the West Antarctic ice sheet, were thinning twice as fast near the coast by 2004 as they had during the 1990s. Warmer seawater erodes the bond between coastal ice and the bedrock below, "like weakening the cork in a bottle," said Robert H. Thomas, a glacier expert for NASA in Wallops Island, Virginia. "You start to let stuff out" (Revkin 2004). In 2004, Thomas and several coauthors wrote in *Science*, "Recent aircraft and satellite laser altimeter surveys of the Amundsen Sea sector of West Antarctica show that local glaciers are discharging about 250 cubic kilometers of ice per year to the ocean, about 60 percent more than is accumulated within their catchment basins" (Thomas et al. 2004, 255). Currently, such a discharge could raise world sea levels about 0.2 millimeters per year—not a startling amount. However, the long-term implications of such ice flow may be more ominous: "Most of these glaciers flow into floating ice shelves over bedrock up to hundreds of meters deeper than previous estimates, providing exit routes for ice from further inland if ice-sheet collapse is underway" (Thomas et al. 2004, 255).

Thinning Accelerating

West Antarctica's ice sheets (areas of floating ice at the edge of the ice shelf) have been thinning at accelerating rates since at least the middle 1990s, increasing the chance that at least some of them will collapse and contribute to worldwide sea level rise, according to data analyzed by Fernando Paolo of the Scripps Institution of Oceanography in San Diego, California, and his colleagues. The ice sheets in some areas have declined by as much as 18 percent in slightly less than 20 years (Paolo et al. 2015). "Eighteen percent over the course of 18 years is really a substantial change," said Paolo. "Overall, we show not only that the total ice shelf volume is decreasing, but we see an acceleration in the last decade" (Aguillera 2015).

The researchers explained the importance of this decline in ice thickness: "The floating ice shelves surrounding the Antarctic ice sheet restrain the grounded ice-sheet flow. Thinning of an ice shelf reduces this effect, leading to an increase in ice discharge to the ocean. . . . West Antarctic losses increased by 70 percent in the last decade, and earlier volume gain by East Antarctic ice shelves ceased" (Paolo et al. 2015). The rapid acceleration in melting after 2003 followed a decade of relative stability. The unstable sectors of the West Antarctic ice sheet could lose half their volume in 200 years at current melting rates.

A major portion of the ice shelves on the southern Antarctic peninsula have destabilized since 2009. As B. Wouters and colleagues reported in *Science*, "Ice mass loss of the marine-terminating glaciers has rapidly accelerated from close to balance in the 2000s to a sustained rate of -56 ± 8 gigatons per year, constituting a major fraction of Antarctica's contribution to rising sea level. The widespread, simultaneous nature of the acceleration, in the absence of a persistent atmospheric forcing, points to an oceanic driving mechanism" (Wouters et al. 2015, 899).

Late in May 2014, two groups of scientists "reported that Thwaites glacier, a keystone holding the massive West Antarctic ice sheet together, is starting to collapse. In the long run, they say, the entire ice sheet is doomed. It would release enough meltwater to raise sea levels by more than three meters" (Sumner 2014, 683). With his colleagues, Anders Leverman of the Potsdam Institute for Climate

Impact Research in Germany asserted "that even if emissions were to stop tomorrow, we have probably locked in several feet of sea level rise over the long term" (Gillis 2013). "The surprises keep coming," said Andrew J. Monaghan, a scientist at the National Center for Atmospheric Research in Boulder, Colorado, a participant in the study. "When you see this type of warming, I think it's alarming" (Gillis 2012).

Cautions and Debates

Eric Rignot of the University of California–Irvine, coauthor of one paper, suggested that one-third of West Antarctica could be gone within 100 to 200 years. Rignot noted that the scientific community "still balks at this"—particularly the 100-year projection—but that he thinks observational studies are showing that ice sheets can melt at a faster pace than model-based projections have considered (Mooney and Warrick 2014).

In 2014, with lead author J. Mouginot and B. Scheuchi, Eric Rignot wrote:

We combine measurements of ice velocity from Landsat feature tracking and satellite radar interferometry, and ice thickness from existing compilations to document 41 years of mass flux from the Amundsen Sea Embayment . . . of West Antarctica. The total ice discharge has increased by 77 [percent] since 1973. Half of the increase occurred between 2003 and 2009. Grounding-line ice speeds of Pine Island glacier stabilized between 2009 and 2013, following a decade of rapid acceleration, but that acceleration reached far inland and occurred at a rate faster than predicted by advective processes. Flow speeds across Thwaites glacier increased rapidly after 2006, following a decade of near stability, leading to a 33[-percent] increase in flux between 2006 and 2013. Haynes, Smith, Pope, and Kohler glaciers all accelerated during the entire study period. The sustained increase in ice discharge is a possible indicator of the development of a marine ice sheet instability in this part of Antarctica. (Mouginot et al. 2014, 1576)

Benjamin Strauss of Climate Central, estimated that "12.8 million Americans live on land less than 10 feet above their local high-tide line" (Mooney and Warrick 2014). Given population trends over the last few centuries, those figures may me much higher by the time the ice melts. Other bodies of ice—in Greenland, East Antarctica, and mountain glaciers—will also be melting at the same time.

Writing in the *Washington Post*, Chris Mooney and Jody Warrick (2014) described other experts' cautions about these studies:

Other scientists urged caution in interpreting the findings, saying it is not clear whether the recent accelerated melting is an anomaly or a persistent phenomenon that will continue into the future. Ocean circulation patterns in the south polar region are still not fully understood, and it is possible that the migration of warmer water into the Amundsen Sea is unrelated to the overall climate warming trend, said Olga Sergienko, a glaciologist Princeton University's Cooperative Institute for Climate Science who was not involved in the studies. "This represents only about 20 years of observation, and on the time scale of ice sheets that's just a blink," said Sergienko, who also is with the National Oceanic and Atmospheric Administration's Geophysical Fluid Dynamics Laboratory in Princeton, N.J. (Mooney and Warrick 2014)

Antarctica: Scientific Issues

Warmth can erode ice in many ways. The most familiar to us is warm air on a sunny day—direct sunshine augmenting the atmosphere's warmth. Direct sunlight plays only a minor role in melting Antarctic ice, however. More often relatively warm water erodes an ice shelf from below. Pools of liquid water also can form on top of ice sheets and work their way into them. Melt water and rainfall also may drain into crevasses in ice, provoking vertical fractures. Heavy ice near the top of a sheet also may break apart, shearing off huge chunks. These last two are being studied as ways in which enough of the East Antarctic ice sheet could have eroded to become the main source for high sea levels during warm periods over the last 25 million years. In some periods, ice has melted relatively quickly, raising sea levels 20 meters or more.

David Pollard and colleagues have studied such periods, and found that:

In response to atmospheric and ocean temperatures typical of past warm periods, floating ice shelves may be drastically reduced or removed completely by increased oceanic melting, and by hydrofracturing due to surface melt draining into crevasses. Ice at deep grounding lines may be weakened by hydrofracturing and reduced buttressing, and may fail structurally if stresses exceed the ice yield strength, producing rapid retreat. Incorporating these mechanisms in our ice-sheet model accelerates the expected collapse of the West Antarctic ice sheet to decadal time scales, and also causes retreat into major East Antarctic subglacial basins, producing ~17 m global sea level rise within a few thousand years. The mechanisms are highly parameterized and should be tested by further process studies. But if accurate, they offer one explanation for past sea level high stands, and suggest that Antarctica may be more vulnerable to warm climates than in most previous studies. (Pollard et al. 2015, 112)

The speed at which Antarctic ice may melt depends not only on how much temperatures rise but also on the ways in which ice moves within the ice cap. Jonathan L. Bamber, David G. Vaughan, and Ian Joughin have been studying these "rivers" of subsurface Antarctic ice. "It has been suggested," they write, "that as much as 90 percent of the discharge from the Antarctic ice sheet is drained through a small number of . . . ice streams and outlet glaciers fed by relatively stable and inactive catchment areas." Their research suggests that "each major drainage basin is fed by complex systems of tributaries that penetrate up to

1,000 kilometers from the grounding line to the interior of the ice sheet" (Bamber et al. 2000, 1248). Such "complex flows" are noted throughout the Antarctic ice sheet by these researchers.

Bamber et al. assert that "this finding has important consequences for the modeled or estimated dynamic response time of past and present ice sheets to climate forcing" (Bamber et al. 2000, 1248). The researchers also find evidence of similar ice-sheet dynamics in Greenland, although they are smaller in scale. "This evidence," they write, "challenges the view that the Antarctic plateau is a slow-moving and homogenous region" (Bamber et al. 2000, 1250). These researchers also contend that the dynamics of large ice flows are too complex for current models to predict, so climate modelers have little idea how global warming will affect the largest of Earth's remaining ice masses.

Increasing Snowfall over Interior Antarctica

Global warming can work in contradictory ways. Witness the fact that sea level rise may be slowed by increasing snowfall over Antarctica as provoked by rising temperatures. The eastern half of Antarctica has been gaining weight, more than 45 billion tons a year, according to one scientific study. Data from satellites bouncing radar signals off the ground show that the surface of eastern Antarctica appears to be slowly growing higher—by 1.8 centimeters a year—as snow and ice pile up (Chang 2005). As temperatures rise, so does the amount of moisture in the air, causing snowfall to increase in cold areas such as inland eastern Antarctica. "It's been long predicted by climate models," said Dr. Curt H. Davis, a professor of electrical and computer engineering at the University of Missouri. In the meantime, however, another study found that changes in snowfall had been insignificant since the 1950s (Monaghan et al. 2006).

Satellite radar altimetry measurements suggest that the East Antarctic ice sheet interior north of 81.6° south increased in mass by 45±7 billion tons per year from 1992 to 2003. Comparisons with contemporaneous meteorological model snowfall estimates suggest that the gain in mass was associated with increased precipitation. A gain of this magnitude is enough to slow sea level rise by 0.12±0.02 millimeters per year (Davis et al. 2005). The accumulation occurring across 2.75 million square miles of eastern Antarctica corresponds to a gain of 45 billion tons of water a year or the removal of the top 0.12 millimeter of the world's oceans. According to Davis, Antarctica "is the only large terrestrial ice body that is likely gaining mass rather than losing it" (Chang 2005).

The data, from two European Space Agency satellites, cover 1992 to 2003, but because the satellites do not pass directly over the South Pole, they did not provide any information for a 1,150-mile-wide circular area there. Assuming that snow was falling there at the same rate seen in the rest of Antarctica, the total gain in snowfall would correspond to a 0.18-millimeter-a-year drop in sea levels (Chang 2005). R.A. Winkelmann and colleagues wrote in *Nature* (2012, 239), "Snowfall and discharge are not independent, but . . . future ice discharge will increase by up to three times as a result of additional snowfall under global warming."

Chemicals, Climate Change, and the Food Chain

As glaciers melt along the fringes of Antarctica, trace amounts of DDT are showing up in Adélie penguins. Although the amounts currently are too little to harm the birds, they suggest, according to one report, that "the presence of the chemical could be an indication that other frozen pollutants will be released because of climate change, says Heidi Geisz, a marine biologist at Virginia Institute of Marine Science in Gloucester" (Callaway 2008). She says that other chemical pollutants such as polychlorinated biphenyls (PCBs) and polybrominated diphenyl ethers (PBDEs) also could be released.

Scientists collected three decades of data indicating that climate change is contributing to declining populations of Antarctic fur seals in the southern Atlantic Ocean by reduced availability of prey, causing significant declines in in birth weight. "Our results provide compelling evidence that selection due to climate change is intensifying, with far-reaching consequences for demography as well as phenotypic and genetic variation," wrote Jaume Forcada and Joseph Ivan Hoffman (2014, 462).

"Antarctica's fate is not as simple as that of an ice cube melting in the sun, scaled up a trillionfold," Jane Qiu wrote in *Science* (2012).

> Changing wind patterns are an unsung force shaping Antarctica's future. Retreating sea ice and stronger winds have caused seawater to mix more deeply, a process that churns sunlight-dependent phytoplankton into the ocean's depths. As a result, phytoplankton biomass has declined by 12 percent over the past 30 years. Higher on the food chain, that means fewer krill and fish larvae. These creatures are also getting hammered by the loss of sea ice, which hides them from predators. The complex interplay between air, sea, and ice has emerged as a central theme underlying climate change in Antarctica. Shifting wind patterns and corresponding ocean changes can explain climate responses across the continent. (Qiu 2012, 879)

Climate, Stratospheric Ozone Levels, and the Circumpolar Vortex

Changes in stratospheric ozone chemistry also may have aided the growth of sea ice around Antarctica since the late 1980s, according to a report in *Geophysical Research Letters* in April 2009 by scientists from British Antarctic Survey (BAS) and NASA (Turner at al. 2009). The research indicates that "the ozone hole delay[ed] the impact of greenhouse gas increases on the climate of the continent" ("Increasing" 2009). This has produced an average increase of 100,000 square kilometers a decade since the 1970s. The increase in sea ice has been used by climate contrarians to refute effects of global warming—that is, as a counterpoint to rapid melting of sea ice in the Arctic.

Professor John Turner of BAS and lead author of the report said,

> Our results show the complexity of climate change across the Earth. While there is increasing evidence that the loss of sea ice in the Arctic has occurred due to human activity, in the Antarctic human influence through the ozone

hole has had the reverse effect and resulted in more ice. Although the ozone hole is in many ways holding back the effects of greenhouse gas increases on the Antarctic, this will not last, as we expect ozone levels to recover by the end of the 21st century. By then there is likely to be around one-third less Antarctic sea ice. ("Increasing" 2009)

Although sea ice has increased slightly around the coast of East Antarctica, it has decreased more rapidly in West Antarctica on the Antarctic peninsula, which has warmed by almost 3°C since the 1960s. Even farther west, sea ice cover over the Ross Sea has increased. Depletion of ozone "has strengthened surface winds around Antarctica and deepened the storms in the South Pacific area of the Southern Ocean that surrounds the continent. This resulted in greater flow of cold air over the Ross Sea (West Antarctica), leading to more ice production in this region," according to this research ("Increasing" 2009).

The work of David W. J. Thompson and Susan Solomon may be "the strongest evidence yet" that a shift in the Antarctic Oscillation "could explain a number of different components of [Antarctic] climate trends," according to David Karoly, a meteorologist at Monash University in Clayton, Australia (Kerr 2002). The researchers linked cooling in the stratosphere induced by depleted ozone levels with acceleration of winds. "During the summer–fall season," Thompson and Solomon have written, "the trend toward stronger circumpolar flow has contributed substantially to the observed warming over the Antarctic peninsula and Patagonia and to the cooling over eastern Antarctica and the Antarctic plateau" (Thompson and Solomon 2002, 895).

Writing in the May 3, 2002 edition of *Science*, Thompson, a professor of atmospheric science at Colorado State University, and Solomon, a senior scientist at the National Oceanic and Atmospheric Administration in Boulder, Colorado, asserted that ozone depletion over the Antarctic may help explain both contradictory trends. "Ozone seems to be capable of tickling the Southern Hemisphere patterns," Thompson said (Chang 2002). Thompson and Solomon assert that a vortex of winds blowing around Antarctica traps cold air at the South Pole and has strengthened in the past few decades, keeping the cold air even more confined. The Antarctic peninsula lies outside the wind vortex and thus escapes the cooling effect. Ozone depletion may be a key causal factor in strengthening the wind pattern, according to Thompson and Solomon. "That's where we speculate," Dr. Thompson said, "and the emphasis is on the word 'may'" (Chang 2002).

Scientists already knew that ozone depletion has cooled the upper atmosphere. Thompson and Solomon's research indicates that parts of Antarctica the troposphere, the lowest six miles of the atmosphere, also has cooled. "It's a lot of food for thought in there," said Dr. John E. Walsh, a professor of atmospheric science at the University of Illinois (Chang 2002). Walsh said the data tying the cooling to stronger winds were convincing. "My one reservation," he said, "is the link to the ozone" (Chang 2002). He noted that the ozone hole was usually largest in November or December but that the greatest cooling had been some six months later. Thompson agreed that ozone depletion could not explain the whole climactic picture and said other influences such as ocean currents probably played important roles, too. "I seriously doubt it's the only player," he said. "I think it's one of many" (Chang 2002).

The idea that stratospheric ozone depletion has been a factor in driving a stronger circumpolar vortex (with cooling inside the vortex and warming outside) has been gaining support. In 2003, Nathan P. Gillett and David W. J. Thompson published results of a modeling study supporting this effect during the spring and summer. "The results," they wrote in *Science*, "provide evidence that anthropogenic emissions of ozone-depleting gases have had a distinct impact on climate not only at stratospheric levels but at Earth's surface as well" (Gillett and Thompson 2003, 273; Karoly 2003).

Mark P. Baldwin and colleagues wrote in *Science*, "The resulting ozone 'hole' leads to a relative reduction in solar heating and a stronger vortex. Observations and recent model simulations show that the strengthening of the polar vortex during spring leads to lower surface temperatures over Antarctica and higher temperatures in the mid-latitudes of the Southern Hemisphere that persist into summer" (Baldwin et al. 2003, 317).

Further Reading

Aguillera, Mario. "Antarctic Ice Shelves Rapidly Thinning; New Study Reveals Accelerating Losses over Two Decades." UC San Diego News Center, March 26, 2015. http://ucsdnews.ucsd.edu/pressrelease/antarctic_ice_shelves_rapidly_thinning.

Baldwin, Mark P., et al. "Weather from the Stratosphere?" *Science* 301 (July 18, 2003): 317–318.

Bamber, Jonathan L., David G. Vaughan, and Ian Joughin. "Widespread Complex Flow in the Interior of the Antarctic Ice Sheet." *Science* 287 (February 18, 2000): 1248–1250.

Barley, Shanta. "Major Antarctic Glacier Is 'Past Its Tipping Point.'" *New Scientist*, January 13, 2010. http://www.newscientist.com/article/dn18383-major-antarctic-glacier-is-past-its-tipping-point.html?DCMP=OTC-rss&nsref=environment.

Britt, Robert Roy. "Antarctic Ice Shelves Falling Apart." Explorezone.com, April 9, 1999. http://www.explorezone.com/archives/99_04/09_antarctic_ice.htm (no longer available).

Callaway, Ewen. "Melting Glaciers Release Toxic Chemical Cocktail." *New Scientist*, May 7, 2008. http://www.newscientist.com/article/dn13848.

Chang, Kenneth. "Ozone Hole Is Now Seen as a Cause for Antarctic Cooling." *The New York Times*, May 3, 2002, A16.

Chang, Kenneth. "Warming Is Blamed for Antarctica's Weight Gain." *The New York Times*, May 20, 2005, A22.

Davis, Curt H., et al. "Snowfall-Driven Growth in East Antarctic Ice Sheet Mitigates Recent Sea-Level Rise." *Science* 308(2) (June 24, 2005): 1898–1901.

Dutrieux,, Pierre, et al. "Strong Sensitivity of Pine Island Ice-Shelf Melting to Climatic Variability." *Science* 343 (January 10, 2014): 174–178.

Feldmann, Johannes, and Anders Levermann. "Collapse of the West Antarctic Ice Sheet after Local Destabilization of the Amundsen Basin." *Proceedings of the National Academy of Sciences*. 112(46) (November 2015): 201512482.

Forcada, Jaume, and Joseph Ivan Hoffman. "Climate Change Selects for Heterozygosity in a Declining Fur Seal Population." *Nature* 511 (July 24, 2014): 462–465.

Fox, Douglas. "Antarctica Undercut," *National Geographic*, January 2012, 35.

Gertner, Jon. "The Secrets in Greenland's Ice Sheet." *New York Times Sunday Magazine*. November 15, 2015. http://www.nytimes.com/2015/11/15/magazine/the-secrets-in-greenlands-ice-sheets.html.

Gillet-Chaulet, Fabien, and Gael Durand. "Ice-Sheet Advance in Antarctica." *Nature* 467 (October 14, 2010): 794–795.

Gillett, Nathan P., and David W. J. Thompson. "Simulation of Recent Southern Hemisphere Climate Change." *Science* 302 (October 10, 2003): 273–275.

Gillis, Justin. "West Antarctica Warming Faster Than Thought, Study Finds." *The New York Times*, December 23, 2012.

Gillis, Justin. "Timing a Rise in Sea Level." *The New York Times*, August 12, 2013. http://www.nytimes.com/2013/08/13/science/timing-a-rise-in-sea-level.html.

"A Growing Rift in Antarctic Ice." NASA Earth Observatory, October 19, 2012. http://earthobservatory.nasa.gov/IOTD/view.php?id=79440&src=eoa-iotd.

"Increasing Antarctic Sea Ice Extent Linked to the Ozone Hole." NASA Earth Observatory and British Antarctic Survey, April 21, 2009. https://www.bas.ac.uk/media-post/increasing-antarctic-sea-ice-extent-linked-to-the-ozone-hole/.

Joughin, Ian, Benjamin E. Smith, and Brooke Medley. "Marine Ice Sheet Collapse Potentially Under Way for the Thwaites Glacier Basin, West Antarctica." *Science* 344 (May 16, 2014): 735–738.

Karoly, David J. "Ozone and Climate Change." *Science* 302 (October 10, 2003): 236–237.

Katz, Richard F., and M. Grae Worster. "Stability of Ice-Sheet Grounding Lines." *Proceedings of the Royal Society A*, January 13, 2010. http://rspa.royalsocietypublishing.org/content/early/2010/01/13/rspa.2009.0434.abstract. doi: 10.1098/rspa.2009.0434.

Kerr, Richard A. "West Antarctica's Weak Underbelly Giving Way?" *Science* 281 (July 24, 1998): 499–500.

Kerr, Richard A. "A Single Climate Mover for Antarctica." *Science* 296 (May 3, 2002): 825–826.

Kerr, Richard A. "Antarctic Glacier Off Its Leash." *Science* 327 (January 22, 2010): 409.

"Major Iceberg Cracks Off Pine Island Glacier." NASA Earth Observatory, November 15, 2013. http://earthobservatory.nasa.gov/IOTD/view.php?id=82392&src=eoa-iotd.

"Melting Antarctic Glacier Could Flood Coastal Areas, Scientists Say." *Los Angeles Times*, July 124, 1998. http://articles.latimes.com/1998/jul/24/news/mn-6808.

Monaghan, Andrew J., et al. "Insignificant Change in Antarctic Snowfall since the International Geophysical Year." *Science* 313 (August 11, 2006). 827–831.

Mooney, Chris, and Jody Warrick. "Research Casts Alarming Light on Decline of West Antarctic Glaciers." *Washington Post*, December 5, 2014. http://www.washingtonpost.com/national/health-science/research-casts-alarming-light-on-decline-of-west-antarctic-ice-sheets/2014/12/04/19efd3e4-7bbe-11e4-84d4-7c896b90abdc_story.html?wpisrc=nl-headlines&wpmm=1.

Mouginot, J. E., E. Rignot, and B. Scheuchl. "Sustained Increase in Ice Discharge from the Amundsen Sea Embayment, West Antarctica, from 1973 to 2013." *Geophysical Research Letters* 41(5) (March 16, 2014): 1576–1584.

"New Ice Island at Pine Island Glacier." NASA Earth Observatory, July 28, 2013, http://earthobservatory.nasa.gov/IOTD/view.php?id=81674&src=eoa-iotd.

Paolo, Fernando S., Helen A. Fricker, and Laurie Padman. "Volume Loss from Antarctic Ice Shelves Is Accelerating." Science (online), March 26, 2015. http://science.sciencemag.org/content/early/2015/03/25/science.aaa0940. doi: 10.1126/science.aaa0940.

Pollard, David, Robert M. DeConto, and Richard B. Alley. "Potential Antarctic Ice Sheet Retreat Driven by Hydrofracturing and Ice Cliff Failure." *Earth and Planetary Science Letters* 412 (February 15, 2015): 112–121.

"Polynyas and the Pine Island Glacier, Antarctica." NASA Earth Observatory. NASA Earth Observatory, November 18, 2011. http://earthobservatory.nasa.gov/IOTD/view.php?id=76437&src=eoa-iotd.

Qiu, Jane. "Winds of Change." *Science* 338 (November 16, 2012): 879–881.

Revkin, Andrew. "Antarctic Glaciers Quicken Pace to Sea; Warming Is Cited." *The New York Times*, September 24, 2004, A24.

Rignot, E., et al. "Widespread, Rapid Grounding Line Retreat of Pine Island, Thwaites, Smith, and Kohler Glaciers, West Antarctica from 1992 to 2011." *Geophysical Research Letters*, May 2014. doi: 10.1002/2014GL060140.

Rignot, E. J. "Fast Recession of a West Antarctic Glacier." *Science* 281 (July 24, 1998): 549–551.

Rignot, E., et al. "Widespread, Rapid Grounding Line Retreat of Pine Island, Thwaites, Smith, and Kohler Glaciers, West Antarctica from 1992 to 2011." *Geophysical Research Letters*, May 2014. doi: 10.1002/2014GL060140.

Rohter, Larry. "Antarctica, Warming, Looks Ever More Vulnerable." *The New York Times*, January 25, 2005. http://www.nytimes.com/2005/01/25/science/earth/25ice.html.

Schneider, Stephen H., and R. S. Chen. "Carbon Dioxide Warming and Coastline Flooding: Physical Factors and Climatic Impact." *American Review of Energy* 5 (1980): 107–140.

Sumner, Thomas. "No Stopping the Collapse of West Antarctic Ice Sheet." *Science* 344 (May 16, 2014): 683.

Thomas, R., et al. "Accelerated Sea-Level Rise from West Antarctica." *Science* 306 (October 8, 2004): 255–258.

Thompson, David W. J., and Susan Solomon. "Interpretation of Recent Southern Hemisphere Climate Change." *Science* 296 (May 3, 2002): 895–899.

Turner J., et al. "Non-Annular Atmospheric Circulation Change Induced by Stratospheric Ozone Depletion and Its Role in the Recent Increase of Antarctic Sea Ice Extent." *Geophysical Research Letters* 36 (April 23, 2009): L08502. doi: 10.1029/2009GL037524.

"Unstable Antarctica: What's Driving Ice Loss?" NASA Earth Observatory. December 15, 2010. http://www.nasa.gov/topics/earth/features/unstable-antarctica.html.

Winkelmann, R., et al. "Increased Future Ice Discharge from Antarctica Owing to Higher Snowfall." *Nature* 492 (December 13, 2012): 239–242. doi:10.1038/nature11616.

Wouters, B., et al. "Dynamic Thinning of Glaciers on the Southern Antarctic Peninsula." *Science* 348 (May 22, 2015): 899–903.

See also: Adaptation, Animal, Amphibians, and Warming Habitats; Animal Life, Antarctic; Ice Melt, Antarctica; Ice Shelves, Antarctic; Inland Cooling, Antarctica; Ocean Circulation; Oceans' Absorption of Heat; Sea Level Rise; Temperatures, Greenhouse Gas Levels and

GLOBAL WARMING, CHINA

China is the wildest card in the world greenhouse deck. On the one hand, the world's most populous country is streamlining energy efficiency and experimenting with new fuel sources. On the other, China is undergoing an industrial revolution with a population of around 1.4 billion, consuming enormous amounts of coal and oil even as its economy becomes more efficient. By 2015, China was consuming half the world's coal and emitting 50 percent more greenhouse gases than the United States, following a decades-long economic boom.

The scale of industrial development in China in the late 20th and early 21st centuries has no parallel in human history. To gauge the scale of the building boom (and greenhouse gas generation) in China, consider the amount of cement manufactured and used there between 2010 and 2013—6.1 gigatons. The United States used 4.4 gigatons of cement during the *entire* 20th century.

In *Carbon Shock*, Mark Schapiro pointed out that although China's share of worldwide greenhouse gas emissions rose rapidly from about 10 percent in 1990 to almost 30 percent in 2013, much of this represents manufacturing exported back to countries that once manufactured basic commodities (such as cement and steel) themselves (Schapiro 2014, 114). Thus, an urban area such as Pittsburgh, Pennsylvania, once a steelmaking hub, is now known as a "green" city with a much smaller carbon footprint than decades ago, even as its residents still consume emission-intensive products from China. Manchester, England (described by Schapiro in detail), the cradle of the fossil fuel manufacturing revolution, has made a point of tracing where the products it now imports come from. As the Chinese become more affluent, of course, more of their industrial output is consumed at home. Much of the emission-rich cement production has gone into Chinese construction.

During this boom, China's greenhouse gas emissions grew rapidly, exceeding those of the United States around 2007. According to the Netherlands Environmental Agency, China's carbon dioxide emissions increased 8 percent in 2007. This increase represented two-thirds of global growth in greenhouse gas emissions. By 2007, China's emissions exceeded those of the United States by 14 percent. Per capita, however, United States emissions were still four times those of China, 19.4 tons to 5.1 tons (Rosenthal 2008).

By 2008, growth in China's carbon dioxide emissions was accelerating at a rate far greater than previous estimates, according to analysis by University of California economists. Maximillian Auffhammer, a University of California–Berkeley assistant professor of agricultural and resource economics, and Richard Carson, University of California–San Diego professor of economics, calculated their estimates from pollution data in China's 30 provinces. During 2008, the carbon emissions of China's electric-power industry jumped by 30 percent, according to the Center for Global Development. The same report expected emissions from power production in China and India to double by 2020. China's emissions leaped from 2.3 billion tons in 2007 to an estimated 3.1 billion in 2008 while U.S. emissions remained stable at 2.8 billion. Paul Ting, an oil analyst, said that China relies on coal for three-quarters of its energy consumption. "They cannot get away from coal," he said (Mufson 2008). Even with China's rapid increases, electricity usage in the United States in 2007 produced roughly 9.5 tons of carbon dioxide per person compared with 2.4 tons per person in China and 0.6 in India.

"Making China and other developing countries an integral part of any future climate agreement is now even more important," said Auffhammer. "What we're finding instead is that the emissions growth rate is surpassing our worst expectations," he said, "and that means the goal of stabilizing atmospheric CO_2 is going to be much, much harder to achieve" ("Growth" 2008). Much of this increase will come from new coal-fired electric power capacity in plants built to last for 40 to 75 years, which all but guarantees high emissions far into the future.

China's growth has been explosive. In 2014, its economy expanded at 7.3 percent—the *slowest* expansion of gross domestic product (GDP) in 25 years. Coal use—and the electricity it produces—grew around 10 percent a year (doubling every 7 years compounded), reflecting a similar rise in the country's GDP. Almost 75 percent of growth in global fossil fuel carbon and cement production between 2010 and 2012 occurred in China. The country's declining growth rate and efforts to install more wind and solar energy by 2015 was causing coal combustion to stabilize, as many coal-fired power plants were operating well below capacity, even as some 150 such plants were still in planning stages or under construction (Wong 2015). "China already has more coal capacity than it will ever need," said Zhang Boting, vice chairman of the China Society for Hydropower Engineering. "A few years down the road we'll see what a waste these plants are" (Wong 2015).

Winter Takes a Vacation

China's winter of 2006–2007 was unusually mild, drawing attention to the warming climate. A popular 1,400-year-old ice festival in Harbin in northeast China, some 400 miles east of the Russian border, literally melted, threatening a tourist attraction that usually draws 5 million people a year. As Edward Cody of the *Washington Post* wrote,

> The hands had melted off a delicately entwined couple of ballet dancers crafted by an ice-sculpting team from Vladivostok. Eaves fashioned from packed snow-drooped into icicles at the Roast Meat Fire House restaurant. Authorities banned people from approaching the ice-cube tower at Ice and Snow World because big chunks kept falling off. . . . Heads are falling from statues and intricately sculpted ice animals are turning into shapeless blobs. (Cody 2007).

In the midst of the nonwinter of 2006–2007, the China Meteorological Administration said that temperatures probably would continue to rise by 7 to 10.8°F by 2100 compared to average temperatures between 1961 and 1990. In Beijing, during the Lunar New Year celebrations, the warmest since authorities began keeping records in 1951, people jogged without jackets in Ritan Park as boys played basketball in T-shirts. On February 5, the temperature rose to a record of 61°F (Cody 2007).

In the meantime, Jiang Yu, a spokeswoman for the foreign ministry, said that China placed primary responsibility for global warming on richer, developed nations. "It must be pointed out that climate change has been caused by the long-term historic emissions of developed countries and their high per capita emissions," she said, adding that developed countries have responsibilities for global warming "that cannot be shirked" (Yardley 2007).

Alarms over global warming also have begun to ring in some of China's official agencies Oceanographers at China's State Oceanic Administration argue that a sea level rise of three feet a century could cause flooding in many of China's coastal areas, home to half of China's large cities and 40 percent of its population.

The Chinese Search for Resources

With its expanding industries, more automobiles, and more affluent population, China scours the world looking for energy in all its forms. China's demand for energy and other natural resources can roil world markets. "China's demand has also provided life support to coal producers suffering from declining use in the United States and other industrialized countries," wrote Clifford Kraus and Keith Bradsher in *The New York Times* (2014). "The dynamic growth of China's economy and energy growth is reshaping global energy markets, and both the economic and strategic implications are still being developed," said Mark J. Finley, BP's general manager for global energy markets and U.S. economics. By 2013, China was burning 10.1 million barrels of oil a day, one-ninth of the world's supply, but was only producing 4.2 million barrels itself.

Before 2015, with Chinese GDP and energy consumption growing at an average of 10 percent a year, global prices for oil and other natural resources soared. When growth slowed to less than 7 percent in 2015, with a lower rate projected for 2016, the prices of commodity futures sank worldwide.

Chinese oil companies have an eye on U.S. production of oil by hydraulic fracturing ("fracking"). "The China market feels that the revolution in shale gas will be coming very soon," Zhang Mi, chair and president of Honghua Group, an exporter of drilling rigs, told *The New York Times* (Kraus and Bradsher 2014). Some in China see shale as a substantial source for natural gas in five to 10 years to relieve some of the burden of low-energy "brown" coal that is so abundant in China. However, much of China's shale-gas deposits lie deep underground in complex geological formations, raising safety and environmental issues.

China and Automobiles

In 2009, China for the first time purchased more cars than did U.S. residents—12.8 million to 10.3 million. China's car sales increased 42 percent in 2009 compared to 2008, 72 percent for sport utility vehicles (SUVs), in a year when most of the world was in recession (Bradsher 2009). Automobile sales in China increased more than 800 percent from 2000 to 2007 (Bradsher 2008). By 2008, more Buicks were being sold in China than in the United States, in a market where car size is closely identified with social and economic status. Some wealthy Chinese pay more than $200,000 for a Hummer.

All new cars, minivans, and SUVs sold in China starting July 1, 2007, had to meet fuel-economy standards stricter than those of the United States'. New construction codes encourage the use of double-glazed windows to reduce air conditioning and heating costs and high-tech lightbulbs that produce more light with fewer watts (Bradsher and Barboza 2006).

China's highway system may soon surpass that of the United States. The 23,000 miles of highway in 2006 had doubled that of 2001. The Chinese government in 2006 announced plans to build 53,000 freeway miles by 2035. The U.S. Interstate Highway System, which is 50 years old, comprises 46,000 miles. As with the U.S. Interstate system. China's goal is to consolidate the nation, and to allow the easy transport of military forces between regions. Policy anticipates that western territories such as Tibet and Xinjiang (meaning "New Frontier") will be fully integrated ethnically and economically (Conover 2006).

The number of passenger cars on the road, about 6 million in 2000, rose to about 20 million in 2006 and more than 35 million by 2015 (Conover, 2006). China accounted for 18 percent of global growth in automobile sales between 2002 and 2012 (Bradsher 2003, 1). During the 1990s, motor vehicle sales in the Chinese countryside rose from 40,000 to almost 500,000 per year (Leggett 2001). Shanghai Automotive Industrial Corp. is planning to license General Motors technology to build a basic pickup truck for China's farmers. The new vehicle, to be called "Combo," will be produced in a nonprofit government car factory. This is one of GM's efforts to tap an auto market of "one billion consumers" and a fast-growing network of national highways (Leggett 2001).

Chinese car culture resembles that of the United States: "City drivers, stuck in ever-growing jams, listen to traffic radio. They buy auto magazines with titles like *The King of Cars, AutoStyle, China Auto Pictorial, Friends of Cars, Whaam* ('The Car—The Street—The Travel—The Racing'). Two-dozen titles now compete for space in kiosks. The McDonald's Corporation said that it expects half of its new outlets in China to include drive-throughs. Whole zones of major cities, like the Asian Games Village area in Beijing, have been given over to car lots and showrooms" (Conover 2006).

As China's fleet of motor vehicles expanded, its consumption of oil also increased from roughly 2.2 million barrels a day in 1988 to 5.2 million barrels a day in 2003, or roughly 150 percent in 15 years (an average of 10 percent a year). Oil use accelerated after that to almost 7 million barrels a day by 2007. The International Energy Agency issued figures from its office in Paris indicating that increases in Chinese greenhouse gas emissions between 2000 and 2030 "will nearly equal the increase from the entire industrialized world" (Bradsher 2003, 1). During the mid-1990s, people in China owned a mere handful of private cars, but private automobile ownership grew by 26 percent between 1996 and 2000—and by 69 percent in 2003 alone (English 2004, 1).

In addition to massive industrial expansion, since 2000 China has been adding 7.5 billion square feet of residential and commercial real estate per year, as much as all existing retail shopping centers and strip malls in the United States, according to the U.S. Energy Information Administration (Kahn and Yardley 2006). An increasing proportion of this space is air-conditioned. In addition, most Chinese buildings, even new ones, have little or no thermal insulation and require twice as much energy to heat or cool as similar floor space in similar U.S. or European climates, according to the World Bank. China has energy-efficiency standards, but most new buildings do not meet them (Kahn and Yardley 2006).

To light, heat, and cool all this new space (as well as industrial plants that produce so many exported goods), China in 2005 alone added 66 gigawatts of electricity—as much as Great Britain's annual demand. In 2006, it added 102 gigawatts, the total demand of France. Two-thirds of this new power is generated using coal. China has built small, inexpensive coal-fired plants that only rarely use the latest more efficient combined-cycle turbines (Kahn and Yardley 2006).

Further Reading

Bradsher, Keith. "China Prospering but Polluting; Dirty Fuels Power Economic Growth." *International Herald-Tribune*, October 22, 2003, 1.

Bradsher, Keith. "With First Car, a New Life in China." *The New York Times*, April 24, 2008. http://www.nytimes.com/2008/04/24/business/worldbusiness/24hrstcar.html.

Bradsher, Keith. "Recession Elsewhere, but It's Booming in China." *The New York Times*, December 10, 2009. http://www.nytimes.com/2009/12/10/business/economy/10consume.html.

Bradsher, Keith, and David Barboza. "Pollution from Chinese Coal Casts a Global Shadow." *The New York Times*, June 11, 2006. http://www.nytimes.com/2006/06/11/business/worldbusiness/11chinacoal.html.

Cody, Edward. "Mild Weather Takes Edge Off Chinese Ice Festival; Residents of Tourist City Blame Global Warming." *Washington Post*, February 25, 2007, A19. http://www.washingtonpost.com/wp-dyn/content/article/2007/02/24/AR2007022401421.html.

Conover, Ted. "Capitalist Roaders." *New York Times Sunday Magazine*, July 2, 2006. http://www.nytimes.com/2006/07/02/magazine/02china.html.

English, Andrew. "Feeding the Dragon: How Western Car-Makers Are Ignoring Ecological Dangers in Their Rush to Exploit a Wide-Open Market." *Daily Telegraph* (London), October 30, 2004, 1.

"Growth in China's CO2 Emissions Double Previous Estimates." Environment News Service, March 11, 2008. http://www.ens-newswire.com/ens/mar2008/2008-03-11-01.asp.

Kahn, Joseph, and Jim Yardley. "As China Roars, Pollution Reaches Deadly Extremes." *The New York Times*, August 26, 2006. http://www.nytimes.com/2007/08/26/world/asia/26china.html.

Kraus, Clifford, and Keith Bradsher. "China's Global Search for Energy." *The New York Times*, May 22, 2014. http://www.nytimes.com/2014/05/22/business/international/chinas-global-search-for-energy.html?hp&_r=0.

Leggett, Karby. "In Rural China, General Motors Sees a Frugal but Huge Market: It Bets Tractor Substitute Will Look Pretty Good to Cold, Wet Farmers." *Wall Street Journal*, January 16, 2001, A19.

Mufson, Steven. "Power-Sector Emissions of China to Top U.S." *Washington Post*, August 27, 2008, D1. http://www.washingtonpost.com/wp-dyn/content/article/2008/08/26/AR2008082603096_pf.html.

Rosenthal, Elisabeth. "China Increases Lead as Biggest Carbon Dioxide Emitter." *The New York Times*, June 14, 2008. http://www.nytimes.com/2008/06/14/world/asia/14china.html.

Schapiro, Mark. *Carbon Shock: A Tale of Risk and Calculus on the Front Lines of the Disrupted Global Economy; How Carbon Is Changing the Cost of Everything.* White River Junction, VT: Chelsea Green, 2014.

Wong, Edward. "Glut of Coal-Fired Plants Casts Doubts on China's Energy Prorities." *The New York Times*, November 12, 2015, A6.

Yardley, Jim. "China Says Rich Countries Should Take Lead on Global Warming." *The New York Times*, February 7, 2007. http://www.nytimes.com/2007/02/07/world/asia/07china.html.

See also: Automobiles; Coal; Deforestation; Desertification; Drought, Worldwide; Extinctions; Floods; Land Use; Sea Level Rise; Solar Power; Wind Gains Power Share

GLOBAL WARMING, GREENLAND

Melting ice has been changing the economy and culture of Greenland as tourists and energy entrepreneurs arrive from lower latitudes. For example, a shrimp factory in Narsaq that was once the town's largest employer has closed because the shrimp departed the nearby warming waters. All but one of the town's eight fishing boats have been retired. The population of Narsaq fell from 3,000 to 1,500 in 10 years, and the suicide rate has risen among residents. "Fishing is the heart of this town," said Hans Kaspersen, 63, a fisherman. "Lots of people have lost their livelihoods" (Rosenthal 2012).

Elsewhere, some of Greenland's 57,000 people are finding that warming temperatures and melting ice offer opportunities. Deposits of gems and minerals have emerged from melting ice, including large deposits of rare earth metals essential for cell phones, electric cars, and wind turbines. Gold, iron, zinc, and offshore oil also are being sought. More than 160 active licenses had been issued by 2014 from Greenland's Bureau of Minerals and Petroleum; at the turn of the century, there were fewer than 20. Greenland also is being pressured to relax its "zero tolerance" policy for uranium mining because melting ice has offered access to deposits that have been known for several decades.

"For me, I wouldn't mind if the whole ice cap disappears," said Ole Christiansen, the chief executive of NunamMinerals, described as "Greenland's largest homegrown mining company." "As it melts, we're seeing new places with very attractive geology" (Rosenthal 2012).

A society mainly comprising fishermen and hunters may be facing large-scale development, as well as foreign workers immigrating to Greenland from around the world. Writing for *The New York Times*, Elisabeth Rosenthal sketched some of the social problems that the people of Greenland may be facing: "An economy supported by corporate mining raises difficult questions. How would Greenland's insular settlements tolerate an influx of thousands of Polish or Chinese construction workers, as has been proposed? Will mining despoil a natural environment essential to Greenland's national identity—the whales and seals, the silent icy fjords, and mythic polar bears? Can fishermen reinvent themselves as miners?" (Rosenthal 2012). "I think mining will be the future, but this is a difficult phase," said

Jens B. Frederiksen, Greenland's housing and infrastructure minister and a deputy premier. "It's a plan that not everyone wants. It's about traditions, the freedom of a boat . . ." (Rosenthal 2012).

Ice Melt Imperils Hunters

Temperatures in Greenland are rising more quickly than almost any other place on Earth—4.5°F since 1987, five times the world average. In northeastern Greenland, a rise of 14 degrees to 21°F is anticipated by some climate models before century's end. Winter pack ice is breaking up, and ice fishing, on which many people depend, has become risky and dangerous.

The erosion of Greenland's ice poses practical dangers for indigenous hunters. DeNeen L. Brown of the *Washington Post* described an Inuit hunter's confrontation with glacial ice in Greenland made more dangerous by a warming climate. Aqqaluk Lynge had been chasing a seal. Fear chilled him when the seal dove under the ice and did not return. Patience is essential when seal hunting, so he waited—but the seal never came back. When an animal begins to act strangely, such as not coming up for air, something out of order is happening in nature, wrote Brown (2002).

The iceberg was at his back. Suddenly it began moving like a monster that was waking up. Lynge . . . looked up in alarm, knowing that these floating mountains . . . for all their frozen beauty, are ruthless and deadly. So he decided to get moving . . . but the engine on his motorboat wouldn't start. Just then he noticed the iceberg moving. If the tip is moving, he knew, it could mean that one end is moving up and the other end is moving down. . . . A friend in another boat nearby quickly gave him a tow. Soon they were speeding away from the iceberg, not waiting to look back. Behind them they heard it turning. "We looked back and saw the whole iceberg was collapsing, exploding almost," he said. "We were so afraid." Then it flipped, creating a great tidal wave that crashed hard onto nearby shorelines. By then the two men were out of the wave's path. "When we were finally far away, we could breathe normally again. We were looking back and seeing nothing was left. It exploded underneath the surface of the sea." (Brown 2002)

Economic Benefits from a Warmer Climate

Some of Greenland's people, most of whom live on the coasts near the edge of the ice cap, are reaping benefits from a warming climate in which average winter temperatures rose about 10°F between 1991 and 2007. New pastures are being used to graze sheep, and new fields grow potatoes. The hay-growing season has been lengthening in southern coastal Greenland, where farmers also have planted Chinese cabbage, several types of flowers, and turnips. Greenland farmers raised 22,000 lambs for local consumption during 2006, another market that is thriving on the warming edges of the retreating ice cap. Some of the lamb has been exported to restaurants in Europe.

By 2007, a few food markets in Greenland were selling local cauliflower, broccoli, and cabbage for the first time. Eight farmers grew potatoes commercially. Five grew vegetables, and home gardeners harvested a few strawberries. Greenland Beer's unique selling point is the purity of its glacier-fed water (Native-grown hops may be next). Ewes were having fatter lambs, and more of them during growing season that extended from the middle of May to about September 15, three weeks more than 10 years previously. "Now spring is coming earlier, and you can have earlier lambings and longer grazing periods," said Eenoraq Frederiksen, 68, a sheep farmer whose farm, near Qassiarsuk, is accessible by a harrowing drive across a rudimentary road plowed in the hillside. "Young people now have a lot of possibilities for the future" (Lyall 2007).

Kim Hoegh-Dam believes that warming coastal waters near Greenland will bring cod that have been abandoning the North Sea and other more southerly waters. He has raised more than $1 million to buy cod trawlers and three processing plants. "Global warming will increase the cod tremendously and will bring other species up from the south," he said (Struck 2007). A government trawler sent to test the cod runs caught 25 tons in one hour, a harvest so large that the crew cut their trip short. Although some conservations voice concerns about overfishing (a factor in past declining cod catches), the mouths of others water at the prospects not only for cod but also other sea creatures that thrive in relatively cold water such as shrimp. "The only limiting factor on human endeavor in Greenland is the temperature," Hoegh said, while bouncing on a fast motorboat past icebergs to visit the agriculture station in this country of few roads. Warm the temperature a bit, and new endeavors pop out like lambs from ewes, he believes (Struck 2007).

Ilulissat, Greenland's third-largest village with 4,500 people, includes one posh hotel, the Arctic, which doubled its size in 2008. From the windows of the Arctic, icebergs move so quickly that they appear to have formed and reformed overnight. "Nobody would have predicted 10 to 15 years ago that Greenland would lose ice that fast," said Konrad Steffen, a glaciologist at the University of Colorado. "That revises all of the textbooks" (Henry 2008).

Farther north along Greenland's coast, fishermen work from boats for longer periods as pack ice forms later and melts earlier. In recent winters, pack ice has failed to form on large areas of the coast, allowing fishing by boat year round for halibut and other species. Records at an ice patrol station, Daneborg said, showed that the water was open for 80 days during 1997. Now it stays ice free for 140 days, said Soren Rysgaard, a researcher for the Greenland Institute of Natural Resources (Struck 2007).

Warmth Changes Traditions

A warming climate also prompts changes in traditional ways of life. For example, Doug Struck reported in the *Washington Post* that Ono Fleischer, one of the most renowned dog sledders in Greenland, took a dog team across the huge island in 19 days last spring to marry his companion, Karo Thomsen, in a village in eastern Greenland. "They intended to sled back," wrote Struck, "but a warm rain put a dangerous glaze on the ice cap. They had to give away their 12 dogs and fly back to Ilulissat" (Struck 2007). Some villages in northern Greenland had to appeal for emergency food when a lack of ice prevented them from hunting for seals. Sled dogs had to eat donated European dog food instead of seal scraps. "If a seal hunter can't hunt, what is he to do?" asked Alfred Jakobsen, Greenland's minister of the environment (Struck 2007).

Business entrepreneurs also are knocking on the door in Greenland, looking for more fossil fuels. Four oil companies have applied to explore offshore as mining companies prospect for uranium and gold. Two aluminum companies want to build smelters using glacial meltwater for hydroelectric power. The U.S. Geological Survey estimates that waters off Greenland's northeastern coast may contain as much as 31 billion barrels of oil and gas. More oil may be found on the west coast, enough to tempt Exxon Mobil, Chevron, Canada's Husky Energy and Cairn Energy, and Sweden's PA Resources. Greenlanders in November 2008 approved a self-rule charter that directs mineral royalties to national development. The idea is to leverage the fruits of global warming to wean Greenland off its annual $680 million subsidy from Denmark.

Further Reading

Brown DeNeen L. "Greenland's Glaciers Crumble; Global Warming Melts Polar Ice Cap into Deadly Icebergs." *Washington Post*, October 13, 2002, A30.

Henry, Tom. "Global Warming Grips Greenland, Leaving Lasting Mark." *Toledo Blade,* October 12, 2008. http://www.toledoblade.com/apps/pbcs.dll/article?AID=2008810109858 (no longer available).

Lyall, Sarah. "Warming Revives Flora and Fauna in Greenland." *The New York Times*, October 28, 2007. http://www.nytimes.com/2007/10/28/world/europe/28greenland.html.

Rosenthal, Elisabeth. "A Melting Greenland: Perils against Potential." *The New York Times*, September 23, 2012. http://www.nytimes.com/2012/09/24/science/earth/melting-green land-weighs-perils-against-potential.html.

Struck, Doug. "Icy Island Warms to Climate Change." *Washington Post*, June 7, 2007, A1. http://www.washingtonpost.com/wp-dyn/content/article/2007/06/06/AR20070 60602783.html?referrer=email.

See also: Animal Life, Arctic; Arctic Hunters; Climate Change, Greenland; Fisheries; Glacial Erosion; Ice Melt, Greenland; Ice Science, Greenland; Inuit; Sea Level Rise

GLOBAL WARMING, SCANDINAVIA

Sweden and Norway have some of the highest liquor taxes in the world, and these have spawned copious smuggling, mainly from Denmark. Until recently, contraband seized at Malmo, directly across the Oresund Sound from Copenhagen, by Tullverket (Swedish customs) was poured down the drain. These days, a million bottles a year of illicit liquor is trucked to a new high-tech plant in Linköping (80 miles south–southwest of Stockholm) that manufactures biogas fuel for automobiles as well as fertilizer.

The plant also accepts human refuse and waste from packing plants. Out of this noxious mix, all of which used to be regarded as useless garbage, comes biofuel to power buses, taxis, garbage trucks, private cars, and a methane-propelled "biogas train" that runs between Linköping and Västervik on the southeast coast. The train's boosters (not squeamish vegetarians, from the sound of it) have figured that the entrails from one dead cow, previously wasted, buys four kilometers (2.5 miles) on the biogas train.

Creative Ways to Replace Oil

Sweden and other Scandinavian countries have found many creative ways to replace oil with what used to be waste. People in these countries are making resources of whatever they have in abundance: geothermal resources and wind in Iceland, wood and organic waste in Sweden, and wind in Denmark. They are using simple, practical solutions that are available now at modest cost. Iceland plans by 2050 to power all of its passenger cars and boats with hydrogen made from electricity drawn from local renewable resources.

In Amsterdam, as much as 40 percent of city travel was on bicycles by 2007. At many traffic lights, bikes go first. Spain is testing a new solar technology called *concentrating solar power* (CSP) that uses mirrors to focus sunlight with a much greater efficiency than photovoltaic cells. This technology is being tested in Seville (and in the United States by Arizona Power). CSP may open the way for large-scale solar-power plants.

The Linköping plant is not alone, although it is unusual for the number of things it can process. The Danish Crown slaughterhouse uses the fat of 50,000 pigs in an average week to generate biogas. In the United States, ConocoPhillips and Tyson Foods have been making "renewable diesel" from beef, pork, and chicken fat in Texas. The entire Danish Crown plant has been redesigned with an eye to saving energy, part of a 30-year Danish effort to eliminate waste, conserve energy, and reduce consumption of fossil fuels. Surplus heat from Danish power plants is piped to nearby homes via insulated pipes as "cogeneration" or "district heating," which required tearing up streets to install pipes. This system now heats almost two-thirds of Danish homes. Power plants have been radically reduced in size and built closer to people's homes and offices to reduce the amount of power lost in transmission and encourage the use of formerly wasted heat. In the mid-1980s, Denmark had 15 large power plants; it now has several hundred small ones.

Danish building codes enacted in 1979 (and tightened several times since) also require thick home insulation and tightly sealed windows. Between 1975 and 2001, Denmark's heating bill fell 20 percent while the amount of heated space rose by 30 percent. Denmark's gross domestic product has doubled on stable energy usage.

Swedish Auto Industry Opposes Oil Dependence

Ulf Perbo, who heads BIL Sweden, the national association for the automobile industry, said that even automakers in that country want to end oil dependency. "Many people have asked [why BIL] is not against the Oil Commission, but it is not in our interest to be dependent on oil, with regard to the production and sales of cars. Oil is not what interests us; cars are. And oil is going to be a limitation [to the production and sales of cars] in the future" (Johansen 2007, 23).

The Swedish government buys environmentally friendly cars, and some cities offer their drivers free parking. Swedish paper and pulp industries use bark that formerly was wasted to produce energy for their manufacturing processes, and sawmills incinerate wood chips and sawdust to generate power.

Sweden's 9 million people have had one of the world's most impressive records on environmental protection for many years. In 2007, according to the government, Sweden's energy use was roughly 35 percent dependent on oil products. The proportion of oil-heated homes in Sweden was down to 8 percent by 2006, because many neighborhoods used hot water from central plants that burn biofuels, often wood-based pellets. In December 2005, all Swedish gasoline stations were required by an act of parliament to offer at least one alternative fuel. Since the beginning of 2006, householders have been paid to replace oil-burning boilers with environmentally friendly heating systems. Such financial incentives already were available to libraries, aquatic facilities, and hospitals that wanted to switch to more efficient renewable energy.

Sweden has experienced some political bumps in the road on this issue, however. The oil independence panel has its critics. For example, in a country that produces more private cars per capita than any other, the proceedings of the commission completely omitted the word *bicycle*. Stockholm, however, is laced with well-used bicycle paths.

During the 1970s, energy interests in Sweden pitched nuclear power as an antidote to oil. Seven new nuclear reactors came into operation between 1972 and 1980. Because of environmental considerations, high production costs, and low world market prices, Sweden's substantial uranium reserves (250,000 to 300,000 tons, or 20 percent of known world reserves) have not been exploited. Even so, in 2006, 45 percent of Sweden's electricity still was being generated by nuclear power and 8 percent from fossil fuels.

Swedish nuclear power sustained a new blow during 2006 because of a near-meltdown at one of the country's reactors. An electricity failure on July 25, 2006, led to the immediate shutdown of the Forsmark 1 reactor after two of four backup generators that supply power to the reactor's cooling system malfunctioned for 20 minutes. This near-accident revived memories of Ukraine's Chernobyl plant and the radioactive fallout over that showered much of Sweden during April 1986. By 2010, all 12 of Sweden's nuclear plants were shut down and replaced by natural gas, a fossil fuel, until other sources were available.

Sweden's Legal Measures

Enacted in 1991, Sweden's carbon tax supplied an early example of how the idea could be applied on a national scale. Finland, Norway, and the Netherlands later enacted similar taxes. Sweden also levies an energy tax on all fossil fuels. Other taxes also are used as incentives to control pollution, including a nitrogen oxide charge, a sulfur tax, and a tax on nuclear energy production. In many cases, energy taxes have replaced some 50 percent of the income tax burden.

The Swedish energy tax varies widely among different products and types of fossil fuels, with by far the highest rates on gasoline. Responding to the new tax system, the use of biofuels (mainly wood-based) for "district" heating soared some 350 percent during the 1990s. Prices of wood-based fuels also fell dramatically as technology improved. Sweden's carbon tax has been credited with reducing fossil fuel emissions by around 20 percent.

Another conservation measure, *congestion charging*, which levies tolls to drive a car in downtown Stockholm, became a controversial issue in the Swedish general election during late summer 2006. The congestion charge of up to $7 a day was narrowly approved by 52 percent in a referendum September 17. Stockholm already had tested the idea. The congestion charges reduced auto traffic 20 percent to 25 percent, while use of trains, buses, and Stockholm's extensive subway system increased. Emissions of carbon dioxide declined 10 percent to 14 percent in the inner city and 2 percent to 3 percent in Stockholm County. The project also increased the number of environmentally friendly cars, which were exempt from congestion taxes. The Stockholm congestion tax became permanent in August 2007.

Per Bolund, one of 19 Green Party members in Sweden's 349-member Riksdag, has been watching Green Party initiatives work their way into the political mainstream for many years. The Green Party favored a congestion charge for Stockholm for decades, only to watch conservative forces block it. Ironically, the congestion later was implemented under a conservative-right coalition.

The commission on oil dependency also was a long-time Green Party initiative that is now embraced across the Swedish political spectrum. "The Social Democrats stole it from us," Borlund said. He then grinned and said that Greens must be prepared to be mimicked by the political mainstream to succeed (Johansen 2007, 23).

The Swedish government recently adopted another Green Party idea: a vehicle tax based on carbon dioxide emissions rather than weight. Borlund pointed out that some cars such as hybrids are heavy but relatively low in emissions.

Bolund said that Sweden's government was emphasizing biofuels too much and underplaying energy efficiency, which is often not profitable but *is* the best way to reduce greenhouse gas emissions. Sweden's metal and heavy industries based on forest products have profited from cheap hydropower, he said. Many homes still are heated by inefficient baseboard electricity, a holdover from the days of cheap hydropower. Plentiful hyrdopower has led to waste, Bolund said.

Bolund said that some of Europe's heavy industries such as automaking consume three times the electricity per unit of production as similar producers in Central Europe. Saab and Volvo are not especially energy efficient, he said. When Swedish officials brag that they have reduced the used of oil in home heating to almost zero, said Bolund, they are ignoring the fact that half that total is from nuclear plants, and the other half from often inefficient hydropower. "Any time you see those statistics you have to say—yes, but. . . ." Do not stereotype Sweden as a green heaven, he said. The same conflicts take place there as in the rest of the world, he emphasized. "We have a long way to go" (Johansen 2007, 23). Since then, Sweden has raised the power share of wind, solar, and biomass and achieved considerable conservation to reduce power demand somewhat. In January 2017, nuclear still provided 40 percent of Swedish electricity.

Further Reading

Johansen, Bruce E. "Scandinavia Gets Serious about Global Warming." *The Progressive,* July 2007, 22–25.

See also: Automobiles

ICE MELT, ANTARCTICA

Scientists have studied the historic climate—the paleoclimate—of Antarctica in search of conditions similar to those forecast for the end of the 21st century as temperatures and atmospheric carbon dioxide levels rise. Similar conditions have been studied in the early Pliocene Epoch, about 3 million years ago, and even further into the past, some 24 million years before the present, when carbon dioxide levels were extremely high compared to long-term averages, producing major expansions and contractions of Antarctica's ice and leading to worldwide sea level changes of as much as 200 feet.

An Analog in the Pliocene?

As noted, for an analog of what may befall the West Antarctic ice sheet during the coming century, scientists are looking to the early Pliocene Epoch when carbon dioxide levels reached 400 parts per million, roughly today's levels. These levels combined with changes in the angle of Earth's axis, which changed the amount of sunlight in that area and produced rapid ice melting within a few thousand years that contributed to several meters of sea level rise worldwide (Naish et al. 2009, 322). T. Naish and colleagues wrote in *Nature*:

> Our data provide direct evidence for orbitally induced oscillations in the WAIS [Western Antarctic ice sheet], which periodically collapsed, resulting in a switch from grounded ice, or ice shelves, to open waters in the Ross embayment when planetary temperatures were up to approx. 3°C warmer than today and atmospheric CO_2 concentration was as high as 400 ppm. The evidence is consistent with a new ice-sheet [and] ice-shelf model that simulates fluctuations in Antarctic ice volume of up to +7 m in equivalent sea level associated with the loss of the WAIS and up to +3 m in equivalent sea level from the East Antarctic ice sheet, in response to ocean-induced melting paced by obliquity. During interglacial times, diatomaceous sediments indicate high surface-water productivity, minimal summer sea ice and air temperatures

above freezing, suggesting an additional influence of surface melt under conditions of elevated CO_2. (Naish et al. 2009, 322)

As David Pollard and Robert M. DeConto wrote in *Nature*, today the West Antarctic ice sheet "is fringed by vulnerable floating ice shelves that buttress the fast flow of inland ice streams. Grounding lines are several hundred meters below sea level and the bed deepens upstream, raising the prospect of runaway retreat" (Pollard and DeConto 2009, 329). Meltwater ponding, a precursor of ice breakup, has been observed in the Ross ice shelf (Archer 2009, 142). Their models indicate that, in the past, this area has experienced "brief but dramatic retreats, leaving only small, isolated ice caps on West Antarctic islands. Transitions between glacial, intermediate, and collapsed states are relatively rapid, taking one to several thousand years."

An Analogous Period 24 Million Years Ago?

T. R. Naish and colleagues have presented sediment data from shallow marine cores in the western Ross Sea that exhibit well-dated cyclic variations that link the extent of the East Antarctic ice sheet directly to orbital cycles during the Oligocene–Miocene transition (24.1 million to 23.7 million years ago) (Naish et al. 2001, 719), a time when planetary temperatures averaged 3 to 4°C warmer than today, and the atmospheric carbon dioxide level was twice as high. These studies, they say, "Should help to provide realistic analogs for their future behavior following the increased levels of atmospheric CO_2 and temperature projected for the end of this century" (Naish et al. 2001, 723).

Anther international research team studied sediment and fossil samples from 3,000 feet below the surface in the Cape Roberts area of the East Antarctica ice sheet, which covers most of the continent. The scientists focused on a 400,000-year period roughly 24 million years ago when carbon dioxide levels were twice as high as today and temperatures were as much as 4°C higher. These conditions are similar to those anticipated for Earth in the next 50 to 100 years because of human-induced rises in greenhouse gas levels.

The team's study was published October 18, 2001, in *Nature* and concluded: "Studies of Antarctic ice sheets during that time should help to provide realistic analogs for their future behavior following the increased levels of atmospheric CO_2 and temperature projected for the end of this century" (Dalton 2001). This report continued:

To find out what we are in for, we had to go back to a time when there were similar conditions. We wanted to find a comparable period in the past when carbon dioxide levels were as high as they are expected to reach. There is every indication that this will be a model for what will happen in the future. We found the ice sheets 24 million years ago were smaller and more dynamic than today. There is good evidence for a lot of instability at the margins of the sheets. . . . We know there are glacial cycles, with the ice altering more when the climate is warmer, but it is now a question of working out how the speed and frequency of these cycles will change. They were probably faster in the past and will probably be faster in the future. (Dalton 2001)

Researchers have identified the West Antarctic ice sheet as the main contributor to a massive and relatively rapid increase in global sea levels at the end of the last ice age. This event provoked a surge of freshwater from melting glacial ice into the world ocean that caused a rise of 20 meters in 500 years. Writing in *Science*, hydrologist Victor Baker of the University of Arizona said that since the 1960s evidence has emerged for "repeated, catastrophic failures of ice dams" (O'Neill 2002). According to Baker, "the southern edge of the Laurentide ice sheet, which covered the central and eastern regions of northern North America, was repeatedly drained by a superflood that had a major influence on global climate during the last glacial period" (O'Neill 2002).

Ice Sheets during a Super Greenhouse?

Paleoclimates are sometimes stories with considerable nuance—in at least one instance some 50 million years ago, for example. This finding, described in *Science* in January 2008, calls into question a linear conviction that a "supergreenhouse" effect will unquestionably melt all glaciers (Bornemann et al. 2008, 189). Either that or glacial cycles can occur with what can only be regarded as awesome speed in geological time—if the scientists' proxies are not misrepresenting what actually happened.

In this study, which was based on proxies—organic molecules in ocean sediments and chemicals in ancient fossil shells—the "super greenhouse climate" was "not a barrier to the formation of large ice sheets, calling into question the common assumption that the poles were always ice-free during past periods of intense global warming," according to the study's authors ("Study Says" 2008).

The glaciers that formed during a 200,000-year period probably formed ice sheets that were two-thirds the size of today's Antarctic ice cap. At the same time, surface temperatures in the west Atlantic Ocean probably were more than 95 to 99°F. The Turonian age was noted for large climatic variations because of changes in Earth's orbit.

In the event climate contrarians jump on this study as reason not to worry about global warming, Thomas Wagner, a German scientist at Newcastle University in England who was among its authors, said "It's difficult to draw a direct relationship between our findings and the current discussion on the climate. The results, however, show that even in a very warm world it is possible, at least temporarily, to build up larger ice caps in cooler regions" ("Study Says" 2008). In addition, some scientists say that the proxies are not ironclad: "We're right at the limit of what our proxies tell us, or a bit beyond," said Timothy Bralower, a paleoceanographer at Pennsylvania State University. Even if the proxies are correct, the research team that "found" the ice maintains that it formed only "under certain rare conditions" (Bornemann et al. 2008, 191).

Ice-Melt Chronicle

Antarctica, which contains 90 percent of Earth's freshwater as ice, has become a land of climatic paradoxes. Ice sheets have been melting around the edges of the continent more quickly than anticipated, as warming temperatures accelerate glacial movement into the surrounding oceans. The Antarctic peninsula is among the most rapidly warming areas on Earth, with large ice shelves crumbling into the surrounding seas in the last several years. Temperatures on the West Antarctic peninsula have risen 8.8°F in winter since 1950 and 4.5°F in summer (Glick 2004, 33). At the same time, sections of Antarctica's interior have experienced a pronounced cooling trend while most other areas of Earth have warmed. Some Antarctic ice shelves are even disintegrating in winter.

Are Antarctic ice sheets thickening or thinning? Is sea ice expanding or contracting? Both questions are open to debate, at least for now. These debates are of great interest for the rest of the world because significant melting of land-based Antarctic ice could raise sea levels and inundate the coastal residences of many hundreds of millions of people. Whatever the outcome of these debates, many observations indicate a breakdown provoked by climate change will affect the Antarctic food chain, which begins with krill and ends with penguins and whales. These problems may be intensified by human-caused declines in stratospheric ozone levels as well.

The speed and quantity of Antarctic ice flowing into southern oceans is the single most important factor in forecasting sea level rise around the world. "The highest thinning rates occur where warm water at depth can access thick ice shelves via submarine troughs crossing the continental shelf," wrote H. D. Pritchard and colleagues in *Nature* (2012, 502). "Wind forcing could explain the dominant patterns of both basal melting and the surface melting and collapse of Antarctic ice shelves, through ocean upwelling. . . . This implies that climate forcing through changing winds influences Antarctic ice-sheet mass balance, and hence global sea level, on annual to decadal time scales."

Ice loss in Antarctica as a whole increased by 75 percent during the decade ending in 2007 as glacier movement to the sea accelerated. By 2007, glacier discharge into the sea there had caught up with the rate in Greenland. Scientists led by Eric Rignot of NASA's Jet Propulsion Laboratory in Pasadena, California, and the University of California–Irvine, detected a rapid increase, doubling the contribution to world sea level rise from 0.3 millimeters (0.01 inches) a year in 1996, to 0.5 millimeters (0.02 inches) a year in 2006 (Rignot et al. 2008, 106; "Antarctic Ice Loss" 2008).

Rignot said the losses, which were primarily concentrated in West Antarctica's Pine Island Bay sector and the northern tip of the Antarctic peninsula, "are caused by ongoing and past acceleration of glaciers into the sea." This is mostly a result of warmer ocean waters that bathe the buttressing floating sections of glaciers, causing them to thin and then collapse. "Changes in Antarctic glacier flow are having a significant, if not dominant, impact on the mass balance of the Antarctic ice sheet," Rignot said (Rignot et al. 2008, 106; "Antarctic Ice Loss" 2008). The team found that the net loss of ice mass from Antarctica increased from 112 gigatons (plus or minus 91) a year in 1996 to 196 gigatons (plus or minus 92) a year in 2006. A gigaton is 1 billion metric tons, or more than 2.2 trillion pounds ("Antarctic Ice Loss" 2008). This estimate of Antarctica's ice mass between 2002 and 2009 says that most of the ice loss is taking place in coastal regions at an estimated 57 gigatons per year ("Unexpected" 2009). The collapse of ice shelves over water does not by itself raise sea levels, but their demise accelerates the melting of land-based ice, which increases sea level rise in the future. "Loss of ice shelves surrounding the continent could have a major effect on the rate of ice flow off the continent," said glaciologist Ted Scambos (Toner 2002). At least five of the ice shelves on the Antarctic peninsula are receding faster than at any time in recent history.

The same study also confirms earlier estimates that West Antarctica is losing some 132 gigatons of ice per year. "While we are seeing a trend of accelerating ice loss in Antarctica, we had considered East Antarctica to be inviolate," said lead author Jianli Chen. "But if it is losing mass, as our data indicate, it may be an indication the state of East Antarctica has changed. Since it's the biggest ice sheet on Earth, ice loss there can have a large impact on global sea level rise in the future" (Chen et al. 2009, 859; "Unexpected" 2009).

Measuring Antarctic Ice Loss

Measurements of the time variable gravity from the Gravity Recovery and Climate Experiment (GRACE) satellites determined mass variations of the Antarctic ice sheet during 2002–2005. Researchers found that the ice sheet mass decreased significantly, at a rate of 152 ± 80 km^3/year of ice, equivalent to 0.4 ± 0.2 mm/year of global sea level rise. Most of this mass loss came from the West Antarctic ice sheet. By 2006, that ice sheet was losing 36 cubic miles of ice a year. Increasing snowfalls in the interior (a product of warmer temperatures) had ceased to balance accelerating melting along the coasts of the continent. These findings imply that global sea levels will rise more rapidly than earlier thought.

These measurements found that the amount of water pouring annually from the ice sheet into the ocean equals the water used by the people of the United States in three months and is causing global sea level to rise by 0.4 millimeters a year (Eilperin 2006). "The ice sheet is losing mass at a significant rate," said Isabella Velicogna, the study's lead author and a research scientist at the University of Colorado–Boulder's Cooperative Institute for Research in Environmental Sciences. "It's a good indicator of how the climate is changing. It tells us we have to pay attention" (Eilperin 2006).

Richard Alley, a Pennsylvania State University glaciologist who has studied the Antarctic ice sheet but was not involved in the GRACE project, said that more research is needed to determine if the shrinkage is a long-term trend because the new report is based on just three years of data. "One person's trend is another person's fluctuation," he said. However, said Alley, the switch in balance from accumulating ice to net melting in Antarctica had not been anticipated this soon by many researchers. "It looks like the ice sheets are ahead of schedule" in terms of melting, Alley said. "That's a wake-up call. We better figure out what's going on" (Eilperin 2006).

Continued Antarctic Melting

In 2007, NASA researchers using 20 years of data from space-based sensors (1987 through 2006) found that the Antarctic ice cap has been melting over time farther inland from the coast. Ice and snow also are melting at higher altitudes. Melting also was increasing on Antarctica's largest (eastern) ice shelf. Snow and ice in Antarctica has been melting as far inland as 500 miles from the coast and as high as 1.2 miles above sea level in the Transantarctic Mountains. The same study also found that melting has been increasing on the Ross ice shelf, both in geographic area and duration.

As in Greenland, satellite sensors found that melting snow and ice on the surface was forming ponds, with meltwater filling small cracks, which cause larger fractures in the ice shelf. "Persistent melting on the Ross ice shelf is something we should not lose sight of because of the ice shelf's role as a 'brake system' for glaciers," said the study's lead author, Marco Tedesco, a research scientist at the Joint Center for Earth Systems Technology, which is cooperatively managed by NASA's Goddard Space Flight Center in Greenbelt, Maryland, and the University of Maryland.

> Ice shelves are thick ice masses covering coastal land with extended areas that float on the sea, keeping warmer marine air at a distance from glaciers and preventing a greater acceleration of melting. The Ross ice shelf acts like a freezer door, separating ice on the inside from warmer air on the outside. So the smaller that door becomes, the less effective it will be at protecting the ice inside from melting and escaping. ("NASA Researchers" 2007)

A team of NASA and university scientists has found clear evidence that extensive areas of snow melted in West Antarctica during January 2005 in response to temperatures that were considerably above average. The area of widespread melting reached to within 300 miles of the South Pole. This is one of a number of exceptional warm episodes noted in polar regions during recent years; others have been described by scientists and residents on Baffin Island and in northern Alaska, western Greenland, and northern Sweden. Andrew Revkin wrote of this report in *The New York Times*: "Balmy air, with a temperature of up to 41 degrees in some places, persisted across three broad swathes of West Antarctica long enough to leave a distinctive signature of melting, a layer of ice in the snow that cloaks the vast ice sheets of the frozen continent. The layer formed the same way a crust of ice can form in a yard in winter when it's a warm day and then a freezing night follow a snowfall, the scientists said" (Revkin 2007).

By 2009, scientists at the U.S. Geological Survey (USGS) and the British Antarctic Survey, with the assistance of the Scott Polar Research Institute at the University of Cambridge and Germany's Bundesamt fur Kartographie und Geodasie, found that Antarctica's Wordie ice shelf had disappeared. The northern part of the Larsen ice shelf also no longer existed. An area of ice three times the size of Rhode Island (more than 8,500 square kilometers) had been shed by the Larsen ice shelf between 198 and 2008. "This continued and often significant glacier retreat is a wakeup call that change is happening in our Earth system and we need to be prepared," said USGS glaciologist Jane Ferrigno, lead author on the study. "Antarctica is of special interest because it holds an estimated 91 percent of the Earth's glacier volume, and change anywhere in the ice sheet poses significant hazards to society" ("Antarctic Ice Shelf Disappears" 2009).

Further Reading

"Antarctic Ice Shelf Disappears, Arctic Melting Rapidly." Environment News Service, April 3, 2009. http://www.ens-newswire.com/ens/apr2009/2009-04-03-01.asp (no longer available).

"Antarctic Ice Loss Speeds Up, Nearly Matches Greenland Loss." NASA Earth Observatory. January 23, 2008. Accessed March 12, 2008. http://www.jpl.nasa.gov/news/news.cfm?release=2008-010.

Archer, David. *The Long Thaw: How Humans Are Changing the Next 100,000 Years of Earth's Climate.* Princeton, NJ: Princeton University Press, 2009.

Bornemann, Andre, et al. "Isotopic Evidence for Glaciation during the Cretaceous Supergreenhouse." *Science* 319 (January 11, 2008): 189–192.

Chen, J. L., et al. "Accelerated Antarctic Ice Loss from Satellite Gravity Measurements." *Nature Geoscience* 2 (November 22, 2009): 859–862.

Dalton, Alastair. "Ice Pack Clue to Climate-change Effects." *The Scotsman* (Edinburgh), October 18, 2001, 7.

Eilperin, Juliet. "Antarctic Ice Sheet Is Melting Rapidly; New Study Warns of Rising Sea Levels." *Washington Post*, March 3, 2006, A1.

Glick, Daniel. "The Heat Is On: Geosigns." *National Geographic*, September 2004, 12–33.

Naish, T., et al. "Orbitally Induced Oscillations in the East Antarctic Ice Sheet at the Oligocene/Miocene Boundary." *Nature* 413 (October 18, 2001): 719–723.

Naish, T., et al. "Obliquity-Paced Pliocene West Antarctic Ice Sheet Oscillations." *Nature* 457 (March 19, 2009): 322–325.

"NASA Researchers Find Snowmelt in Antarctica Creeping Inland." NASA Earth Observatory, September 20, 2007. https://www.nasa.gov/centers/goddard/news/topstory/2007/antarctic_snowmelt.html.

O'Neill, Graeme. "The Heat Is On." *Sunday Herald-Sun* (Sydney, Australia), March 31, 2002.

Pollard, David, and Robert M. DeConto. "Modelling West Antarctic Ice Sheet Growth and Collapse through the Past Five Million Years." *Nature* 457 (March 19, 2009): 329–332.

Pritchard, H. D., et al. "Antarctic Ice-Sheet Loss Driven by Basal Melting of Ice Shelves." *Nature* 484 (April 26, 2012): 502–505.

Revkin, Andrew C. "Analysis Finds Large Antarctic Area Has Melted." *The New York Times*, May 16, 2007. http://www.nytimes.com/2007/05/16/science/earth/16melt.html.

Rignot, Eric, et al. "Recent Antarctic Ice Mass Loss from Radar Interferometry and Regional Climate Modelling." *Nature Geoscience* 1 (2008): 106–110.

"Study Says Glaciers Formed during a Very Warm Period." *The New York Times*, January 11, 2008. http://www.nytimes.com/2008/01/11/world/europe/11glacier.html.

Toner, Mike. "Huge Ice Chunk Breaks Off Antarctica." *Atlanta Journal-Constitution*, March 20, 2002, A1.

"Unexpected Ice Loss Detected in East Antarctica." NASA Earth Observatory, November 24, 2009. http://earthobservatory.nasa.gov/Newsroom/view.php?id=41455&src=eoa-nnews (no longer available).

See also: Animal Life, Antarctic; Global Warming, Antarctica; Ice Shelves, Antarctic; Inland Cooling, Antarctica; Ocean Circulation; Oceans' Absorption of Heat; Sea Level Rise; Temperatures, Greenhouse Gas Levels and

ICE MELT, ARCTIC

Speaking at the annual meeting of the American Geophysical Union in San Francisco in December 2007, Richard Alley of Pennsylvania State University surveyed the major—and accelerating—effects that a relatively small amount of warming (compared to what is anticipated for the rest of the century) has already had on the melting of ice in the Arctic and Antarctic. "If a very small warming makes such a difference," asked Alley, "it raises the question of what happens when more warming occurs." At the same meeting, Josefino Comiso of NASA's Goddard Space Flight Center in Greenbelt, Maryland, said, "The tipping point for perennial sea ice has likely already been reached" (Kerr 2008). "When the ice thins to a vulnerable state, the bottom will drop out and we may quickly move into a new, seasonally ice-free state of the Arctic," said glaciologist Mark Serreze. "I think there is some evidence that we may have reached that tipping point, and the impacts will not be confined to the Arctic region" ("As Arctic Sea Ice" 2008).

Shrinking sea ice in the Arctic reinforces conditions that provoke further warming, one important example of positive climatic feedbacks that make global warming dangerous in the long term. This process is called *polar amplification* and was described thusly by James A. Scree and Ian Simmonds:

> Polar areas warm faster than the tropics. . . . The rise in Arctic near-surface air temperatures has been almost twice as large as the global average in recent decades.. A feature known as "Arctic amplification" increased concentrations of atmospheric greenhouse gases have driven Arctic and global average warming. . . . A better understanding of the processes responsible for the recent amplified warming is essential for assessing the likelihood, and impacts, of future rapid Arctic warming and sea ice loss. Here we show that the Arctic warming is strongest at the surface during most of the year and is primarily consistent with reductions in sea ice cover. Changes in cloud cover, in contrast, have not contributed strongly to recent warming. Increases in atmospheric water vapor content, partly in response to reduced sea ice cover, may have enhanced warming in the lower part of the atmosphere during summer and early autumn. We conclude that diminishing sea ice has had a leading role in recent Arctic temperature amplification. The findings reinforce suggestions that strong positive ice-temperature feedbacks have emerged in the Arctic, increasing the chances of further rapid warming and sea ice loss, and will probably affect polar ecosystems, ice-sheet mass balance and human activities in the Arctic. (Screen and Simmonds 2010, 1334)

Ice Cover and Climate Change

Over several decades, the extent of snow cover in the Arctic has changed significantly and has had consequent impacts on climate. As snow covers the land surface for less time, reflectivity (albedo) changes, allowing less reflection and more absorption of solar radiation at the surface. Between 1979 (when satellite mapping began) and 2012, the extent of snow cover at high latitudes has been declining an average of 17.6 percent per decade. Because of this, "The snow-cover study authors . . . found an overall decline in snow cover from 1967 through 2012, and also detected an acceleration of snow loss after the year 2003. Between June 2008 and June 2012, North America experienced three record-low snow cover extents. In Eurasia, each successive June from 2008 to 2012 set a new record for the lowest snow cover extent yet recorded for that month" (Derksen and Brown 2012; "Snow Cover" 2013).

The British scientific journal *Nature* reported in 2014 that snow thickness had decreased by around 37 percent in the western Arctic and by 56 percent in the Beaufort area, comparing observations from 1954 to 1991 from Soviet ice stations with radar surveys (2009–2013) by satellite that had been partially verified by reports from the surface by Melinda Webster of the University of Washington and her colleagues ("Arctic Snowpack" 2014).

Other research has identified a link between rising air temperatures and shrinking snow cover, according to NASA's Earth Observatory ("Snow Cover" 2013). The Earth Observatory's report continued: "Snow has very high albedo, reflecting up to 90 percent of the sunlight it receives. As snow cover declines, dark soils and vegetation absorb more of the sun's energy. . . . The uppermost layer of permafrost that thaws each summer. When organic material in thawing permafrost decomposes, it can release methane, a potent greenhouse gas when released to the atmosphere."

During November 2010, extreme warmth in northeastern Canada was related to an ice-free Hudson Bay, which is unusual. The contrast of temperatures at coastal stations in years with and without sea ice cover on the neighboring water body is useful for illustrating the dramatic effect of sea ice on surface air temperature. Sea ice insulates the atmosphere from ocean water warmth, allowing surface air to achieve temperatures much lower than that of the ocean. It is for this reason that some of the largest positive temperature temperatures on the planet have occurred in the Arctic because sea ice area has decreased in recent years (Hansen et al. 2010).

Weather patterns can influence the way in which Arctic ice melts and reforms, providing variations from year to year that obscure long-term trends. For example, weather patterns (especially a severe cyclonic storm in August 2012, the most intense August storm to hit the region since satellite monitoring began in 1979) helped to destroy ice, whereas the patterns of 2013 inhibited ice breakup. The vast difference between the two melt seasons had researchers investigating the ways in which cyclones can either exacerbate or dampen the effects of climate warming on sea ice.

"As global warming continues and cyclones become more intense, many researchers worry that summers like that of 2012 could become the norm," Lauren Morello wrote in *Nature* (2013). In the 2012 cyclone, "wind and wave action driven by the tempest caused a massive chunk of ice . . . to separate from the main pack. . . . The sheared-off portion eventually melted, depleting the overall ice cover and leaving the main pack more vulnerable to erosion from wind, waves, and warm water stirred up by the storm" (Morello 2013). The area of disturbed ice has been variously estimated at between 150,000 and 400,000 square kilometers. In 2013, cyclones brought in cold air and aided ice formation, blocking sunshine late in the melt season. "Yet scientists say that future summer storms are more likely to accelerate ice loss than to slow it, as rising temperatures continue to deplete the Arctic's ice cover." Morello continued. "Thick, 'multiyear' ice that has survived more than one summer thaw once covered large expanses of the Arctic Ocean. Now those waters are dominated by thin, 'first-year' ice—which forms in autumn and melts the next summer, and is more vulnerable to the fierce winds and waves that the cyclones bring."

The Ice May No Longer Recover

In the past, low ice years often were followed by recovery the next year when cold winters favored accumulation or cool summers kept ice from melting. That kind of balancing cycle stopped after 2002. "If you look at these last few years, the loss of ice we've seen, well, the decline is rather remarkable," said Mark Serreze (Human 2004).

A NASA analysis of satellite data for the first time in 2010 measured the pace at which stronger, older "multiyear" ice is melting in the Arctic Ocean and being replaced by thinner, more fragile ice that usually melts each year. Some scientists suspected that losses in ice coverage over the Arctic Ocean have been mainly from wind pushing the ice, movement that scientists call "export." In this study, Ron Kwok and Glenn Cunningham at NASA's Jet Propulsion Laboratory in Pasadena, California, used satellite data to show the role of export versus actual melting. within the Arctic Ocean basin. Kwok and Cunningham showed that from 1993 to 2009, 1,400 cubic kilometers (336 cubic miles) of ice was lost because of melt, not export. "The paper shows that there is indeed melt of old ice within the Arctic basin and the melt area has been increasing over the past several years," Kwok said. "The story is always more complicated—there is melt as well as export—but this is another step in calculating the mass and area balance of the Arctic ice cover" ("NASA Study" 2010).

One study found that "multiyear" ice, which had been assumed to be more durable than the thinner single-year ice that forms after melting, may be at least as vulnerable. NASA scientist Josefino C. Comiso found that "Arctic multiyear ice 'extent'—which includes all areas where at least 15 percent of the ocean surface is covered by multiyear ice—has been vanishing at a rate of about 15.1 percent per decade. . . . Over the same period, the 'area' covered by multiyear ice—which discards open water among ice floes and focuses exclusively on regions that are completely covered—has been shrinking by about 17.2 percent per decade" (Comiso 2012; "Oldest Arctic" 2012).

"The Arctic sea ice cover is getting thinner because it's rapidly losing its thick component," Comiso said. "At the same time, the surface temperature in the Arctic is going up, which results in a shorter ice-forming season. It would take a persistent cold spell for multiyear sea ice to grow thick enough again to be able to survive the summer melt season and reverse the trend" (Comiso 2012). "We're seeing more melting of multiyear ice in the summer," said Julienne Stroeve, senior research scientist with the National Snow and Ice Data Center in Boulder, Colorado. "We may soon reach a threshold beyond which the sea ice can no longer recover." "We have already witnessed major losses in sea ice, but our research suggests that the decrease over the next few decades could be far more dramatic than anything that has happened so far," said Marika Holland of the National Center for Atmospheric Research, also in Boulder. "These changes are surprisingly rapid" ("As Arctic Sea Ice" 2008).

"It's hard even for people like me to believe, to see that climate change is actually doing what our worst fears dictated," said Jennifer A. Francis, a Rutgers University scientist who studies the effect of sea ice on weather patterns. "It's starting to give me chills, to tell you the truth" (Gillis and Foster 2012).

Robert F. Spielhagen and colleagues wrote in *Science* early in 2011 that relatively warm water flowing from the Atlantic Ocean into the Arctic is strongly affecting the coverage and distribution of ice. "Early 21st-century temperatures of Atlantic Water entering the Arctic Ocean are unprecedented over the past 2,000 years and are presumably linked to the Arctic amplification of global warming," they wrote. "Records of its natural variability are critical for the understanding of feedback mechanisms and the future of the Arctic climate system, but continuous historical records reach back only about 150 years" (Spielhagen et al. 2011, 450).

Arctic Ice Melt Outpaces Models

How well do observations and models agree on Arctic ice melt? Julienne Stroeve and colleagues compared more than a dozen models and found that nearly all of them underestimated the speed of ice melt—in many cases by large amounts. "These findings have two important implications," said a summary of this study in *Science*. "First, that the effect of rising greenhouse gases may have been more important than has been believed; and second, that future loss of Arctic sea ice may be more rapid and extensive than predicted" ("Melting Faster" 2007). "Climate models have predicted a retreat of the Arctic sea ice; but the actual retreat has proven to be much more rapid than the predictions," said Claire Parkinson, a climate researcher at NASA Goddard. "There continues to be considerable interannual variability in the sea ice cover, but the long-term retreat is quite apparent" ("Visualizing" 2012).

In addition to sea ice, land ice in the northern and southern Canadian Arctic archipelago declined sharply between 2004 and 2009, according to a study in *Nature* as reported by a team led by Alex Gardner of the University of Michigan. In six years (2004–2009), the islands lost an average of 61 gigatons (61 billion tons of ice) per year. The scientists also found that the rate of loss is accelerating. From 2004 to 2006, the average mass loss was roughly 31 gigatons per year; from 2007 to 2009, the loss increased to 92 gigatons per year ("Ice Loss" 2011).

Sea ice across the Arctic Ocean has been melting much more quickly than earlier anticipated even by the most recent computer models. Latest projections indicate that the Arctic Ocean may be without summer ice by 2020. Even projections made by the Intergovernmental Panel on Climate Change (IPCC) in its 2007 assessments (forecasting an ice-free Arctic summer between 2050 and 2100) were out of date weeks after they were made public, according to reports by scientists at the National Center for Atmospheric Research and the University of Colorado's National

Snow and Ice Data Center. The study, "Arctic Sea Ice Decline: Faster Than Forecast?" was published in May 2007 in the online edition of *Geophysical Research Letters*. This study was led by Julienne Stroeve of the National Snow and Ice Data Center and funded by the National Science Foundation and NASA.

In a separate study released in March 2006, Stroeve and her team showed that dwindling Arctic sea ice may have reached "a tipping point that could trigger a cascade of climate change reaching into Earth's temperate regions" ("Arctic Ice Retreating" 2007). "This suggests that current model projections may in fact provide a conservative estimate of future Arctic change, and that the summer Arctic sea ice may disappear considerably earlier than IPCC projections," said Stroeve ("Arctic Ice Retreating" 2007).

The pronounced thinning of Arctic sea ice has made the ice pack more brittle and susceptible to wind drift. The volume of Arctic sea ice decreased by one-third during 2007–2011 compared with the 1979–2006 mean. As Richard Kerr wrote in *Science* (2012), "Thinning of Arctic sea ice has made it more brittle, and more likely to break up because of wind drift, according to a model simulation by Jinlun Zhang at the University of Washington in Seattle and his colleagues" ("Thinning Ice" 2012). "The volume of Arctic sea ice decreased by one-third during 2007–2011 compared with the 1979–2006 mean."

Scientists were surprised by the sudden retreat of the Arctic sea ice during the summer of 2007, which was much more extensive than their models had anticipated. "The Arctic is often cited as the canary in the coal mine for climate warming," said Jay Zwally, a climate expert at NASA. "Now as a sign of climate warming, the canary has died" (Kolbert 2007, 44). "We could very well be in that quick slide downward in terms of passing a tipping point," said Mark Serreze, a senior scientist at the National Snow and Ice Data Center, in Boulder, Colorado. "It's tipping now. We're seeing it happen now" ("As Arctic Sea Ice" 2008). Bob Corell, who headed a multinational Arctic assessment, said, "We're moving beyond a point of no return" "As Arctic Sea Ice" 2008).

By 2007, Baffin Island in the Canadian Arctic had lost half its ice in 50 years as glaciers on its northern mountain ranges eroded. In another 50 years, what remains may be gone, according to research by geological sciences Professor Gifford Miller of the University of Colorado–Boulder's Institute of Arctic and Alpine Research and colleagues ("Baffin Island" 2008). "Even with no additional warming, our study indicates these ice caps will be gone in 50 years or less," he said (Anderson et al. 2008). Many observations combined with some climatic perspective present scientists with an alarming forecast of global ice-melting patterns to come. Bear in mind that because of thermal inertia ice that is melting now in the Arctic, Antarctic, and mountain ranges of the Earth reflects greenhouse-gas emissions of roughly 50 years ago—that is, in this instance, the mid-to-late 20th century. Add the fact, explained elsewhere in this work, that ice melts much more quickly than it forms. Combine that with knowledge that oceans rise not only from actual rise of water but also from expansion of its volume with warmth. The geophysical facts present an alarming warning of accelerating rises in sea levels in the coming century and beyond. Taking a world-wide perspective, combining it with historical perspective and an understanding of climate science illustrates why scientists who understand all of this have been sounding alarms.

Five years after the record melt of 2007, another massive decline in 2012 again forced scientists to question their models. The director of the National Snow and Ice Data Center (NSIDC), Mark Serreze, said that what remained of the Arctic ice pack was thinner and frailer than before made up of first-year ice. "We have entered a new regime," he said. "The sea ice is in such poor health in spring that large parts of it can't survive the summer melt season, even without boosts from extreme weather" (Schiermeier 2012, 185)

As Quirin Schiermeier wrote in *Nature* (2012),

Computer models that simulate how the ice will respond to a warming climate project that the Arctic will be seasonally "ice free" (definitions of this vary) some time between 2040 and the end of the century. But the observed downward trend in sea ice cover suggests that summer sea ice could disappear completely as early as 2030, something that none of the models used for the next report by the Intergovernmental Panel on Climate Change comes close to forecasting. (Schiermeier 2012, 185)

Arctic Warming and Midlatitude Climate Changes

Changes in upper-air circulation patterns over the Arctic have created long-lasting weather patterns that have been locked into place by atmospheric blocking across large parts of the Northern Hemisphere. Research into this phenomenon has increased because of their effects on the lives of people in the midlatitudes, said Bob Henson, a meteorologist and acting spokesperson for the University Corporation for Atmospheric Research in Boulder, Colorado. By the time the blocks ease, they can leave behind a "stack of broken records," he said (Gaarder 2013).

These patterns can shift at a given location, given shifting positions of the block, as Nancy Gaarder of the *Omaha World-Herald* wrote in 2013 in describing extremes in the upper Midwest:

Daffodils and crab apples were putting on their show. Golf courses and ball fields were packed. Planting was under way in backyard gardens. Such was the record warmth of March 2012. Fast forward a year: The soil's been too cold for planting, trees and flowers have yet to bud, and cabin fever has taken hold. "A tale of extremes," said Barbara Mayes, a National Weather Service meteorologist. (Gaarder 2013)

Writing in *Geophysical Research Letters*, Jennifer Francis, an atmospheric scientist based at Rutgers University, and atmospheric scientist Stephen Varvus of the University of Wisconsin–Madison have proposed a theory that associates rising temperatures in the Arctic with changes in weather conditions (and the persistence of particular patterns) at lower latitudes.

As Francis and Vavrus wrote,

These effects are particularly evident in autumn and winter consistent with sea-ice loss, but are also apparent in summer, possibly related to earlier snow melt on high-latitude land. Slower progression of upper-level waves would cause associated weather patterns in mid-latitudes to be more persistent, which may lead to an increased probability of extreme weather events that result from prolonged conditions, such as drought, flooding, cold spells, and heat waves. (Francis and Vavrus 2012)

"The Arctic is warming at two to three times the rate of the rest of the globe," Francis said.

As Nancy Gaarder also wrote,

As it warms, there's less contrast between the temperature of Arctic air and the atmosphere farther south. As a result, the jet stream weakens. A strong jet stream tends to flow fairly directly, west to east. A weakened jet meanders at a slower pace, looping north and south. The consequences: A weakened jet is more likely to form atmospheric blocks, which tend to create "stuck" weather patterns. The meandering allows Arctic air to plunge southward or warm air to surge northward. Combined, these two factors stack the odds in favor of prolonged hot or cold spells and contribute to stalled storm systems. (Gaarder 2013)

Jennifer Francis proposed her theory in 2011—just before record melting of sea ice shocked scientists in 2012. Within three years, she had become a major subject of controversy, as Eli Kintisch reported in *Science*: "The idea has drawn extensive attention from the media, the public, and influential policy makers, such as White House science adviser John Holdren. Many researchers, however, are skeptical, and some have been vociferously critical. But Francis, whose life and work has been shaped by two round-the-world sailing adventures, has remained calm throughout the storm" (Kintisch 2014, 250).

"The question really is not whether the loss of the sea ice can be affecting the atmospheric circulation on a large scale," said Francis. "The question is, how can it not be, and what are the mechanisms?" (Gillis and Foster 2012). At its basis, Francis theorizes that a warming Arctic and melting sea ice changes the behavior of the jet stream, a high-atmosphere river of air that steers storms and air masses, generally from west to east—except when it slows down and moves north and south as well. These kinks in the upper-level airflow can have major impacts on weather in any given location, serving up heat or cold, drought or deluge. "This means that whatever weather you have today—be it wet, hot, dry, or snowy—is more likely to last longer than it used to," said Francis. "If conditions hang around long enough, the chances increase for an extreme heat wave, drought or cold spell to occur," she said, but the weather can change rapidly once the kink in the jet stream moves along (Gillis and Foster 2012).

With his colleagues, climate scientist Ralf Jaiser at the Alfred Wegener Institute for Polar and Marine Research in Potsdam, Germany, wrote in *Tellus* (2012), a meteorological journal published in Stockholm, "Our analysis suggests that Arctic sea ice concentration changes exert a remote impact on the large-scale atmospheric circulation during winter, exhibiting a barotropic structure with similar patterns of pressure anomalies at the surface and in the midtroposphere. These are connected to pronounced planetary wave train changes notably over the North Pacific."

Arctic Warming and Extreme Weather Worldwide

Mounting evidence indicates that changes in the Arctic play a role in driving extreme weather at lower latitudes by altering the jet stream. This alteration has made for summer heat waves in North America, Europe, and Asia.

> The now clearly accelerating decline of summer ice—punctuated by exceptional losses in 2007 and now in 2012—has persuaded everyone that summer Arctic sea ice will be a goner far sooner than the end of the century, as current models predict. So the full knock-on effects of an ice-free Arctic Ocean—from the loss of polar bear habitat to possible increases of weather extremes at midlatitudes—could be here in many people's lifetimes. How far wrong the models might be, however, is still very much in dispute. (Kerr 2012, 1591)

As tundra and permafrost thaw, they release stored carbon dioxide and methane that further raise the atmosphere's level of greenhouse gases. During Alaska's unusually warm, dry winter of 2000–2001, the snowless tundra caught fire along Norton Sound, a possible precursor of larger, smoldering fires that could further accelerate global warming. There exists a palatable fear among climate scientists that such feedbacks in the oceans and on land could release large amounts of carbon dioxide and methane currently stored in frozen earth.

In 2009, scientists reported that lightning strikes in the Arctic had increased to 20 times their levels earlier in the 20th century, igniting tundra fires that added more "natural" carbon dioxide to the atmosphere (Jarvis 2009, n.p.). Additional evidence that Earth's frozen regions are already releasing additional greenhouse gases into the air has been provided by M. L. Goulden, et al., who studied boreal forests, finding " clear evidence that carbon dioxide locked into permafrost several hundred to 7,000 years ago is now being given off to the atmosphere as warming climate melts the permafrost" (Davis 2001, 270). The same is true in many cases for methane. According to Neil Davis, author of *Permafrost: A Guide to Frozen Ground in Transition* (2001), this "relict" carbon dioxide

> represents a massive source since it is estimated that the carbon dioxide contained in the seasonally and perennially frozen soils of boreal forests is 200 [billion] to 500 billion metric tons, enough if all released to increase the atmosphere's concentration of carbon dioxide by 50 percent. Hence, it is possible that the release of carbon dioxide from melting permafrost during warming, or locking it into newly frozen soil during cooling, may accelerate climate change. (Davis 2001, 270)

Nineteenth-Century Cooling Trend Quashed

Temperatures are now warmer in the Arctic than at any point in the last 2,000 years and at a time when Earth is receiving less heat from the sun because of changes in its orbit. The difference, suggest Darrell S. Kaufman and colleagues in *Science* in 2009 is greenhouse gas emissions from human activity. A cooling trend during the late 19th century "was caused by the steady orbitally driven reduction in summer insolation. . . . [It] was reversed during the 20th century, with four of the five warmest decades of our 2000-year-long reconstruction occurring between 1950 and 2000" (Kaufman et al. 2009, 1236). This study also provides fresh evidence for indications that human-generated warming could interrupt the natural ice age cycle. "The slow cooling trend is trivial compared to the warming that's been happening and that's in the pipeline," said Kaufman, lead author of the study (Revkin 2009). After a cooling of less than 0.5°F per millennium, the Arctic has warmed 2.2°F since 1900. "The fast rate of recent warming is the scary part," said coauthor Jonathan T. Overpeck. "It means that major impacts on Arctic ecosystems and global sea level might not be that far off unless we act fast to slow global warming" (Revkin 2009). "It's basically saying that greenhouse gas emissions are overwhelming the system," said coauthor David Schneider ("Warming Past" 2009).

According to glaciologist Mark Serreze,

> Climate models tell us that it is the Arctic sea ice cover that declines first, and that Antarctic ice extent falls only later, and may even (as observed) temporarily increase in response to changing patterns of atmospheric circulation. In other words, events are unfolding pretty much as expected. Finally, the statement that there was "substantial recovery" this year in the Arctic is simply rubbish. Ice extent at the end of the melt season in the Arctic [in 2008, compared to 2007] was second lowest on record and ice extent is still (as of early January) well below normal. (Revkin 2009)

The Greening of the Arctic

As temperatures warm in the Arctic and ice melts in the Arctic Ocean, forests of spruce trees and shrubs are spreading over northern Canada's tundra at a rate much faster than anticipated by many climate models. A study in the *Journal of Ecology*, which used tree rings to date the year of establishment and death of spruce trees and reconstruct changes in tree line vegetation, analyzed the density and altitude of tree line forests in southwestern Yukon over the past three centuries ("Canadian Tundra" 2007).

Tough Sledding on the Iditarod

The Iditarod, Alaska's annual celebration of Alaska's original form of indigenous transportation (now known as the "Super Bowl of dogsled racing"), has been running into tough sledding in a globally warmed world. In 2003, for the first time, the Iditarod was postponed for lack of snow. Snow was shipped in for the race's start in March 2003, and the race route was revised to avoid areas short on snow. As in 2002, unusual rain soaked parts of the race route. The route was finally changed as Alaska experienced its mildest winter in more than a century of record keeping.

In 2015, weather was so mild in Alaska that the Iditarod's opening ceremony was held March 9 in Anchorage with trucked-in snow, but the actual starting point was moved 225 miles north to Fairbanks. Parts of the traditional trail south of Fairbanks contained exposed grass, rocks, and gravel that made racing impossible. By 2016, trucked-in snow was becoming part of the annual ritual. That year, it arrived in railroad cars idled by a slump in coal transport. The seven rail cars brought about four inches of dirty snow with the consistency of sand, imported from Fairbanks, for a parade route that had been shortened from 11 miles to three (Johnson 2016).

Weather during the Iditarod (the first half of March) seems to be getting more variable and irregularly warmer, ranging from record snows during 2012 to a paucity of snow and drizzly mildness in 2013 that forced race organizers to import snow for the largely ceremonial start of the race. After the faux start in Anchorage, dogs, sleds, and mushers were loaded onto trucks and driven north to the village of Willow for the start of the race. By the time racers reached McGrath, on their way to Nome, the temperature had broken 50°F for a day. As one veteran onlooker quipped, "I've never sunbathed on the Iditarod Trail before!" (McGrath 2013, 94).

Mike Williams, a Yupiaq from the small village of Akiak on western Alaska's lower Kuskokwim River, has run the Iditarod for 15 years. He described how snow cover has declined in an irregular fashion year by year. More recently, a large number of mushers and dogs have been injured by rough terrain. Several mushers had to quit the race due to "broken sleds and broken spirits" ("Where's the Snow" 2014). Unusually warm weather in 2014 also made training for the race difficult. "We trained in ice and it rained most of the time and we only had half a day of good snow." Mike recounted. "All winter long back in southwest Alaska, in our home, it was the worst training conditions we have seen. The ice was thin and we could not set our fish traps underneath the ice as was usual" ("Where's the Snow" 2014).

The sloppy Iditarod was just a forewarning. By early May, the ice was the thinnest Williams had ever seen, making hunting dangerous. "Our people are suffering for it," he said. "We are anticipating some erosion due to lack of permafrost and its continuously thawing. If we don't have normal winters anymore. . . ." The lack of snow and cold upsets the natural cycle. Anchorage in 2014–2015 experienced its warmest winter on record and hundreds of dump trucks again imported snow to the racecourse to get the Iditarod started in early March.

Further Reading

Johnson, Kirk. "As Alaska Warms, the Iditarod Adapts." *The New York Times*, March 6, 2016. http://www.nytimes.com/2016/03/07/us/as-alaska-warms-the-iditarod-adapts .html.

McGrath, Ben. "The White Wall: In the Iditarod, Alaska's Mushing Dynasties Confront the Elements—and a Generational Divide." *The New Yorker,* April 22, 2013, 81–95.

"Where's the Snow? Mike Williams, First Stewards Board Member, Talks about Climate Change and Its Culture-Changing Impacts to the Native Communities of Alaska." First Stewards, May 8, 2014. http://www.firststewards.org/2014-press .html.

"The conventional thinking on tree line dynamics has been that advances are very slow because conditions are so harsh at these high latitudes and altitudes," explained study author Ryan Danby, a biologist with the University of Alberta. "But what our data indicate is that there was an upslope surge of trees in response to warmer temperatures. It's [as if] it waited until conditions were just right and then it decided to get up and run, not just walk" ("Canadian Tundra" 2007).

"The tundra is becoming greener with the growth of more shrubs," said Vladimir E. Romanovsky, a professor at the geophysical institute of the University of Alaska. This development is causing problems in some areas such as where herds of reindeer migrate. At the same time, there is some decrease in the greening of the northern forest areas, probably because of drought. The glaciers are continuing to shrink, and river discharge into the Arctic Ocean is rising, Romanovsky said (Schmid 2006).

The Arctic Sea—A Stagnant Soup?

The Transpolar Drift is a strong Arctic current that runs from central Siberia to Greenland and then into the Atlantic and disperses pollutants. Could it stagnate in coming years as waters warm, leaving the area a polluted sea? This cold surface current was first discovered in 1893 by Norwegian explorer Fridtjof Nansen, who tried unsuccessfully to use it to sail to the North Pole. Together with the Beaufort Gyre, the Transpolar Drift keeps Arctic waters well mixed and ensures that pollution never lingers for long.

Without this current, the Arctic would retain such nuclear testing residues as strontium-90 and caesium-137, as well as pesticides and petroleum residue. Examining a business-as-usual greenhouse gas scenario, with carbon dioxide levels doubling by 2070, Y. Gao and colleagues (2009, 375) found that the Transpolar Drift stops and the Beaufort Gyre, Greenland Current, and Gulf Stream weaken considerably. "One reason for this sluggish behaviour is a change in wind patterns driven by global warming and rapid melting of the Arctic sea ice" (Ravilious 2009).

Further Reading

Anderson, Rebecca K., et al. "A Millennial Perspective on Arctic Warming from 14-C in Quartz and Plants Emerging from Beneath Ice Caps." *Geophysical Research Letters* 35 (2008): L01502. doi: 10.1029/2007/GL032057.

"Arctic Ice Retreating 30 Years Ahead of Projections." Environment News Service, April 30, 2007. http://www.ens-newswire.com/ens/apr2007/2007-04-30-04.asp (no longer available).

"Arctic Snowpack Thins." *Nature* 512 (August 14, 2014): 115.

"As Arctic Sea Ice Melts, Experts Expect New Low." *The New York Times*, August 28, 2008. http://www.nytimes.com/2008/08/28/science/earth/28seaice.html.

"Baffin Island Ice Caps Shrink by Half in 50 Years." Environment News Service, January 28, 2008. http://www.ens-newswire.com/ens/jan2008/2008-01-28-02.asp.

"Canadian Tundra Turning Green." Environment News Service, March 3, 2007. http://www.ens-newswire.com/ens/mar2007/2007-03-06-02.asp (no longer available).

Comiso, Josefino C. "Large Decadal Decline of the Arctic Multiyear Ice Cover." *Journal of Climate* 25 (2012): 1176–1193. doi: http://dx.doi.org/10.1175/JCLI-D-11-00113.1.

Davis, Neil. *Permafrost: A Guide to Frozen Ground in Transition.* Fairbanks: University of Alaska Press, 2001.

Derksen, C., and R. Brown. "Spring Snow Cover Extent Reductions in the 2008–2012 Period Exceeding Climate Model Projections." *Geophysical Research Letters* 39(19) (October 16, 2012). http://onlinelibrary.wiley.com/doi/10.1029/2012GL053387/abstract.

Francis, Jennifer A., and Stephen J. Vavrus. "Evidence Linking Arctic Amplification to Extreme Weather in Mid-latitudes." *Geophysical Research Letters,* 39 (March 17, 2012): L06801. http://marine.rutgers.edu/~francis/pres/Francis_Vavrus_2012GL051000_pub.pdf.

Gaarder, Nancy. "Arctic Warming Blamed for Last Year's Heat and This Year's Chill." *Omaha World-Herald*, March 30, 2013. http://www.omaha.com/article/20130330/NEWS/130339998/1685#arctic-warming-blamed-for-last-year-s-heat-and-this-year-s-chill (no longer available).

Gao, Y., et al. "Sources and Pathways of 90Sr in the North Atlantic-Arctic Region: Present Day and Global Warming." *Journal of Environmental Radioactivity*, 5 (May 2009): 375–395.

Gillis, Justin, and Joanna M. Foster. "Weather Runs Hot and Cold, So Scientists Look to the Ice." *The New York Times,* March 28, 2012. http://www.nytimes.com/2012/03/29/science/earth/arctic-sea-ice-eyed-for-clues-to-weather-extremes.html.

Hansen, James E., et al. "Global Temperature and Europe's Frigid Air." Draft paper, December 10, 2010. 20101211_TemperatureandEurope(3).pdf.

Human, Katy. "Disappearing Arctic Ice Chills Scientists; A University of Colorado Expert on Ice Worries That the Massive Melting Will Trigger Dramatic Changes in the World's Weather." *Denver Post*, October 5, 2004, B2.

"Ice Loss in the Canadian Arctic Archipelago." NASA Earth Observatory, June 1, 2011. http://earthobservatory.nasa.gov/IOTD/view.php?id=50726&src=eoa-iotd.

Jaiser, R., et al. "Impact of Sea Ice Cover Changes on the Northern Hemisphere Atmospheric Winter Circulation." *Tellus* 64 (2012): 11595. http://www.tellusa.net/index.php/tellusa/article/view/11595.

Jarvis, Lisa. "Kindling for Climate Change." *Chemical and Engineering News*, August 17, 2009.

Kaufman, Darrell S., et al. "Recent Warming Reverses Long-Term Arctic Cooling." *Science* 325 (September 4, 2009): 1236–1239.

Kerr, Richard A. "Climate Tipping Points Come in from the Cold." *Science* 319 (January 11, 2008): 153.

Kerr, Richard A. "Ice-Free Arctic Sea May Be Years, Not Decades, Away." *Science* 337 (September 28, 2012): 1591.

Kintisch, Eli. "Into the Maelstrom." *Science* 344 (April 18, 2014): 250–253.

Kolbert, Elizabeth. "Testing the Climate." (Talk of the Town) *The New Yorker*, December 24 and 31, 2007, 43–44.

"Melting Faster." *Science* 316(1) (May 18, 2007): 955.

Morello, Lauren. "Summer Storms Bolster Arctic Ice." *Nature* 500 (August 29, 2013): 512.

"NASA Study Shows Role of Melt in Arctic Sea Ice Loss." NASA press release, November 9, 2010. http://www.jpl.nasa.gov/news/news.php?feature=2811.

"Oldest Arctic Sea Ice Is Disappearing." NASA Earth Observatory. March 1, 2012. http://earthobservatory.nasa.gov/IOTD/view.php?id=77270&src=eoa-iotd.

Ravilious, Kate. "The Arctic Sea a Stagnant Soup?" *New Scientist*, August 6, 2009. http://www.newscientist.com/article/mg20327204.800-arctic-ocean-may-be-polluted-soup-by-2070.html.

Revkin, Andrew C. "Gore, Will, Climate, and Complexity." *The New York Times*, February 25, 2009. http://dotearth.blogs.nytimes.com/2009/02/25/gore-and-will-and-climate-and-the-press.

Schiermeier, Quirin. "Ice Loss Shifts Arctic Cycles." *Nature* 489 (September 13, 2012): 185–186.

Schmid, Randolph E. "State of the Arctic Warming, with Widespread Melting." Associated Press, November 16, 2006 (LEXIS).

Screen, James A., and Ian Simmonds. "The Central Role of Diminishing Sea Ice in Recent Arctic Temperature Amplification." *Nature* 464 (April 29, 2010): 1334–1337.

"Snow Cover Extent Declines in the Arctic." NASA Earth Observatory, January 8, 2013. http://earthobservatory.nasa.gov/IOTD/view.php?id=80102&src=eoa-iotd.

Spielhagen, Robert F., et al. "Enhanced Modern Heat Transfer to the Arctic by Warm Atlantic Water." *Science* 331 (January 28, 2011): 450–453.

"Thinning Ice More Fragile and Mobile." *Nature* 491 (November 15, 2012): 304.

"Visualizing the 2012 Sea Ice Minimum." NASA Earth Observatory, September 27, 2012. http://earthobservatory.nasa.gov/IOTD/view.php?id=79256&src=eoa-iotd.

"Warming Past, Present." *Omaha World-Herald*, September 4, 2009, 5A.

See also: Adaptation, Animals, Lizards, and Warming Habitats; Animal Life, Arctic; Arctic Hunters; Atmospheric Circulation; Climate Change, Abrupt Nature of; El Niño and La Niña; Inuit; Permafrost; Sea Ice, Arctic; Summer Ice, Arctic; Temperatures, Global; Temperatures, Greenhouse Gas Levels and

ICE MELT, GREENLAND

The largest mass of ice in the Northern Hemisphere resides atop Greenland and constitutes approximately 10 percent of the world's frozen water. This ice is being measured and monitored as never before by satellites, aircraft, and dozens of down-clad scientists braving temperatures as low as 30 degrees below zero and deadly snow-cloaked crevasses that corrugate the slumping edges of the ice cap (Revkin 2004). By 2014, Greenland was averaging almost 300 billion tons of ice loss a year, or more than 1 trillion tons in four years, according to surveys carried out by NASA and an independent team of scientists. Over time, the pace of loss has accelerated and now amounts to 9 trillion tons in a century (McMillan et al. 2016).

In 2016, Greenland's ice-melting season began surprisingly early in April. Scientists receiving the first reports of ice melting first thought their instruments had malfunctioned. Between 2012 (when melting was first detected at the highest point of the ice sheet) and 2016, 1 trillion tons of ice melted across Greenland, enough to cover an area the size of New York state to a depth of 23 feet (Kolbert 2016, 50).

Ice melt in Greenland has accelerated significantly since 1990, according to a report in the *Journal of Climate* (January 15, 2008) coauthored by Konrad Steffen, professor of geography at the University of Colorado (Steffen 2007). A scientific team surveyed the rate of summer melting between 1958 and 2006 and found that the five largest melting years had all occurred after 1995. Greenland's ice has continued to erode since then. The year 1998 was the biggest loss (109 cubic miles) followed by 2002, 2003, 2007, 2010, and 2012, which exceeded all previous years to that time. "Ice is moving faster into the ocean, and that will add to the sea level rise," said Steffen ("Ice Sheet" 2008). The erosion of Greenland's ice cap tends to accelerate during El Niño years (such as 1998 and 2015) when temperatures rise worldwide.

Satellite monitoring of Greenland's ice sheet began in the late 1970s. Since then, ice extent has eroded steadily. Records were set and reset in 2005, 2007, 2010, and 2012. According to J. Harper and colleagues, writing in *Nature*,

> Much of the increased surface melt is occurring in the percolation zone, a region of the accumulation area that is perennially covered by snow and firn (partly compacted snow). The fate of melt water in the percolation zone is poorly constrained: some may travel away from its point of origin and eventually influence the ice sheet's flow dynamics and mass balance and the global sea level, whereas some may simply infiltrate into cold snow or firn and refreeze with none of these effects. (Harper et al. 2012, 240)

Steady Ice Erosion

In 20 years, the loss of ice mass from the Greenland ice sheet has quadrupled and has accounted for some 25 percent of global sea level rise during that time. Much of this melting takes place as relatively warm ocean water erodes outlet glaciers from below. According to a report in *Nature*,

> Recent evidence suggests that an anomalous inflow of subtropical waters driven by atmospheric changes, multidecadal natural ocean variability and a long-term increase in the North Atlantic's upper ocean heat content since the 1950s all contributed to a warming of the sub-polar North Atlantic. This led, in conjunction with increased runoff, to enhanced submarine glacier melting. Future climate projections raise the potential for continued increases in

warming and ice-mass loss, with implications for sea level and climate. (Straneo and Heimbach 2013, 36)

Ice loss has been migrating north and west across Greenland since 2005, according to satellite-borne sensors migration of ice loss across the length of coastal Greenland, enlarging an area in the island's southeast that has been losing ice for many years. An analysis in 2010 described "an ongoing northward migration of increasing mass loss" along the western coast from the southern tip to the far north (Cox 2010). The analysis from Shfaqat Abbas Khan of the National Space Institute of Denmark was published in late March 2010 in *Geophysical Research Letters*. The study surveyed ice loss between February 2003 and June 2009. "When we look at the monthly values from GRACE, the ice mass loss has been very dramatic along the northwest coast of Greenland," said coauthor John Wahr, a physicist at University of Colorado–Boulder. "This is a phenomenon that was undocumented before this study. Our speculation is that some of the big glaciers in this region are sliding downhill faster and dumping more ice in the ocean" (Cox 2010; Khan et al. 2010).

"Having such extreme melting so far north, where it is usually colder than the southern regions is extremely interesting," said Marco Tedesco, director of the Cryospheric Processes Laboratory at the City College of New York, who is leader of a project studying variables that affect Greenland's ice melt. "In 2007, the record occurred in southern Greenland, mostly at high elevation areas where in 2008 extreme snowmelt occurred along the northern coast" ("Satellite Data" 2008). Melting in northern Greenland lasted as many as 18 days longer than in previous years. Some of the most rapid melting took place near Ellesmere Island, including the Petermann glacier, which lost 29 square kilometers in July 2008.

Greenland's ice is only a fraction of Antarctica's, but it is melting more rapidly, in part because summers are warmer, allowing for more rapid runoff. Greenland's southern tip is no farther north than Juneau or Stockholm. The persistence of the ice cap results from its mass, the fact that ice makes its own climate. The ice itself deflects sunlight, heat, and weather systems from the south. The elevation of the ice sheet also helps to keep it cold. As it erodes, these advantages diminish (Appenzeller 2007, 68).

Greenland Ice Melt Past and Future

Paleoclimatic data reveals that Greenland's ice can melt relatively quickly. Philippe Huybrechts, a glaciologist and ice-sheet modeler at the Free University of Brussels, has modeled the behavior of Greenland's ice sheet and found that with an anticipated annual temperature increase of 8°C, "the ice sheet would shrink to a small glaciated area far inland and sea level would rise by six meters" (Schiermeier 2004, 114). Over the course of several centuries, if global warming indeed leads to Greenland's ice cap melting, "removal of the Greenland ice sheet due to a prolonged climatic warming would be irreversible," according to climate models run by several scientists (Toniazzo et al. 2004, 21).

In Ilulissat (a name meaning "iceberg") on the west coast of Greenland, rain fell during December 2007 and January 2008. "Twenty years ago, if I had told the people of Ilulissat that it would rain at Christmas 2007, they would have just laughed at me. Today it is a reality," said Konrad Steffen, director of the Cooperative Institute for Research in Environmental Sciences at the University of Colorado (Friedman 2008). Melting of the Greenland ice sheet increased 30 percent between 1979 and 2007. By that time, Greenland was losing 200 cubic kilometers of ice per year—from actual melting and ice sliding into the ocean from outlet glaciers along its edges—which far exceeds the volume of all the ice in the European Alps, Steffen added. "Everything is happening faster than anticipated," he said (Friedman 2008). Air temperatures on the Greenland ice sheet increased by some 7°F between 1991 and 2008.

"The amount of ice lost by Greenland over the last year [2007] is the equivalent of two times all the ice in the Alps, or a layer of water more than one-half mile deep covering Washington, D.C.," according Konrad Steffen (Kolbert 2007, 43). Ice is melting most rapidly at the edges of the ice sheet. Although Greenland has been gaining ice mass at higher elevations due to increases in snowfall (warmer air holds more moisture, thus more snow), the gain is more than offset by an accelerating mass loss, primarily from rapidly thinning and accelerating outlet glaciers, Steffen said.

Greenland's ice sheet melted in an unusual area during 2008—its northern fringes—according to NASA scientist Marco Tedesco and colleagues. The average temperature between June and August 2008 was as much as 3°C above average, with new record temperatures at many ground-based weather stations in this area ("Melting" 2009; Tedesco et al. 2008).

By 2009, Greenland's ice sheet was losing an average of 50 cubic miles of ice a year (Folger 2010, 52). The coastal waters of northeastern and northwestern Greenland, now ice free more than six months per year, are estimated to hold 10 billion barrels of offshore oil and gas (Folger 2010, 63).

The length of the melting season for Greenland ice reached a new record in 2010. "This past melt season [2010] was exceptional, with melting in some areas stretching up to 50 days longer than average," said Tedesco ("New Melt" 2011; Tedesco et al. 2011). Melting during 2010 stretched from late April to mid-September. According to Tedesco, summer temperatures in Greenland during 2010 were as much as 3°C above the average. The capital, Nuuk, experienced its warmest spring and summer since records began in 1873. A longer ice-free season reduced albedo and increased warming.

Manhattan-Sized Ice Chunks Break Off

A 2.7 square mile piece of ice broke off Greenland's Jakobshavn Isbrae glacier on July 6 and 7, 2010. This event pushed the calving front—where the ice sheet meets the ocean—inland almost a mile in one day to a position farther inland than any on record. Warming Arctic waters were a major factor in the breakup. "While there have been ice breakouts of this magnitude from Jakobshavn and other glaciers in the past, this event is unusual because it occurs on the heels of a warm winter that saw no sea ice form in the surrounding bay," said Thomas Wagner, cryospheric program scientist at NASA headquarters in Washington, D.C. "While the exact relationship between these events is being determined, it lends credence to the theory that warming of the oceans is responsible for the ice loss observed throughout Greenland and Antarctica" ("Big Chunk" 2010).

"Bare ice is much darker than snow and absorbs more solar radiation," said Marco Tedesco. "Other ice melting feedback loops that we are examining include the impact of lakes on the glacial surface, of dust and soot deposited over the ice sheet and how surface melt water affects the flow of the ice toward the ocean" ("New Melt" 2011).

In 2012, scientists said that "unprecedented" melting took place over a larger area than had been detected in three decades of satellite observation (Slivka 2012). Melting even occurred at Greenland's coldest and highest place, Summit Station. NASA said that nearly the entire Greenland ice sheet—from thin, low-lying coastal glaciers to the two-mile-thick mass at its center—experienced some degree of surface melting in 2012. "When we see melt in places that we haven't seen before, at least in a long period of time, it makes you sit up and ask what's happening," NASA chief scientist Waleed Abdalati said. "It's a big signal, the meaning of which we're going to sort out for years to come" ("Retreat" 2014).

According to NASA's Earth Observatory, "Jakobshavn is Greenland's fastest-moving glacier, and the flow rate is variable with spurts of speed in the summer and additional variation from year to year. Jakobshavn receded more than 40 kilometers (25 miles) between 1850 and 2010. Retreat is accelerating. During the summer of 2012, the glacier surged into the sea at 17 kilometers (10 miles) in one year." Another very large calving event (shedding of ice into the ocean) took place at the end of May 2014. "Sometime between May 9, and June 1 . . . the glacier shed kilometers of ice from its front," NASA reported ("Retreat" 2014).

By mid-July 2012, after an exceptional warm spell, Greenland's usually frozen surface had, according to one account, "turned into a colossal puddle. Even the coldest parts of the world's largest island saw ice thaw and rainfall, fueling concerns over the future of glaciers that hold enough water to raise global sea levels by around seven meters" (Schiermeier 2013, 459). In 20 years, from 1992 to 2011, Greenland and Antarctica lost significant amounts of ice—2,700 billion and 1,350 billion tons, respectively—and raised world sea levels by 0.6 millimeters per year (Schiermeier 2013, 459). Studies of paleoclimate indicate that ice loss is not linear but proceeds in an irregular manner.

On July 16, in the midst of Greenland's record ice melt of summer 2012, ice twice the size of Manhattan Island calved from Petermann glacier on the island's Northwest coast, a breakup that scientists attributed to warmer than usual sea temperatures. In August 2010 (another summer notable for then-record ice melt in Greenland), the same glacier had calved 97 square miles at once, the largest breakoff of its kind on record in Greenland and an area twice as large as 2012's calving.

Ice: "Changing before Our Eyes"

According to Andreas Muenchow, an associate professor of physical ocean science and engineering at the University of Delaware, "The Greenland ice sheet is changing rapidly before our eyes." Muenchow added that although "no individual glacier will be the canary in the coal mine," recent warming has transformed the overall ice sheet (Eilperin and Samenow 2012). The glacier accelerated 10 percent to 20 percent after the 2010 calving

By the summer of 2012, nearly the entire Greenland ice sheet was melting in midsummer. A scientific team found that the accelerated melting rate was being aided by "low-level clouds consisting of liquid water droplets ('liquid clouds'), via their radiative effects, [which] played a key part in this melt event by increasing near-surface temperatures" (Bennartz et al. 2013, 83). In the summer of 2012, melting occurred across 98.6 percent of the Greenland ice sheet. Until that time, it had averaged roughly half ("The Rapid Melt" 2012).

Writing in *Science*, Twila Moon and colleagues said they expect an acceleration of ice melt:

Earlier observations on several of Greenland's outlet glaciers, starting near the turn of the 21st century, indicated rapid (annual-scale) and large (>100%) increases in glacier velocity. Combining data from several satellites, we produce a decade-long (2000 to 2010) record documenting the ongoing velocity evolution of nearly all (200+) of Greenland's major outlet glaciers, revealing complex spatial and temporal patterns. Changes on fast-flow marine-terminating glaciers contrast with steady velocities on ice-shelf–terminating glaciers and slow speeds on land-terminating glaciers. Regionally, glaciers in the northwest accelerated steadily, with more variability in the southeast and relatively steady flow elsewhere. Intraregional variability shows a complex response to regional and local forcing. Observed acceleration indicates that sea level rise from Greenland may fall well below proposed upper bounds. (Moon et al. 2012)

Observers have been using archival photographs from the 1980s that allow a comparison to the current ice loss in Greenland. Using this record, Kurt H. Kjaer and colleagues wrote in *Science* (2012, 569),

We reveal two independent dynamic ice loss events on the northwestern Greenland ice sheet margin: from 1985 to 1993 and 2005 to 2010, which were separated by limited mass changes. Our results suggest that the ice mass changes in this sector were primarily caused by short-lived dynamic ice loss events rather than changes in the surface mass balance. This finding challenges predictions about the future response of the Greenland ice sheet to increasing global temperatures.

Further Reading

Appenzeller, Tim. "The Big Thaw." *National Geographic*, June 2007, 56–71.
Bennartz, R., et al. "July 2012 Greenland Melt Extent Enhanced by Low-Level Liquid Clouds." *Nature* 496 (April 4, 2013): 83–86.

"Big Chunk of Ice Breaks Off of Greenland Glacier." LiveScience, July 12, 2010. http://www
.ouramazingplanet.com/big-chunk-of-ice-breaks-off-of-greenland-glacier-0334/.

Cox, John D. "More Greenland Ice Melting Faster." Discovery News, March 24, 2010. http://
news.discovery.com/earth/more-greenland-ice-melting-faster.html.

Eilperin, Juliet, and Jason Samenow. "Greenland Glacier Loses Large Mass of Ice." *Washington Post*, July 17, 2012. http://www.washingtonpost.com/national/health-science/gree
nland-glacier-loses-large-mass-of-ice/2012/07/17/gJQAf5CQsW_print.html.

Folger, Tim. "Viking Weather: The Changing Face of Greenland." *National Geographic*,
June 2010, 48–63.

Friedman, Thomas. "Learning to Speak Climate." *The New York Times*, August 6, 2008.
http://www.nytimes.com/2008/08/06/opinion/06friedman.html.

Harper, J., et al. "Greenland Ice-Sheet Contribution to Sea-Level Rise Buffered by Meltwater Storage in Firn." *Nature* 491 (November 8, 2012): 240–243.

"Ice Sheet of Greenland Melting Away at Faster Pace." *Omaha World-Herald*, January 20,
2008, 18A.

Khan, S. A., et al. "Spread of Ice Mass Loss into Northwest Greenland Observed by GRACE
and GPS." *Geophysical Research Letters* 37 (March 23, 2010): L06501. http://www.agu
.org/pubs/crossref/2010/2010GL042460.shtml (no longer available). doi: 10.1029/20
10GL042460.

Kjær, Kurt H., et al. "Aerial Photographs Reveal Late–20th-Century Dynamic Ice Loss in
Northwestern Greenland." *Science* 337(August 3, 2012): 569–573.

Kolbert, Elizabeth. "Testing the Climate." *The New Yorker*, December 31, 2007, 43–44.

Kolbert, Elizabeth. "A Song of Ice: What Happens When a Country Starts to Melt?" *The
New Yorker*, October 24, 2016, 50–61.

McMillan, Malcolm, et al. "A High-Resolution Record of Greenland Mass Balance." *Geophysical Research Letters* 43 (July 9, 2016): 7002–7010.

"Melting on the Greenland Ice Cap, 2008." NASA Earth Observatory, February 24, 2009.
http://earthobservatory.nasa.gov/IOTD/view.php?id=37215.

Moon, T., et al. "Twenty-First-Century Evolution of Greenland Outlet Glacier Velocities."
Science 336 (May 4, 2012): 576–578.

"New Melt Record for Greenland Ice Sheet; CCNY's Marco Tedesco Says 'Exceptional' Season Stretched up to 50 Days Longer Than Average." NASA Earth Observatory. January 21,
2011. http://www.eurekalert.org/pub_releases/2011-01/ccon-nmr012111.php.

"The Rapid Melt of Greenland" *Nature* 491 (November 8, 2012): 163.

"Retreat of Jakobshavn Glacier, Greenland." NASA Earth Observatory, June 11, 2014. http://
earthobservatory.nasa.gov/IOTD/view.php?id=83837&src=eoa-iotd.

Revkin, Andrew. "An Icy Riddle as Big as Greenland." *The New York Times*, June 8, 2004.
www.nytimes.com/learning/teachers/featured_articles/20040609wednesday.html.

"Satellite Data Reveals Extreme Summer Snowmelt in Northern Greenland, CCNY Professor Says." NASA Earth Observatory, October 8, 2008 (no longer available). http://
earthobservatory.nasa.gov/Newsroom/MediaAlerts/2008/2008100827658.html.

Schiermeier, Quirin. "A Rising Tide: The Ice Covering Greenland Holds Enough Water to
Raise the Oceans Six Meters—and It's Starting to Melt." *Nature* 428 (March 11, 2004):
114–115.

Schiermeier, Quirin. "Greenland Defied Ancient Warming." *Nature* 493 (January 24, 2013):
459–460. http://www.nature.com/news/greenland-defied-ancient-warming-1.12265.

Slivka, Kelly. "Burst of Melting in Greenland's Ice Sheet Gives Scientists a Rare Climatic Snapshot." *The New York Times*, July 25, 2012, A4.

Steffen, Konrad. "Greenland Melt Accelerating, According to Colorado University-Boulder Study." Press release, University of Colorado, December 11, 2007 via NASA Earth Observatory, December 25, 2007. http://www.news.uiuc.edu/news/07/1210nitrogen.html (no longer available).

Straneo, Fiammetta, and Patrick Heimbach. "North Atlantic Warming and the Retreat of Greenland's Outlet Glaciers." *Nature* 504 (December 5, 2013): 36–43.

Tedesco, M., et al. "Extreme Snowmelt in Northern Greenland during Summer 2008." *Eos* 89(41) (October 7, 2008): 391.

Tedesco, M., et al. "The Role of Albedo and Accumulation in the 2010 Melting Record in Greenland." *Environmental Research Letters* 6(1) (January 21, 2011), n.p. http://iopscience.iop.org/article/10.1088/1748-9326/6/1/014005/meta;jsessionid=110ADC707B6345F40B3BE3E2C9A259C8.ip-10-40-1-105.

Toniazzo, T., J. M. Gregory, and P. Huybrechts. "Climatic Impact of a Greenland Deglaciation and Its Possible Irreversibility." *Journal of Climate* 17(1) (January 1, 2004): 21–33.

See also: Animal Life, Arctic; Arctic Hunters; Climate Change, Greenland; Fisheries; Glacial Erosion; Global Warming, Greenland; Ice Science, Greenland; Inuit; Sea Level Rise

ICE SCIENCE, GREENLAND

Until recently, many climate scientists anticipated the impacts of melting polar ice sheets in the far future—a matter of centuries. Then they learned that their models were not squaring with real-world ice-melt patterns, which generally have been faster, often in unpredictable and nonlinear ways. "When you look at the ice sheet, the models didn't work, which puts us on shaky ground," said Richard Alley, a geosciences professor at Pennsylvania State University (Rudolf 2007).

Scientists have now admitted they have no accurate way to predict the melting of polar ice. Thus, estimates of how much ice may melt and how quickly cover a wide range. In addition, the actual melting takes place perhaps a century or two after a given quantity of greenhouse gases have been consumed and released into the air because of thermal inertia. The major question becomes not how much ice may melt by a given date, but how much more melting is "in the pipeline," guaranteed by greenhouse gases already burned for which effects have not been fully realized.

In July 2006, researchers floating in a small boat studied melting dynamics of a mile-wide glacial lake in Greenland. Ten days later, in just 90 minutes, the lake was sucked through a crack in the ice with a force equaling Niagara Falls. This event was studied by several scientists led by Ian Joughin of the University of Washington and Sarah Das of the Woods Hole Oceanographic Institution in Woods Hole, Massachusetts, to construct a unique description of Greenland ice sheet "plumbing" that shows how summertime melting accelerates ice loss. Their results, published in *Science Express* on April 17, 2008, concluded that summer melt is only one factor influencing the erosion of Greenland's ice (Hansen 2008).

"For years, people have said that the increasing length and intensity of the melt season in Greenland could yield an increase in ice discharge," said Joughin, lead author on the paper in *Science* Express. "Greater melt in future summers would cause ice to flow faster toward the coast and draw down more of the ice sheet" ("Researchers Warm" 2008). This study found that violent draining of surface lakes had a short-lived influence on the ice sheet's local movement. Water is then quickly distributed to the ice sheet's base, which contributes to glaciers' movement toward the sea. "If you're really going to get a lot of ice out of Greenland, that would have to occur through outlet glaciers, but those are not being affected very much by seasonal melt," Joughin said. "The outlet glaciers are more affected by the removal of their shelves and grounded ice in their fjords, which decreases resistance to ice flow" ("Researchers Warm" 2008).

Breaking Up from Below

Greenland's coastal ice often melts with a boost from the sea as warm water upwells from offshore and erodes areas that have been locked in ice for millennia. "Changes in the ocean eat the ice sheet from underneath," said Sarah Das, a glaciologist at Woods Hole. "Warmer water causes the glaciers to calve and melt back more quickly" (Faris 2008).

NASA's Earth Observatory described the role that melting water plays in eroding Greenland's ice sheet: "Each spring and summer, as the air warms up and the sunlight beats down on the Greenland ice sheet, sapphire-colored ponds spring up like swimming pools in a suburban neighborhood. As snow and ice melt atop the glaciers, the water flows in channels and streams and collects in depressions on the surface. . . . These melt ponds and lakes sometimes sometimes disappear quickly," sucked into sub-surface rivers through deep cracks called "moulins" ("Greenland Melt" 2013). "Moulins . . . act like plumbing drains and pipes moving melted water from the surface to the bottom of the ice sheet and eventually to the coast. Greenland's glaciers have been accelerating their movement and calving in recent decades, and some scientists have theorized that the melt water lubricates the interface between the ice sheet and rocky landscape, allowing glaciers to move faster" ("Greenland Melt" 2013).

By 2009, some glaciers along Greenland's west coast were melting 100 times faster from below than on their surfaces as they entered the ocean, suggesting that the role of relatively warm ocean water is playing a critically important role in eroding Greenland's coastal glaciers. Eric Rignot and Isabella Velicogna, both with NASA's Jet Propulsion Laboratory in Pasadena, California, and the University of California–Irvine with colleague Michele Koppes of the University of British Columbia, Vancouver, Canada, measured underwater melting rates of four glaciers in central west Greenland during summer 2008. Using equipment lowered into glacier fjords, they took samples of water at various depths to measure temperature and salinity and trace ocean currents.

Rignot said the new study complements other recent research on the effects of ocean conditions in Greenland fjords. "Our study fills the gap by actually looking at these submarine melt rates, something that had never been done before in Greenland," he said. "The results indicate rather large values that have vast implications for the evolution of the glaciers if ocean waters within these fjords continue to war" ("NASA Finds" 2010).

"All major Greenland glaciers end up in the ocean, and tidewater glaciers control 90 percent of the ice discharged by Greenland into the sea," Rignot said. "Submarine melting may therefore have a large indirect impact on the ice mass budget of the entire Greenland ice sheet. If we are to determine the future of the Greenland ice sheet more reliably in a changing climate, more complete and detailed studies of the interactions between ice and ocean at the ice sheet's margins are essential" ("NASA Finds" 2010).

The power of warm-water ice erosion from below is illustrated in a *Harper's* magazine account by Gretel Ehrlich (2015) from Qaanaaq and Siorapaluk, the two northernmost villages in Greenland, where hunters relayed descriptions of "rotten ice"—melted by currents from below to a point where ice that had been several feet thick was reduced to a few inches, even at temperatures of −30°F. In large part because of such decay, by 2014 Greenland's ice sheet was shedding ice five times faster than it had during the 1990s. "It is breaking up from below," one hunter told Ehrlich. "If the ice goes, it will be a disaster. Without ice we are nothing" (Ehrlich 2015, 41).

Rotting ice destroys lives in Greenland and elsewhere in the Arctic. As Ehrlich described,

Jens [a hunter] explained that only the shore-fast ice was strong enough for a dogsled, that hunting had been impossible all winter. Despondent, he left. I heard rifle shots. What was that? I asked. "Some of the hunters are shooting their dogs because they have nothing to feed them," I was told. A fifty-pound bag of dog food costs more than the equivalent of fifty U.S. dollars. One bag lasts two dogs for ten days. We stared at the rotting ice. It was down there that a modern shaman named Panippaq, who was said to be capable of heaping up mounds of fish at will, had committed suicide as he watched the sea ice decline. (Ehrlich 2015, 46, 47)

"Additional global warming of 2 to 3 [°C] under a business-as-usual scenario would be expected to yield about 5 [°C] additional warming over Greenland," said James E. Hansen, retired director of NASA's Goddard Institute for Space Studies. "Such a level of additional warming would spread summer melt over practically the entire ice sheet and considerably lengthen the melt season. It is inconceivable that the ice sheet could long withstand such increased meltwater before entering a period of rapid disintegration, but it is very difficult to predict when such a period of large rapid change would begin" (Hansen 2006, 20).

Scientists have been learning that ice melts in nonlinear sways—and often much more quickly than it accumulates. Greenland's Jakobshavn glacier (which probably delivered the iceberg that sunk the *Titanic*) is four miles wide and 1,000 feet thick and shunts more ice into the sea than any other—and that pace has been accelerating. For a decade beginning in the 1990s, it doubled its speed, delivering an average of 11 cubic miles of ice to the sea each year. The glacier's ice tongue, the point at which the glacier meets the ocean, also has retreated four miles since 2000 (Appenzeller 2007, 61).

Glacier Calving and Earthquakes

Greenland's mile-thick ice sheet has been sloughing off chunks of ice so large that the act of calving into the ocean (the source of nearly half of Greenland's ice loss) causes earthquakes that average 5.0 on the Richter scale. By 2014, Greenland was experiencing seven times as many of these earthquakes as in 1990. T. Murray and colleagues, writing in *Science*

> show that calving at Greenland's Helheim glacier causes a minutes-long reversal of the glacier's horizontal flow and a downward deflection of its terminus. The reverse motion results from the horizontal force caused as icebergs capsize and accelerate away from the glacier front. The downward motion results from a hydrodynamic pressure drop behind the capsizing berg, which also causes an upward force on the solid Earth. These forces are the source of glacial earthquakes, globally detectable seismic events whose proper interpretation will allow remote sensing of calving processes occurring at increasing numbers of outlet glaciers in Greenland and Antarctica. (Murray et al. 2015)

Each of these incidents involves approximately a gigaton of ice that may be 4 kilometers (more than two miles) long. By 2014, Greenland was losing an average of 378 gigatons of ice a year, enough to raise world sea levels by one millimeter. That is a miniscule portion of Greenland's ice, but melting is increasing. Melt it all, and world seas would rise some 6 meters or 20 feet.

Elevation Influences Melting Rates

Greenland's ice also melts at an accelerating rate because of a temperature–elevation feedback. As ice melts, its elevation declines into relatively warmer air, which increases the speed of melting. Alexander Robinson, a scientist at the Complutense University of Madrid, told Jon Gertner of *The New York Times* (2015) that once the effects of lower elevation are combined with changes in albedo (as dust and dirt concentrate on the surface of the ice, it becomes darker as it melts, which increases its absorption of heat), "You're imposing a process that won't stop, that becomes irreversible, unless you really cool the climate in the other direction. And I think we're getting close to temperatures where it becomes a danger that Greenland becomes auto-sustainable, that it can just melt itself." Robinson thinks we will cross the threshold within the next few decades.

Ice Melt in Winter

By 2005, some ice melted in Greenland even during December. "We have never seen that," glaciologist Konrad Steffen said. "It is significantly warmer now, and it happened quite suddenly. This year [2005], the temperatures were warmer than I have ever experienced [before]" (Hotz 2006).

Melting ice is changing the geography of Greenland, with islands never previously mapped emerging from the melting ice cap. "We are already in a new era of geography," said Arctic explorer Will Steger. "This phenomenon—of an island all of a sudden appearing out of nowhere and the ice melting around it—is a real common phenomenon now." With 27,555 miles of coastline and thousands of fjords, inlets, bays and straits, Greenland's geography is becoming obsolete almost as soon as new maps are created. Routes once pioneered on a dogsled are routinely paddled in a kayak now. Many features such as the Ward Hunt ice shelf in Greenland's northwest have disappeared for good. In August 2006, Steger discovered a new island himself off the coast of Svalbard. "We saw it ourselves up there, just how fast the ice is going," he said (Rudolf 2007).

Scientists who think they are looking at solid ice now often find mazes of tunnels, and cracks below the surface, melting ice at work. "We are witnessing enormous changes, and it will take some time before we understand how it happened, although it is clearly a result of warming around the glaciers," said Eric Rignot, a scientist at California Institute of Technology's Jet Propulsion Laboratory (Vedantam 2006).

Greenland's Ice Melt Reinforces Itself

The nature of how ice melts is also accelerating the speed of erosion of parts of Greenland's ice cap. As Horst Machguth of the Geological Survey of Iceland and Denmark and colleagues explained in *Nature Climate Change* (2016):

> Approximately half of Greenland's current annual mass loss is attributed to runoff from surface melt. At higher elevations, however, melt does not necessarily equal runoff, because meltwater can refreeze in the porous near-surface snow and firn. Two recent studies suggest that all or most of Greenland's firn pore space is available for meltwater storage, making the firn an important buffer against contribution to sea level rise for decades to come. Here, we employ in situ observations and historical legacy data to demonstrate that surface runoff begins to dominate over meltwater storage well before firn pore space has been completely filled. Our observations frame the recent exceptional melt summers in 2010 and 2012, revealing significant changes in firn structure at different elevations caused by successive intensive melt events. In the upper regions (more than about 1,900 m above sea level), firn has undergone substantial densification, while at lower elevations, where melt is most abundant, porous firn has lost most of its capability to retain meltwater. Here, the formation of near-surface ice layers renders deep pore space difficult to access, forcing meltwater to enter an efficient surface discharge system and intensifying ice sheet mass loss earlier than previously suggested. (Machguth et al. 2016)

Scientists' understanding of how the melting of ice accelerates is still developing. "Until recently," Chelsea Harvey wrote in the *Washington Post* (2016),

> many scientists have assumed that most of Greenland's firn space is still available for trapping meltwater. But the new research shows that this is likely no longer the case. Through on-the-ground observations, the scientists have shown that the recent formation of dense ice layers near the ice sheet's surface are making it more difficult for liquid water to percolate into the firn— meaning it's forced to run off instead. (Harvey 2016)

"I think the most notable result of our study is showing that the firn reacts faster to an atmospheric warming than expected," Machguth said. By examining the cores, the researchers found that the deluge of meltwater in recent years had trickled into the firn and frozen into chunks called *ice lenses*. These lenses then began to hinder any additional liquid water from trickling down through the firn, meaning the meltwater began to accumulate and freeze near the surface, increasing the number and thickness of the existing lenses in a kind of vicious cycle (Harvey 2016).

These changes to the firn are probably irreversible. "While new firn can form as more snow falls and accumulates on Greenland's surface, the process can take decades—and might not be able to occur at all in a warming climate," Harvey wrote (2016).

Arctic Ice Loss and Greenland Melting

The melting of ice in Greenland may be accelerated by loss of the Arctic ice cap. Melting ice in the Arctic Ocean may be associated with the summer formation of long-lived, nearly stationary high-pressure systems in the upper atmosphere known as *blocking highs*. Such a pressure pattern formed north of Greenland during the summer of 2015 as northern Greenland experienced record melting with summer temperatures rising as high as 19°C (66°F). According to Eli Kintisch writing in *Science* (2016),

> Such atmospheric blocks are expected to result from melting sea ice, some researchers say—a claim that has added fuel to a contentious dispute over the global influence of the warming Arctic. Until now, the dispute has focused on how disappearing sea ice might be favoring extreme mid-latitude weather, such as floods in Texas or heat waves in Russia. The new study . . . [has] expanded the debate to the melting of Greenland.

Further Reading

Appenzeller, Tim. "The Big Thaw." *National Geographic*, June 2007, 56–71.
Ehrlich, Greterl. "Rotten Ice: Traveling by Dogsled in the Melting Arctic." *Harper's*, April 2015, 41–50.
Faris, Stephan. "Ice Free." *New York Times Sunday Magazine*, July 27, 2008. http://www.nytimes.com/2008/07/27/magazine/27wwln-phenom-t.html.

Gertner, Jon. "The Secrets in Greenland's Ice Sheet." *New York Times Sunday Magazine,* November 15, 2015. http://www.nytimes.com/2015/11/15/magazine/the-secrets-in -greenlands-ice-sheets.html.

"Greenland Melt Ponds." NASA Earth Observatory. March 21, 2013. http://earthobservatory .nasa.gov/IOTD/view.php?id=80677&src=eoa-iotd.

Hansen, James E. "Declaration of James E. Hansen." *Green Mountain Chrysler-Plymouth-Dodge-Jeep, et al., Plaintiffs v. Thomas W. Torti, Secretary of the Vermont Agency of Natural Resources, et al., Defendants.* Case Nos. 2:05-CV-302 and 2:05-CV-304, Consolidated. United States District Court for the District of Vermont. August 14, 2006. http://www .giss.nasa.gov/~dcain/recent_papers_proofs/vermont_14aug20061_textwfigs.pdf (no longer available).

Hansen, Kathryn. "Researchers Warm Up to Melt's Role in Greenland Ice Loss." April 27, 2008. NASA.gov. https://www.nasa.gov/centers/goddard/news/topstory/2008/greenland _speedup.html.

Harvey, Chelsea. "What Scientists Just Discovered in Greenland Could Be Making Sea-Level Rise Even Worse." *Washington Post*, January 4, 2016. https://www.washingtonpost.com /news/energy-environment/wp/2016/01/04/what-scientists-discovered-in-greenland could-be-making-sea-level-rise-even-worse/?wpmm=1&wpisrc=nl_headlines.

Hotz, Robert Lee. "Greenland's Ice Sheet Is Slip, Sliding Away." *Los Angeles Times*, June 24, 2006, n.p.

Kintisch, Eli. "Sea Ice Retreat Said to Accelerate Greenland Melting." *Science* 352 (June 17, 2016): 1377.

Machguth, Horst, et al. "Greenland Meltwater Storage in Firn Limited by Near-Surface Ice Formation." *Nature Climate Change* (2016). http://www.nature.com/nclimate/journal/vaop /ncurrent/full/nclimate2899.html. doi: 10.1038/nclimate2899.

Murray, T., et al. "Reverse Glacier Motion during Iceberg Calving and the Cause of Glacial Earthquakes." *Science* (online), June 26, 2015. http://science.sciencemag.org/content /early/2015/06/24/science.aab0460. doi: 10.1126/science.aab0460.

"NASA Finds Warmer Ocean Speeding Greenland Glacier Melt." NASA Earth Observatory, February 16, 2009. http://earthobservatory.nasa.gov/Newsroom/view.php?id=42840&src =eoa-nnews (no longer available).

"Researchers Warm Up to Melt's Role in Greenland Ice Loss." NASA Earth Observatory, April 17, 2008. http://earthobservatory.nasa.gov/Newsroom/NasaNews/2008/200804 1726522.html (no longer available).

Rudolf, John Collins. "The Warming of Greenland." *The New York Times*, January 16, 2007. http://www.nytimes.com/2007/01/16/science/earth/16gree.html.

Vedantam, Shankar. "Glacier Melt Could Signal Faster Rise in Ocean Levels." *Washington Post*, February 17, 2006, A1.

See also: Animal Life, Arctic; Arctic Hunters; Climate Change, Greenland; Fisheries; Glacial Erosion; Global Warming, Greenland; Inuit; Sea Level Rise

ICE SCIENCE, ANTARCTIC

Measurements of 244 marine glaciers fronts on the Antarctic peninsula that have been draining into the sea (with records dating to 1945) indicate that 87 percent of them are retreating and that "a clear boundary between mean advance and retreat has migrated progressively southward" (Cook et al. 2005, 544). The same study indicates, according to the authors, that "increased drainage of the Antarctic peninsula is more widespread than previously thought" (Cook et al. 2005, 544).

Breakup of the Larsen Ice Shelves

Near the northernmost tip of the Antarctic peninsula, the Larsen ice shelves (known by scientists north–south as "A," "B," and "C") have been disintegrating since 1995, when the A shelf fell apart. Larsen B followed, losing 1,000 square miles over four years. A 1,250-square-mile section of the Larsen B ice shelf disintegrated in just 35 days, setting thousands of icebergs adrift in the Weddell Sea, and leading to its eventual collapse in 2015. At the same time, several other ice shelves in the area eroded and one, Wilkins, on the western side of the peninsula, also collapsed.

In early 1995, the Larsen A ice shelf disintegrated during a single storm after years of shrinking gradually. "The speed of the final breakup was unprecedented, and followed several of the warmest summers on record for this portion of the Antarctic," said Ted Scambos of the National Snow and Ice Data Center at the University of Colorado–Boulder ("Satellite Images" 1998). "Ice shelves appear to be good bellwethers for climate change, since they respond to change within decades, rather than the . . . centuries sometimes typical of other climate systems," said Scambos ("Satellite Images" 1998).

During 1998 and 1999, several reports described the retreat of ice along the shores of the Antarctic peninsula. Reports indicated that an additional 1,100 square miles of ice had melted during 1998. At about the same time, David Vaughan, a researcher with the British Antarctic Survey, and Scambos reported that the Larsen B and Wilkins shelves were in "full retreat" (Britt 1999). These ice sheets had been retreating slowly for some 50 years, losing around 2,700 square miles during that period. A loss of 1,100 square miles in one year (1998) thus represented a major acceleration of the ice sheets' erosion. During that year as much ice was lost as had melted during the preceding four decades.

"This may be the beginning of the end for the Larsen ice shelf," said Scambos as the ice sheet crumbled in 1999. "This is the biggest ice shelf yet to be threatened. . . . The total size of the Larsen B ice shelf is more than all the previous ice that has been lost from Antarctic ice sheets in the past two decades" ("Satellite Images" 1998). Until its dissolution, the Larsen B was the northernmost ice shelf in Antarctica, and therefore "on the front line of the warming trend," said Scambos ("Satellite Images" 1998).

Eric Rignot of the Jet Propulsion Laboratory in Pasadena, California, and Scambos of the National Snow and Ice Data Center (NSIDC) at the University of Colorado–Boulder, described the dynamics of the ice shelves' collapse. Their work indicated a two- to sixfold increase in "centerline speed" for glaciers that would be left exposed by the collapse of the Larsen B ice shelf. One glacier, Hektoria, lost 38 meters of height within six months (Scambos et al. 2004; Rignot et al. 2004). Their observations provided a glimpse of what could happen on a larger scale if other large ice shelves in Antarctica—for example, the huge Ross ice shelf—break up. "These glaciers are themselves too small to contribute any significant amount of ice to the sea-level problem," Rignot said, "but the Ross ice shelf buttresses much larger glaciers. If these are released, the impact would be much greater" (Nesmith 2004).

An analysis by Ted Scambos, glaciologist at the NSIDC, and several colleagues (2013, 441) indicated that after major Antarctic ice shelves—Larsen A and parts of B—collapsed, the remaining ice shelves continued to lose ice. Warming ocean waters driven by strong winds eroded the ice on the Antarctic peninsula and the edges of the West Antarctic ice sheet.

"We knew what was left of the ice shelf would collapse eventually, but this is staggering," said David Vaughan, a British glaciologist. "It's just broken apart. It fell over like a wall and has broken as if into hundreds of thousands of bricks" (Vidal 2002). "This is the largest single event in a series of retreats by ice shelves in the peninsula during the previous 30 years. Satellite images indicated that during 2002 another massive iceberg, larger in area than Delaware, broke away from the Thwaites ice tongue, a sheet of glacial ice that extends into the Amundsen Sea nearly a thousand miles from the Larsen ice sheet. The collapse of the Larsen B ice shelf is unprecedented during the Holocene"—that is, during the last 10,000 years—according to a scientific team writing in *Nature* (Domack et al. 2005, 681). An analysis of the ice shelves' collapse indicated, "The sudden drainage of thousands of small lakes on the surface of Antarctic glaciers seems to have triggered the spectacular collapse of the Larsen B ice shelf in March 2012" (Scambos 2013).

By 2015, 13 years after Larsen B's initial partial collapse, scientists were preparing for its total demise by 2020. The NASA Earth Observatory reported in June 2015 that a "team led by Ala Khazendar of NASA's Jet Propulsion Laboratory found that the remnant of the Larsen B ice shelf is flowing faster, becoming increasingly fragmented, and developing large cracks. Two of its tributary glaciers also are flowing faster and thinning rapidly" (Khazendar et al. 2015). "These are warning signs that the remnant is disintegrating," Khazendar said.

"Although it's fascinating scientifically to have a front-row seat to watch the ice shelf becoming unstable and breaking up, it's bad news for our planet. This ice shelf has existed for at least 10,000 years, and soon it will be gone." The free-floating remnant will shatter into hundreds of icebergs that will drift away, and the glaciers will speed up their unhindered move to the sea "What is really surprising about Larsen B is how quickly the changes are taking place," Khazendar said. "Change has been relentless" ("Last Call" 2015). "A large rift only 12 km downstream from the grounding line is currently traversing the stagnant part of the ice shelf, defining the likely front of the next large calving event. We propose that the flow acceleration, ice front retreat and enhanced fracture of the remnant Larsen B ice shelf presage its approaching demise," Khazendar and colleagues wrote (2015).

Larsen B Melting Linked to Human Activity

Scientists during 2006 reported direct evidence associating 2002's collapse of an Antarctic ice shelf to global warming. The researchers found that stronger westerly winds in the northern Antarctic peninsula, provoked mainly by rising levels of human-induced greenhouse gases, were a major contributor to dramatic summer warming that led to the retreat and collapse of the Larsen B ice shelf. "This is the first time that anyone has been able to demonstrate a physical process directly linking the breakup of the Larsen ice shelf to human activity," said lead author Gareth Marshall from the British Antarctic Survey ("Antarctic Ice Shelf" 2006). The study concluded that global warming as well as human-induced stratospheric ozone loss has changed Antarctic weather, forcing warm winds eastward and over the natural barrier created by a mile-high mountain chain on the Antarctic peninsula. The warm winds sometimes raise temperatures to 5 to 10°C ("Antarctic Ice Shelf" 2006).

These warming winds created conditions that allowed drainage of meltwater into crevasses on the Larsen ice shelf, a key process that led to its breakup. "Climate change does not impact our planet evenly—it changes weather patterns in a complex way that takes detailed research and computer modeling techniques to unravel," Marshall said. "What we've observed at one of the planet's more remote regions is a regional amplifying mechanism that led to the dramatic climate change we see over the Antarctic peninsula" ("Antarctic Ice Shelf" 2006).

Wilkins Ice Shelf Breaks Apart in Winter

In mid-2008, satellite photos from NASA showed that Antarctica's Wilkins ice shelf was close to breaking off from the Antarctica peninsula, and experts said the effects of warming there looked irreversible. In an unusual twist, this ice shelf disintegrated during the *winter*, after enduring summertime surface warmth and sunshine, which usually lead to the disintegration of ice sheets. The winter melting was probably the result of relatively warm ocean water that upwelled from under the ice.

By the end of May 2008, long, thin blocks had broken off the Wilkins ice shelf and were moving away toward the southwest. By late June 28, the portion of the ice shelf connecting to Charcot Island had narrowed, assuming an almost hourglass shape. Immediately northeast of this skinny stretch of shelf, the darker parts of the ice mélange appeared to be melting. Farther northeast, the large blocks of ice began to drift apart. A satellite image taken from space in mid-July showed the continued breakup of the ice shelf. Large, relatively intact plates of ice drifted toward the northeast from the thin piece of shelf that still stretched toward the nearby island. Large blocks of ice in the northeast continued their northward drift, some separated by areas of open water.

Another Ice Shelf's Collapse

In March 2008, another chunk of ice seven times the size of Manhattan Island collapsed into the sea near the base of the Antarctic peninsula. The 160-square mile piece of ice near the Wilkins ice shelf was holding back a larger mass of ice—about the size of Connecticut—that began to move toward the ocean. This particular chunk of ice was only one of several ice shelves adjacent to the Antarctic peninsula that collapsed into the ocean, becoming spectacular poster images for global warming advocates. Writing in *Science*, Kevin Krajick remarked that glaciologists in Antarctica "are keeping an eye on an alarming trend: sudden, explosive calving [of icebergs] in parts of Antarctica. The fear is that if this continues, it may hasten the death of glaciers at an unanticipated rate" (Krajick 2001, 2245).

Ice Melt in Antarctica, Generally

The East Antarctic ice sheet is 2.5 miles (4 kilometers) thick in places and covers 11 million square kilometers, three-quarters of Antarctica. Recent studies challenge previous assumptions about the paleoclimatic melting rates of this ice mass: "Its glaciers were thought to sit mostly above sea level, protecting them from the type of ocean-induced losses that are affecting the West Antarctic ice sheet," Douglas Fox wrote in *Science*. "Combined with several other new lines of evidence, they support the idea that parts of East Antarctica could indeed be more prone to melting than expected" (Fox 2010, 1630). NASA and the German Aerospace Center's Gravity Recovery and Climate Experiment (GRACE) mission published a study in *Nature Geoscience* (Chen et al. 2009) indicating that areas of East Antarctica (with 90 percent of Earth's ice) that had been considered stable were now beginning to lose mass.

While attention has focused on the instability of the West Antarctic ice sheet, an even larger body of ice in East Antarctica (with a drainage, or catchment, area the size of California) is being surveyed for its instability. If it melted, the Totten glacier could raise world oceans by 13 feet. This ice mass melted during the Pliocene Epoch, 3 million years ago, when atmospheric carbon dioxide was at 400 parts per million—its current level (Aitken et al. 2016, 385).

The massive East Antarctic ice sheet, with more frozen water than the rest of Earth combined, has thus far been slow to react to the world's warming climate, possibly because of thermal inertia. Its huge mass and generally elevated altitude have kept this huge climatically pivotal mass of ice largely intact into the early 21st century. A team of scientists writing in *Nature* in 2013 asserted that this stable state of affairs may not last much longer. As ice melt accelerates in West Antarctica, Greenland, and mountain glaciers worldwide, thinning has also been reported along the margin of the East Antarctic ice sheet. B. W. J. Miles and colleagues summarized their research:

> Here we present multidecadal trends in the terminus position of 175 ocean-terminating outlet glaciers along 5,400 kilometers of the margin of the East Antarctic ice sheet, and reveal widespread and synchronous changes. Despite large fluctuations between glaciers—linked to their size—three epochal patterns emerged: 63 percent of glaciers retreated from 1974 to 1990, 72 percent advanced from 1990 to 2000, and 58 percent advanced from 2000 to 2010. These trends were most pronounced along the warmer western South Pacific coast, whereas glaciers along the cooler Ross Sea coast experienced no significant changes. We find that glacier change along the Pacific coast is consistent with a rapid and coherent response to air temperature and sea-ice trends, linked through the dominant mode of atmospheric variability (the Southern Annular Mode). We conclude that parts of the world's largest ice sheet may

be more vulnerable to external forcing than recognized previously. (Miles et al. 2013, 563)

Rapid Sea Level Surges

Even without human intervention, sea levels sometimes have changed rapidly during glacial cycles throughout time. P. U. Clark and colleagues investigated sea level changes roughly 14,200 years ago that resulted in sea level surges of some 40 millimeters a year over 500 years, much more rapid than the 1-millimeter to 2-millimeter sea level rise of the 20th century (Clark et al. 2002, 2438; Sabadini 2002, 2376). These "meltwater pulses" probably originated in Antarctica, mostly from ice-sheet disintegration.

Earth's history reveals periods during which cataclysmic collapse of the West Antarctic ice sheet caused sea level rise at an average rate of one meter every 20 years for centuries. Within such periods there must have been times when the rate of sea level change was even faster. Because hundreds of large cities today are located on coastlines around the world, such abrupt ice-sheet collapse could have devastating consequences. It is impossible to specify the exact level of global warming necessary to cause such an ice-sheet collapse, but it is nearly certain that West Antarctica and Greenland's ice would disintegrate at some point if global warming approaches 3°C (Hansen 2006, 32–33).

Arctic and Antarctic Sea Ice Compared

Paradoxically, as Arctic sea ice has repeatedly set new record lows during the early 21st century, sea ice around Antarctica has been expanding to record surface area, giving joy to many climate-change contrarians. In 2012, for example, the National Snow and Ice Data Center reported that Antarctic sea ice covered 19.44 million square kilometers (7.51 million square miles); the previous record of 19.39 million kilometers (7.49 million square miles) was set in 2006.

The NASA Earth Observatory described a study by sea ice scientists Claire Parkinson and Donald Cavalieri of NASA's Goddard Space Flight Center that surveyed an Antarctic sea ice increase averaging 17,100 square kilometers per year from 1979 to 2010 (Parkinson and Cavalieri 2012). "Much of the increase . . . occurred in the Ross Sea, with smaller increases in Weddell Sea and Indian Ocean. At the same time, the Bellinghausen and Amundsen Seas have lost ice," the Earth Observatory said. "The strong pattern of decreasing ice coverage in the Bellingshausen [and] Amundsen Seas region and increasing ice coverage in the Ross Sea region is suggestive of changes in atmospheric circulation," they noted ("Antarctic Sea Ice" 2012).

Parkinson added, "Both hemispheres have considerable interannual variability, so that in either hemisphere, next year could have either more or less sea ice than this year. Still, the long-term trends are clear, but not equal: the magnitude of the ice losses in the Arctic considerably exceed the magnitude of the ice gains in the Antarctic" ("Antarctic Sea Ice" 2012).

Scientists from the University of Colorado wrote,

Comparing winter and summer sea ice trends for the two poles is problematic since different processes are in effect. During summer, surface melt and ice-albedo feedbacks are in effect; winter processes include snowfall on the sea ice, and wind. Small changes in winter extent may be a more mixed signal than the loss of summer sea ice extent. An expansion of winter Antarctic ice could be due to cooling, winds, or snowfall, whereas Arctic summer sea ice decline is more closely linked to decadal climate warming. ("Antarctic Sea Ice" 2012)

The Scope of Ice Loss

In 1998 alone, Antarctica's ice shelves lost 3,000 square miles of surface area. During March 2000, one of the largest icebergs ever observed broke off the Ross ice shelf near Roosevelt Island. Designated B-15, its initial 4,250 square mile (11,007 square kilometer) area was almost as large as the state of Connecticut.

In mid-May 2002, another massive iceberg broke off the Ross ice shelf, according to the National Ice Center in Suitland, Maryland. Named C-19 to indicate its location in the western Ross Sea, the new iceberg was the second to break from the Ross ice shelf in two weeks. On May 5, researchers spotted a new floating ice mass named C-18. It measured some 41 nautical miles long and four nautical miles wide. An iceberg 200 kilometers (120 miles) long, 32 kilometers (20 miles) wide, and 200 meters (660 feet) thick calved from the Ross ice shelf during late October 2002. The iceberg, designated C-19, is one of the biggest observed in recent years, said David Vaughan of the British Antarctic Survey ("Monster Iceberg" 2002).

Vaughan said he believed that the emergence of several extremely large icebergs from the Ross ice shelf in such a short time should not be a cause for worry about global warming.

That's the normal, natural cycle of the ice shelf. There are areas where we've seen ice shelves retreating over the last 50 years, and we think that is a response to climate change, but this is not one of those areas. The Antarctic peninsula is where climate has been changing most rapidly and where the ice shelves have been retreating. But they don't retreat like this, producing one big iceberg every now and again. They retreat by year-on-year production of lots of little icebergs, a kind of constant retreat. ("Monster Iceberg" 2002)

Further Reading

Aitken, A. R. A., et al. "Repeated Large-Scale Retreat and Advance of Totten Glacier Indicated by Inland Bed Erosion." *Nature* 533 (May 19, 2016): 385–389. http://www.nature.com/nature/journal/v533/n7603/full/nature17447.html.

"Antarctic Ice Shelf Collapse Tied to Global Warming." Environment News Service, October 16, 2006. http://www.ens-newswire.com/.

"Antarctic Sea Ice Reaches New Maximum Extent." NASA Earth Observatory, October 11, 2012. http://earthobservatory.nasa.gov/IOTD/view.php?id=79369&src=eoa-iotd.

Britt, Robert Roy. "Antarctic Ice Shelves Falling Apart." Explorezone.com, April 9, 1999. http://www.explorezone.com/archives/99_04/09_antarctic_ice.htm (no longer available).

Chen, J., et al. "Accelerated Antarctic Ice Loss from Satellite Gravity Measurements." *Nature Geoscience* 2 (November 22, 2009): 859–862.

Clark, P. U., et al. "Sea-Level Fingerprinting as a Direct Test for the Source of Global Meltwater Pulse." *Science* 295 (March 29, 2002): 2438–2441.

Cook, A. J., et al. "Retreating Glacier Fronts on the Antarctic Peninsula over the Past Half-Century." *Science* 308 (April 22, 2005): 541–544.

Domack, Eugene, et al. "Stability of the Larsen B Ice Shelf on the Antarctic Peninsula during the Holocene Epoch." *Nature* 436 (August 4, 2005): 681–685.

Fox, Douglas. "Could East Antarctica Be Headed for Big Melt?" *Science* 328 (June 25, 2010): 1630–1631.

Hansen, James E. "Declaration of James E. Hansen." *Green Mountain Chrysler-Plymouth-Dodge-Jeep, et al., Plaintiffs v. Thomas W. Torti, Secretary of the Vermont Agency of Natural Resources, et al., Defendants.* Case Nos. 2:05-CV-302 and 2:05-CV-304, Consolidated. United States District Court for the District of Vermont. August 14, 2006. http://www.giss.nasa.gov/~dcain/recent_papers_proofs/vermont_14aug20061_textwfigs.pdf (no longer available).

Khazendar, Ala, et al. "The Evolving Instability of the Remnant Larsen B Ice Shelf and Its Tributary Glaciers." *Earth and Planetary Science Letters* 419(1) (June. 2015): 199–210. http://www.sciencedirect.com/science/article/pii/S0012821X1500151X.

Krajick, Kevin. "Tracing Icebergs for Clues to Climate Change." *Science* 292 (June 22, 2001): 2244–2245.

"Last Call for Larsen B." NASA Earth Observatory, June 9, 2015. http://earthobservatory.nasa.gov/IOTD/view.php?id=86002&src=eoa-iotd.

Miles, B. W. J., et al. "Rapid, Climate-Driven Changes in Outlet Glaciers on the Pacific Coast of East Antarctica." *Nature* 500 (August 29, 2013): 563–566.

"Monster Iceberg Heads into Antarctic Waters." Agence France Presse, October 22, 2002 (LEXIS).

Nesmith, Jeff. "Antarctic Glacier Melt Increases Dramatically." *Atlanta Journal-Constitution,* September 22, 2004, 9A.

Parkinson, C. L., and D. J. Cavalieri. "Antarctic Sea Ice Variability and Trends, 1979–2010." *The Cryosphere* 6 (August 15, 2012): 871–880.

Rignot, E., et al. "Accelerated Ice Discharge from the Antarctic Peninsula Following the Collapse of Larsen B Ice Shelf." *Geophysical Research Letters* 31(18) (September 22, 2004). doi: 10.1029/2004GL020697.

Sabadini, Roberto. "Ice Sheet Collapse and Sea Level Change." *Science* 295 (March 29, 2002): 2376–2377.

"Satellite Images Show Chunk of Broken Antarctic Ice Shelf." April 16, 1998. http://www.eurekalert.org/releases/brkantartice.html. (no longer available).

Scambos, T. A., et al. "Glacier Acceleration and Thinning after Ice Shelf Collapse in the Larsen B Embayment, Antarctica." *Geophysical Research Letters* 31(18) (September 22, 2004). doi: 10.1029/2004GL020670.

Scambos, Ted. "Anatomy of an Ice Shelf's Demise." *Nature* 503 (November 28, 2013): 441.

Vidal, John. "Antarctica Sends Warning of the Effects of Global Warming: Scientists Stunned as Ice Shelf Falls Apart in a Month." *The Guardian* (U.K.), March 20, 2002, 3.

INLAND COOLING, ANTARCTICA

Global warming is not linear, and not all regions warm evenly over time. Parts of Antarctica have cooled as others have experienced record high temperatures. Episodes of cooling, especially in some of Antarctica's inland valleys, have been fodder for political critics of global warming who insist that these isolated phenomena are "proof" that a steady rise in greenhouse gas levels worldwide cannot be associated with human actions.

Cooling Has Been Limited

Detailed analyses of Antarctic temperature records for half a century indicate that cooling there has been limited. These studies maintain that assertions of inland warming have been "substantially incomplete owing to the sparseness and short duration of the observations" (Steig et al. 2009, 459). Eric J. Steig and colleagues writing in *Nature*, have instead found that "[significant] warming extends well beyond the Antarctic peninsula to cover most of West Antarctica, an area of warming much larger than previously reported. West Antarctic warming exceeds 0.1°C per decade over the past 50 years, and is strongest in winter and spring" (Steig et al. 2009, 459).

A later study indicated more pronounced warming. According to David H. Bromwich and colleagues, writing in *Nature Geoscience* (2012), reports from Byrd Station in central West Antarctica reveal

> a linear increase in annual temperature between 1958 and 2010 by 2.4 ± 1.2°C (4.4°F) establishing central West Antarctica as one of the fastest-warming regions globally. . . . In contrast to previous studies, we report statistically significant warming during austral summer, particularly in December–January, the peak of the melting season. A continued rise in summer temperatures could lead to more frequent and extensive episodes of surface melting of the West Antarctic ice sheet. These results argue for a robust long-term meteorological observation network in the region. (Bromwich 2012)

The scientists included satellite measurements to interpolate temperatures in the vast areas between Antarctic weather stations. "We now see warming is taking place on all seven of the earth's continents in accord with what models predict as a response to greenhouse gases," said Steig, a professor of earth and space sciences at the University of Washington in Seattle (Chang 2009). The cooling that has been recorded in parts of Antarctica, however, cannot be traced directly to a more general climatic pattern, according to Steig and colleagues. Writing in *Nature*, they said, "Instead, regional changes in atmospheric circulation and associated changes in sea surface temperature and sea ice are required to explain the enhanced warming in West Antarctica" (Steig et al. 2009, 459).

Cooling Inland Valleys

In an apparent contradiction of temperature trends across most of the world, some areas of interior Antarctica have cooled steadily for more than two decades at the same time areas on the fringes of the continent have warmed rapidly. Data assembled by Peter Doran, an associate professor of earth and environmental sciences at the University of Illinois–Chicago, found that temperatures in the McMurdo Dry Valleys of East Antarctica have declined at a rate of 1.2°F per decade since 1986. Similar trends have been observed across the continent's interior since 1978.

The apparent cooling of inland Antarctica has been used by climate skeptics to refute global warming as an idea, much to the consternation of scientists involved in research. Doran and other members of the National Science Foundation's Long-Term Ecological Research Team assembled temperature data in the Dry Valleys near McMurdo Sound, a snow-free mountainous desert of chill and arid soils, bleak bedrock outcroppings, and ice-covered lakes that is home to many microscopic invertebrates, mostly nematodes (Gugliotta 2002).

Doran stressed that although scientists could not explain the falling temperatures, his research "does not change the fact that the planet has warmed up on the whole. The findings simply point out that Antarctica is not responding as expected" (Gugliotta 2002). Doran also warned, "You don't want to overstate the effects of the cooling trend" (Gugliotta 2002).

Measuring Temperatures in the Dry Valleys

The fragile ecosystem of the Dry Valleys requires four to six weeks of above-freezing temperatures during the southern summer, Doran said. This period of relative warmth causes meltwater from hillside glaciers to cascade downward in seasonal arroyos that feed life in local lakes. Researchers have found that temperatures had been dropping, not rising, since 1986, with the most pronounced declines in summer and autumn. Glacial ice has not been melting, so the streams have not been flowing, lakes have been shrinking, and microorganisms have been disappearing (Doran et al. 2002, 517).

Doran said his team also studied data collected since 1966 from permanent installations throughout the Antarctic. Previous studies had indicated overall warming, but the researchers found that these calculations relied disproportionately on readings from the Antarctic peninsula. When the researchers corrected for this distortion, they found that some parts of Antarctica had become colder. "Temperatures were rising between 1966 and 1978," Doran said, but then they started to fall and have continued falling ever since (Doran et al. 2002, 517).

> Our spatial analysis of Antarctic meteorological data demonstrates a net cooling on the Antarctic continent between 1966 and 2000, particularly during summer and autumn. The McMurdo Dry Valleys have cooled by 0.7°C per decade between 1986 and 2000, with similar pronounced seasonal trends. Summer cooling is particularly important to Antarctic terrestrial ecosystems that are poised on the interface of ice and water. . . . Continental Antarctic

cooling, especially the seasonality of cooling, poses challenges to models of climate and ecosystem change. (Doran et al. 2002, 517)

Doran and colleagues did not venture speculation on causes of the temperature decline. They do know that temperatures in the Dry Valleys rise when the wind blows and clouds cover the sky. Doran explained that as winds roll downhill off the Antarctic plateau into the Dry Valleys, the air compresses and heats up as a result, an effect similar to the Chinook winds of the western United States or the dry and warm Santa Ana winds of Southern California (Gugliotta 2002). At the same time, relatively warm summer winds gathering speed over the ocean bring warmer air in from the coast to promote the thaw, he said. Wind generally brings clouds that add to the warming, Doran added (Gugliotta 2002). Recently, however, "We're getting a decrease in winds from both directions." He said that perhaps as a consequence, temperatures in the Dry Valleys are dropping. "It's clearly connected to the winds, but what's controlling the decrease in the winds is not clear" (Gugliotta 2002).

Once the results from Doran and colleagues were made public, several newspaper reports rushed to simplify them into a worldwide cooling trend, taking this news as a reason to refute global warming. The authors of the studies expressed caution. One of the scientists involved in studies that indicated that the Ross ice shelf was thickening, Slawek Tulaczyk of the University of California–Santa Cruz, said that press misinterpretations left him increasingly frustrated by sometimes careless media coverage of the global warming issue.

Data Distortion by Contrarians

When Tulaczyk and Ian Joughin of the NASA's Jet Propulsion Laboratory in Pasadena, California, reported in *Science* that the movement of glacial ice streams appeared to be slowing on the Ross ice shelf, which was allowing the ice to thicken, a headline over an editorial in the *San Diego Union-Tribune* minced no words: "Scientific Findings Run Counter to Theory of Global Warming." The editorial sarcastically asked, "Oh dear. What will the doomsayers say now? How will they explain away yet two more scientific studies that clearly contradict the global warming orthodoxy?" (Davidson 2002). A headline in the *National Post*, a generally right-wing Canadian newspaper, declared, "Antarctic Ice Sheet Has Stopped Melting, Study Finds" (Davidson 2002). "Is Another Ice-Age on the Way?" asked an editorial in the *Rocky Mountain News* (Davidson 2002). Analyzing these reports in the *San Francisco Chronicle*, Keay Davidson commented, "Some media mistakenly equated the phenomenon studied by Joughin and Tulaczyk—a change in ice flow rates—with ice melting rates. The mistake contributed to the erroneous belief that the studies constituted, as it were, scientific 'tests' of the global warming theory" (Davidson 2002).

Contrary to some news reports, "the ice-sheet growth that we have documented in our study area has absolutely nothing to do with any recent climate trends," Tulaczyk said, emphasizing those words in an e-mail to the *San Francisco Chronicle*

(Davidson 2002). The thickening of Antarctic ice in certain regions—especially "Ice Stream C" of the Whillans ice stream, adjacent to the Ross ice shelf—results from the complex internal dynamics of the ice itself. These particular ice-flow changes were unrelated to global warming caused by combustion of fossil fuels; such changes occurred for many millennia before the Industrial Revolution boosted atmospheric levels of heat-trapping gases. The area with the greatest ice thickening is on an ice stream that stopped flowing some 150 years ago (Davidson 2002).

"I keep repeating to journalists that climate science is much like economics. Both deal with complex systems," Tulaczyk observed. "Just as a single stock going up or down cannot be interpreted as a reliable indicator of economic recovery or collapse, we have to accept the occurrence of contradictory trends in the global climate" (Davidson 2002). Contrary to some reports attributed to his research, "global warming is real and happening right now," asserted Doran. The cooling trend in Antarctica, he continued, appears to be a surprising, regional exception to the overall planetary warming (Davidson 2002).

Antarctic Inland Cooling in Global Context

Doran emphasized: "Our paper does not change the global [temperature] average in any significant way. . . . Although we have said that more area of the continent is cooling than warming, one just has to look at the paper itself . . . to see that it is a close call" (Davidson 2002). "Our analysis suggests that about two-thirds of the main continent has been cooling in the last 35 years," Doran continued. "But there is one-third of the continent that has been warming if you remove the [Antarctic] peninsula. And with the peninsula included, it shrinks to 58 percent cooling" (Davidson 2002). Doran bluntly advised the public: "If you want the facts, you have to go to the original scientific peer-reviewed literature, and avoid the broken-telephone effect of the popular press" (Davidson 2002).

Doran made further comments on the op-ed page of *The New York Times*:

[Misinterpretation] had already become legend, and in the four and half years since, it has only grown. Our results have been misused as "evidence" against global warming by Michael Crichton in his novel *State of Fear* and by Ann Coulter in her latest book, *Godless: The Church of Liberalism*. Search my name on the Web, and you will find pages of links to everything from climate discussion groups to Senate policy committee documents—all citing my 2002 study as reason to doubt that the Earth is warming. One recent Web column even put words in my mouth. I have never said that "the unexpected colder climate in Antarctica may possibly be signaling a lessening of the current global warming cycle." I have never thought such a thing either. (Doran 2006)

His data, Doran said, did find that 58 percent of Antarctica cooled from 1966 to 2000. During that period, however, the rest of the continent was warming. Climate models created after Doran's study was released "have suggested a link between the lack of significant warming in Antarctica and the ozone hole over that continent.

These models, conspicuously missing from the warming-skeptic literature, suggest that as the ozone hole heals—thanks to worldwide bans on ozone-destroying chemicals—all of Antarctica is likely to warm with the rest of the planet. An inconvenient truth?" (Doran 2006).

An Unusual Warm Spell

In 2007, scientists from NASA's Jet Propulsion Laboratory and the Cooperative Institute for Research in Environmental Sciences at the University of Colorado–Boulder announced that an unusual warm spell had traversed parts of Antarctica two years earlier, covering an area the size of California. Using its QuikSCAT satellite, the team measured snowfall accumulation and melting in Antarctica and Greenland from July 1999 through July 2005. The research was led by Son Nghiem of JPL in Pasadena and Konrad Steffen, director of the Cooperative Institute for Research in Environmental Sciences.

Steffen further noted that Antarctica as a whole has shown little warming (even slight cooling in some areas) in the recent past with the exception of the Antarctic peninsula (which is warming rapidly). By 2007, however, large regions of the continent were showing the first signs of warming's impacts as interpreted by this satellite analysis. "Increases in snowmelt, such as this in 2005, definitely could have an impact on larger-scale melting of Antarctica's ice sheets if they were severe or sustained over time," he said ("NASA" 2007).

The observed melting occurred in several regions, some far inland at high latitudes (in one instance within 310 miles of the South Pole) and at elevations as high as 6,100 feet, where melt had heretofore never been seen. Temperatures in the affected areas rose above the freezing point and in some cases remained there for a week or so, judging from the satellite's survey of snowfall that had turned to ice from melting and refreezing.

Maximum air temperatures at the time of the melting were unusually high, reaching more than 5°C (41°F) in one affected area. They remained above melting for around a week.

Further Reading

Bromwich, David H., et al. "Central West Antarctica among the Most Rapidly Warming Regions on Earth." *Nature Geoscience*, December 23, 2012. doi: 10.1038/ngeo1671.

Chang, Kenneth. "Study Finds New Evidence of Warming in Antarctica." *The New York Times*, January 22, 2009. http://www.nytimes.com/2009/01/22/science/earth/22climate.html.

Davidson, Keay. "Media Goofed on Antarctic Data; Global Warming Interpretation Irks Scientists." *San Francisco Chronicle*, February 4, 2002, A8.

Doran, Peter T., et al. "Antarctic Climate Cooling and Terrestrial Ecosystem Response." *Nature* 415 (January 30, 2002): 517–520.

Doran, Peter. "Cold, Hard Facts." *The New York Times*, July 27, 2006. http://www.nytimes.com/2006/07/27/opinion/27doran.html.

Gugliotta, Guy. "In Antarctica, No Warming Trend; Scientists Find Temperatures Have Gotten Colder in Past Two Decades." *Washington Post*, January 14, 2002, A2.

"NASA: Vast Areas of West Antarctica Melted in 2005." Environment News Service, May 28, 2007. http://www.ens-newswire.com/ens/may2007/2007-05-28-09.asp#anchor1 (no longer available).

Steig, Eric J., et al. "Warming of the Antarctic Ice-Sheet Surface Since the 1957 International Geophysical Year." *Nature* 457 (January 22, 2009): 459–462.

See also: Atmospheric Circulation; Ice Melt, Antarctica; Global Warming, Antarctica; Ice Shelves, Antarctic; Ocean Circulation; Oceans' Absorption of Heat; Sea Level Rise; Temperatures, Greenhouse Gas Levels and

NORTHEAST AND NORTHWEST PASSAGES

European mariners, including Sir Francis Drake and Captain James Cook, sought and failed to find the fabled Northwest Passage beginning in 1497, when English King Henry VII sent Italian explorer John Cabot to look for a route from Europe to the Orient that would avoid the southern tip of Africa. According to a report released in 2002, Gary Brass, director of the U.S. Arctic Research Commission, said that both the Northwest Passage and the Northern Sea Route (Northeast Passage) could be open soon to vessels lacking reinforcement against the ice for at least a month in the summer.

Both routes soon may be open for the entire summer, according to this report (Kerr 2002, 1490). German explorers asserted in October 2002 that global warming and unusual wind patterns had enabled them to become the first navigators to sail unaided in a yacht through the usually icebound route along Russia's Arctic coast. A 12-man team led by explorer Arved Fuchs, 49, made the trip on its fourth attempt west-to-east during the summer in a 60-foot sail-driven, wooden trawler that they had converted into a yacht.

NASA's Advanced Microwave Scanning Radiometer aboard the Aqua satellite observed open water along nearly the entire route on August 22, 2007. "Although nearly open, the Northwest Passage was not necessarily easy to navigate in August 2007," NASA noted. "Located 800 kilometers (500 miles) north of the Arctic Circle and less than 1,930 kilometers (1,200 miles) from the North Pole, this sea route remains a significant challenge, best met with a strong icebreaker ship backed by a good insurance policy" ("Northwest Passage" 2007). By 2011, both routes became a summer reality as the Arctic ice cap steadily melted. Russian tugboats were traversing the country's northern coast rather routinely, as described by Andrew E. Kramer of *The New York Times*:

Rounding the northernmost tip of Russia in his oceangoing tugboat this summer, Capt. Vladimir V. Bozanov saw plenty of walruses, some pods of beluga whales and in the distance a few icebergs. One thing Captain Bozanov did not encounter while towing an industrial barge 2,300 miles across the Arctic Ocean was solid ice blocking his path anywhere along the route. Ten years ago, he said, an ice-free passage, even at the peak of summer, was exceptionally rare. (Kramer 2011)

As ice melted, these shipping channels widened. With the tugboats came explorers for oil and other resources that long had been locked under the ice at all seasons. ExxonMobil signed a contract to explore the Russian sector of the Arctic Ocean. Rosneft, the Russian state oil company, signed a partnership with ExxonMobil. The irony here is that melting ice caused by combustion of fossil fuels creates an opportunity to find and burn more of them. "It is paradoxical that new opportunities are opening for our nations at the same time we understand that the threat of carbon emissions have become imminent," said Iceland's president, Olafur Ragnar Grimsson (Kramer 2011).

At the same time, Russian Prime Minister Vladimir V. Putin heartily endorsed the new opportunities, regardless of the climatic consequences. "The Arctic is the shortcut between the largest markets of Europe and the Asia-Pacific region," he said. "It is an excellent opportunity to optimize costs" (Kramer 2011). The Northeast Passage cuts 4,000 to 5,000 miles off a more southerly route between Western Europe and Japanese or Chinese ports, with attendant savings in time, payroll, and fuel, thus saving shippers hundreds of thousands and perhaps millions of dollars per trip. Ships also use the route without contracting expensive services from icebreakers.

The Russians have been occasionally cutting a path along this route for a century by searching for favorable summer conditions and remaining as close to land as possible because it is the only route across trackless northern Siberia. Ships have competed with each other to break the speed record for traversing the northeast passage. During the summer of 2011, a tanker carrying natural gas condensate did it in six and a half days, besting the previous record of eight. After the Soviet Union collapsed in 1991, the route opened to freight haulers and prospectors from several nations. Larger ships with deeper drafts began to ply the route as more ice receded year by year. In 2009, two commercial cargo ships made the trip, and 18 did so the next year. After that, the route became rather routine.

Cruise ships have been offering Arctic views on routes from Murmansk to Anadyr, another Russian port not far from Alaska on the Bering Sea. "The voyage offered attractions such as abandoned Russian polar stations," the Australian operator, Aurora Expeditions, advertised (Kramer 2011).

The Northwest Passage

During late August 2007, for the first time, a northwest passage from Baffin Bay to northern Alaska opened during a season of record ice melt for the Arctic ice cap. As ice continued to melt during ensuing summers, this journey became more and more routine. In September 2008, shipping passages opened briefly on both sides of the Arctic ice cap, a Northwest *and* a Northeast Passage, at the same time, the centuries old shoppers' dream came true.

In July 2009, two German ships left South Korea bound for the Netherlands carrying 3,500 tons of construction materials. They were scheduled to arrive in Rotterdam in mid-September by traversing the Arctic via the Northeast Passage, the first-ever commercial traffic over the top of the world. Two tankers (one Finnish and one Latvian) also used the route. "It is global warming that enables us to think about using that route," said Verena Beckhusen, a spokeswoman for shipping company Beluga Group, headquartered in Bremen, Germany (Kramer and Revkin 2009).

By 2010, because of melting ice, the Kodiak-Kenai Cable Co. was making plans to lay fiber-optic cable across the Arctic between London and Tokyo. The route, called ArcticLink, is half the length of others, cutting transmission time almost in half from 140 to 88 milliseconds (Loss 2010). The cable was completed in 2016.

Industrialists Enjoy Open Water

Some industrialists are gleefully anticipating the melting of polar ice so they can open new shipping lanes and ports, drill for oil, or engage in other profitable activities. Bad news for polar bears, for example, may be good for Pat Broe, a Denver entrepreneur, who bought the derelict port of Churchill, Manitoba, on Hudson's Bay, from the Canadian government in 1997 for around $10 Canadian. By Broe's calculations, once the Arctic ice cap melts, Churchill could bring in as much as $100 million a year as a port on Arctic shipping lanes between Europe, Asia, and the Americas—the fabled Northwest Passage—which would be shorter by thousands of miles than current more southerly routes (Krauss et al. 2005).

Oil companies have been pushing into the frigid Barents Sea seeking undersea oil and gas fields made accessible not only by melting ice but also by advances in technology. But now, as thinning ice stands to simplify construction of drilling rigs, exploration is likely to move even farther north (Krauss et al. 2005). In 2004, scientists found evidence of oil in samples taken from the floor of the Arctic Ocean only 200 miles from the North Pole. All told, one-quarter of the world's undiscovered oil and gas resources lies in the Arctic, according to the U.S. Geological Survey (Krauss et al. 2005).

Tourists have also been swarming to the Arctic. One day during the summer of 2005, residents of Pangnirtung on the east coast of Baffin Island were greeted by a surprise: a 400-foot European cruise ship that had dropped anchor unannounced and sent several hundred tourists ashore in small boats.

Further Reading

Kerr, Richard A. "A Warmer Arctic Means Change for All." *Science* 297 (August 30, 2002): 1490–1492.

Kramer, Andrew E. "Warming Revives Dream of Sea Route in Russian Arctic." *The New York Times*, October 18, 2011. http://www.nytimes.com/2011/10/18/business/global/warming-revives-old-dream-of-sea-route-in-russian-arctic.html.

Kramer, Andrew E., and Andrew C. Revkin. "Arctic Shortcut Beckons Shippers as Ice Thaws." *The New York Times*, September 11, 2009. http://www.nytimes.com/2009/09/11/science/earth/11passage.html.

Kraus, Clifford, et al. "As Polar Ice Turns to Water, Dreams of Treasure Abound." *The New York Times*, October 10, 2005. http://www.nytimes.com/2005/10/10/science/10arctic.html.

"Loss of Arctic Ice Opens New Cable Route." *Omaha World-Herald,* January 22, 2010, A4.
"Northwest Passage Nearly Open." NASA Earth Observatory, August 27, 2007. https://archive.org/details/npseaice_amsre_2007234.

See also: Carbon Cycle Feedbacks; Climate Change, Greenland; Ice Melt, Arctic; Sea Ice, Arctic; Summer Ice, Arctic

OZONE DEPLETION AND WARMING, ANTARCTIC

When chlorofluorocarbons (CFCs) were banned in the late 1980s, most experts expected ozone depletion over the Antarctic to be corrected rather quickly. By 2005, however, ozone depletion remained a major problem, and an ozone "hole" was beginning to open over the Arctic as well. The nature of science has evolved since then to explain how the capture of heat near Earth's surface by greenhouse gases speeds cooling in the stratosphere and plays an important role in continuing ozone depletion at that level. Although a single atom of chlorine from a CFC can destroy more than 100,000 ozone molecules (Flannery 2005, 216), these human-created chemicals do more than destroy stratospheric ozone. They also act as greenhouse gases, with several thousand times the per-molecule greenhouse potential of carbon dioxide.

Ozone Depletion and Skin Cancer

In early 2009, a study by Professor Darryn Waugh of the Department of Earth and Planetary Sciences at Johns Hopkins University and his colleagues at the NASA Goddard Space Flight Center in Greenbelt, Maryland, confirmed that increasing retention of heat near Earth's surface would delay healing of stratospheric ozone, in part by distorting wind circulation in the lower stratosphere of the tropical and southern midlatitudes, causing rising incidence of skin cancers in Australia and South America. Ozone levels over these areas may never return to safe levels until excess greenhouse forcing near the surface abates, Waugh and colleagues said.

In tropical and southern midlatitudes, Waugh said "[global] warming causes changes in the speed that the air is transported into and through the lower stratosphere. You're moving the air through it quicker, so less ozone gets formed" ("Global Warming Forecast" 2009). The scientists used computer simulations called the Goddard Earth Observing System Data and Information System (EOSDIS) for research published in *Geophysical Research Letters* in February 2009.

"The risk of skin cancer for fair-skinned populations living in countries like Australia and New Zealand, and probably in Chile and Argentina too, will be greater in the 21st century than it was during the 20th century," said Dan Lubin, an atmospheric scientist at Scripps Institution of Oceanography in La

Jolla, California, who did not participate in the research ("Global Warming Forecast" 2009).

The study suggests that in northern polar regions and northern midlatitudes, ozone in the lower stratosphere escaped most of the impact from increasing greenhouse gases.

Further Reading

"Global Warming Forecast to Delay Ozone Layer Recovery." Environment News Service, February 6, 2009. http://www.ens-newswire.com/ens/feb2009/2009-02-06 -02.asp (no longer available).

Thus, the healing of the stratospheric ozone layer depends, to some degree, on reduction of greenhouse gas levels in the lower atmosphere. The ozone shield protects plant and animal life on land from the sun's ultraviolet rays, which can cause skin cancer, cataracts, and damage to the immune systems of human beings and other animals. Thinning of the ozone layer also may alter the DNA of plants and animals.

Ozone depletion over the Antarctic reaches its height in late winter and early spring as the sun rises after the midwinter night. Solar radiation triggers reactions between ozone in the stratosphere and chemicals containing chlorine or bromine. These chemical reactions occur most quickly on the surface of ice particles in clouds, at temperatures less than $-80°C$ ($-107°F$). In September 2008, the Antarctic ozone hole reached 27 million square kilometers, its maximum size for the year, an area larger than North America.

Warm at the Surface, Cold Aloft

An increasing level of carbon dioxide near Earth's surface "acts as a blanket," said NASA research scientist Katja Drdla. "It is trapping the heat. If the heat stays near the surface, it is not getting up to these higher levels" (Borenstein 2000). Deprived of emitted warmth, the stratosphere cools, which aggravates the depletion of ozone. As levels of greenhouse gases rise, the middle and upper atmosphere are expected to continue cooling, with attendant consequences for ozone depletion. Because of this relationship, problems with ozone depletion depend, in a fundamental way, on mitigation of greenhouse warming.

"The chemical reactions responsible for stratospheric ozone depletion are extremely sensitive to temperature," wrote Drew T. Shindell and colleagues in *Nature* in 1998. "Greenhouse gases warm the Earth's surface but cool the stratosphere radiatively, and therefore affect ozone depletion." Shindell et al. then expected ozone loses in the Arctic to peak in the decade 2010 to 2019 at two-thirds of the "ozone column," or roughly the same ozone loss observed in Antarctica during the early 1990s. "The severity and duration of the Antarctic ozone hole are also expected to increase because of greenhouse gas–induced stratospheric cooling over the coming decades," Shindellet al. asserted (1998, 589).

History of the "Ozone Hole"

CFCs initially raised no environmental questions when they were first marketed by DuPont Chemical during the 1930s under the trade name Freon. Few if any environmental questions were asked then. At about the same time, asbestos was being proposed as a high-fashion material for clothing, and radioactive radium was being built into timepieces so that they would glow in the dark.

Discovery of the "Ozone Hole"

In 1985, a team of scientists working with the British Antarctic Survey reported a startling decline in "column ozone values" above an observation station near Halley Bay (Farman et al. 1985). Ozone depletion had been suspected theoretically as the cause beginning in the early 1970s, and actual ozone densities had been declining over the Antarctic since 1977. The size of the decline in 1985 was a shocking surprise, however, because theorists had expected stratospheric ozone amounts to fall relatively evenly over the entire Earth.

Sherwood Rowland and Mario Molina, the first scientists to discover the ozone "hole," had expected a largely uniform decline of 1 percent to 5 percent (Rowland and Molina 1974). Scientists did not realize then how ozone depletion is related to temperature in the stratosphere. The seasonal variability of the decline was another surprise because existing theoretical models made no allowance for it. Ozone values over Antarctica tended to decline rapidly just as the sun was rising after winter. During the middle 1980s, the cause of dramatic falls in ozone density over the Antarctic was open to debate. Some scientists suspected variability in the sun's radiational output, and others suspected changes in atmospheric circulation. A growing minority began to suspect CFCs. These chemicals were not proven suspects when a majority of the world's national governments signed the Montreal Protocol to eliminate CFCs in 1987.

Definite proof of CFCs' role in ozone depletion developed shortly thereafter when J. G. Anderson and colleagues implicated the chemistry of chlorine and explained a chain of chemical reactions (later broadened to bromides as bit players), the "smoking gun" that explained why ozone depletion was so sharp and limited to specific geographic areas at a specific time of the year (Anderson et al. 1989, 11465). The temperature of the stratosphere became a key ingredient in the mix—the colder the stratosphere, the more active the chlorine chemistry that devoured ozone. By 2000, according to Maureen

Christie, ozone depletion was "significantly affecting ozone levels throughout the Southern Hemisphere" (Christie 2001, 86).

Further Reading

Anderson, J. G., W. H. Brune, and M. H. Proffitt. "Ozone Destruction by Chlorine Radicals within the Antarctic Vortex: The Spatial and Temporal Evolution of ClO/ O_3, Anticorrelation Based on In Situ ER-2 Data." *Journal of Geophysical Research* 94 (1989): 11465–11479.

Christie, Maureen. *The Ozone Layer: A Philosophy of Science Perspective.* Cambridge, UK: Cambridge University Press, 2001.

Farman, J. C., B. G. Gardiner, and J. D. Shanklin. "Large Losses of Total Ozone Reveal Seasonal ClO_x/NO_x Interaction." *Nature* 315 (1985): 207–210.

Rowland, Sherwood, and Mario Molina. "Stratospheric Sink for Chlorofluoromethanes: Chlorine Atom-Catalyzed Destruction of Ozone." *Nature* 249 (June 28, 1974): 810–812.

By 1976, U.S. manufacturers were producing 750 million pounds of CFCs a year and finding all sorts of uses for them—from propellants in aerosol sprays to solvents used to clean silicon chips, automobile air conditioning, and blowing agents for polystyrene cups, egg cartons, and containers for fast food. "They were amazingly useful," wrote Anita Gordon and David Suzuki. "Cheap to manufacture, non-toxic, non-inflammable, and chemically stable" (Gordon and Suzuki 1991, 24). By the time scientists discovered in the 1980s that CFCs were thinning the ozone layer over the Antarctic, they found themselves taking on industries with combined sales of $28 billion a year.

By the time CFC manufacturing was banned internationally in the late 1980s, the chemicals had been used in roughly 90 million car and truck air conditioners, 100 million refrigerators, 30 million freezers, and 45 million air conditioners in homes and other buildings. Because CFCs remain in the stratosphere for as long as 100 years, they will deplete ozone long after industrial production of the chemicals ceased.

In 1998, the Antarctic ozone hole reached a new record size—roughly that of the continental United States. Some researchers came to the conclusion that, as Richard A. Kerr describes in *Science*, "Unprecedented stratospheric cold is driving the extreme ozone destruction. . . . Some of the high-altitude chill . . . may be a counterintuitive effect of the accumulating greenhouse gases that seem to be warming the lower atmosphere. The colder the stratosphere, the greater the destruction of ozone by CFCs" (Kerr 1998).

The area affected by ozone depletion over the Antarctic roughly stabilized after 2000, increasing or decreasing year by year depending on other factors. Overall, according to NASA's Earth Observatory, the Antarctic ozone "hole" has recovered slightly from 2006, its worst year:

In 1979—when scientists were just coming to understand that atmospheric ozone could be depleted—the area of ozone depletion over Antarctica grew to 1.1 million square kilometers, with a minimum ozone concentration of 194 Dobson units. In 1987, as the Montreal Protocol was being signed, the area of the hole reached 22.4 million square kilometers and ozone concentrations dropped to 109 DU. By 2006, the worst year for ozone depletion to date, the numbers were 29.6 million square kilometers and just 84 DU. By 2011, the most recent year with a complete data set, the hole stretched 26 million square kilometers and dropped to 95 DU. ("Watching" 2012).

Antarctic Ozone Depletion Endures

The Antarctic ozone hole formed earlier and endured longer during September and October 2000 than ever before—and by a significant amount. Figures from NASA satellite measurements showed that the hole covered an area of approximately 29 million square kilometers in early September, exceeding the previous record from 1998. These record sizes persisted for several days.

Ozone levels are measured in Dobson units (DUs), which describe the columnar density of a trace gas in Earth's atmosphere. Now ozone levels fell below 100 DUs for the first time ever recorded. The area cold enough to produce ozone depletion also grew by 10 percent to 20 percent more surface area than any other year. The ozone-depletion zone was coming closer to New Zealand, where usual springtime ozone levels average around 350 DUs. During the spring of 2000, ozone levels reached as low as 260 DUs when atmospheric circulation patterns nudged the Antarctic zone northward. Scientists usually regard an area of the stratosphere as ozone depleted when DU levels fall below 220.

The decline in stratospheric ozone is striking when viewed on a graph with any sense of historical proportion. As little as a decade or two will do. Until the 1990s, springtime ozone levels in the Arctic averaged around 500 Dobson units. By 2001, they were averaging 200 to 300. In the Antarctic, in the days before the "ozone hole" (circa 1980), DU values ranged from 250 to 350; by 2000, they ranged from 100 to 200. During 2000, the ozone-depleted area over Antarctica grew to a maximum size equal to Africa. Although the polar reaches of Earth have suffered the most dramatic declines in ozone density, ozone measurements over most of the planet also have declined roughly 15 percent since the middle 1980s.

During early September 2003, early indications were that the area of depleted ozone over Antarctica was approaching near-record size again. By the end of the month, the area of severely depleted ozone was the second-largest on record at about the size of North America. The Antarctic ozone hole swelled to an area of 10 million square kilometers by late August 2005. Only the ozone holes of 1996 and 2000 had been larger (Schiermeier 2005).

The coupling of global warming near the surface with declining stratospheric temperatures continued in 2006 as ozone loss over Antarctica reached a new record, according to scientists with the European Space Agency (ESA). "Such significant ozone loss requires very low temperatures in the stratosphere combined with sunlight," said ESA atmospheric engineer Claus Zehner. "This year's extreme loss of ozone can be explained by the temperatures above Antarctica reaching the lowest recorded in the area since 1979," the beginning of record keeping ("Ozone Loss" 2006).

"There was a lot of Antarctic ozone depletion in 2013," one Earth Observatory analysis noted. "Because of above average temperatures in the Antarctic lower stratosphere, the hole was a bit below average compared to ozone holes observed since 1990," said Paul Newman, an atmospheric scientist at NASA's Goddard Space Flight Center ("Ozone Hole" 2013).

The atmospheric chemistry of stratospheric ozone continued to retard the breakdown of ozone-destroying chemicals until at least 2015, when the zone of depleted ozone above and around Antarctica was the fourth worst since 1979, the year detailed satellite measurements were first taken. That year, according to NASA's Earth Observatory, the ozone hole "spanned 28.2 million square kilometers (10.9 million square miles). . . . The largest single-day ozone hole recorded by satellite was 29.9 million square kilometers (11.5 million square miles) on September 9, 2000" ("Ozone Hole" 2015).

"In 2015," said an Earth Observatory analysis, "the hole started slowly but then quickly expanded to cover a large area. The average size in September–October 2015 was 25.6 million square kilometers (9.9 million square miles)—also the fourth largest since the start of the satellite record." The largest September–October average on record was 26.6 million square kilometers (10.3 million square miles) in 2006 ("Ozone Hole" 2015). According to NASA atmospheric scientist Paul Newman, "There are still plenty of ozone-depleting chlorine and bromine compounds present in the stratosphere. Moreover, the lower stratosphere was colder than in previous years, which creates favorable conditions for ozone-depleting chemical reactions" ("Ozone Hole" 2015).

Do Higher Ozone Levels Equal More Warming?

Restoration of stratospheric ozone may even become more closely linked to greenhouse warming as temperatures continue to rise near Earth's surface. Guy P. Brasseur and colleagues modeled the response of the middle atmosphere to a doubling of carbon dioxide levels near the surface and found that their models indicated a "cooling of about 8 degrees Kelvin is predicted at 50 kilometers during summer. During winter, the temperature is reduced up to 14 degrees K at 60 kilometers in the polar region" (Brasseur et al. 2000, 16). Increasing levels of methane also add to this effect. In addition to its properties as a greenhouse gas, "methane oxidation leads to higher water and OH concentrations in the stratosphere and mesosphere, and hence to less ozone at these altitudes" (Brasseur et al. 2000, 16).

In the mid-1990s, scientists were beginning to model a relationship between global warming and ozone depletion. A team led by Drew Shindell at the NASA Goddard Institute for Space Studies created the first atmospheric simulation to include ozone chemistry. The team found that the greenhouse effect was responsible not only for heating the lower atmosphere but also for cooling the upper atmosphere. The cooling poses problems for ozone molecules, which are most unstable at low temperatures. Based on the team's model, the buildup of greenhouse gases could chill the high atmosphere near the poles by as much as 8 to 10°C. The model predicted that maximum ozone loss would occur between 2010 and 2019 (Shindell et al. 1998, 589).

Healing ozone depletion over Antarctica may aggravate Southern Hemisphere warming, according to a study by scientists at Great Britain's University of Leeds. High-speed winds spurred by depleted ozone aided formation of bright summertime clouds that reflected sunlight. Strong winds over Antarctica toss sea spray, containing droplets that form reflective clouds. Stronger winds spurred by ozone depletion have formed thicker, brighter clouds. As the ozone levels rise, these winds are expected to subside, increasing warming at the surface.

"These clouds have acted like a mirror to the sun's rays, reflecting the sun's heat away from the surface to the extent that warming from rising carbon emissions has effectively been cancelled out in this region during the summertime," said Professor Ken Carslaw, professor of atmospheric science in the School of Earth and Environment, University of Leeds, and coauthor of the study. "If, as seems likely, these winds die down, rising CO_2 emissions could then cause the warming of the Southern Hemisphere to accelerate, which would have an impact on future climate predictions," he added (Isom 2010; Korhonen et al. 2010).

According to Carslaw, sea spray influx provoked an increase in cloud droplet concentration as much as 46 percent in parts of the Southern Hemisphere. Judith Perlwitz, NOAA research scientist and professor at the University of Colorado–Boulder, said that rises in greenhouse gas levels may sustain wind speeds and maintain cloud cover, contrary to Carslaw's forecast. "The question is whether the wind is really going to slow down, and that I doubt," she said. "The future is not just determined by the recovery of the ozone hole," she said. "We're also increasing our use of greenhouse gases, which increases the speed of the winds all year long" (Bhanoo 2010).

Much of the summer warming in southern Africa over recent decades seems to have been a result of the ozone hole over Antarctica. Desmond Manatsa of Bindura University of Science in Zimbabwe and colleagues compared climatic regimes before and after ozone depletion began. Significant warming at Earth's surface strongly correlated with changes in the position and intensity of high- and low-pressure systems in the atmosphere, enhancing the flow of tropical air southward in Africa. According to an account in *Nature*, "These shifts are often attributed to ozone depletion in the upper atmosphere. The expected closure of the ozone hole by 2050 may help to mitigate climate warming in southern Africa, the authors conclude" ("Ozone Hole Fans" 2013).

Further Reading

Bhanoo, Sinya N. "The Ozone Hole Is Mending. Now for the 'But.'" *The New York Times*, January 26, 2010. http://www.nytimes.com/2010/01/26/science/earth/26ozone.html.

Borenstein, Seth. "Arctic Lost 60 percent of Ozone Layer; Global Warming Suspected." Knight-Ritter News Service, April 6, 2000 (LEXIS).

Brasseur, Guy P., et al. "Natural and Human-Induced Perturbations in the Middle Atmosphere: A Short Tutorial." In David E. Siskind et al., *Atmospheric Science Across the Stratopause*. Washington, DC: American Geophysical Union, 2000.

Flannery, Tim. *The Weather Makers: How Man Is Changing the Climate and What It Means for Life on Earth*. New York: Atlantic Monthly Press, 2005.

Gordon, Anita, and David Suzuki. *It's a Matter of Survival*. Cambridge, MA: Harvard University Press, 1991.

Isom, Hannah. "Ozone Hole Healing Could Cause Further Climate Warming." University of Leeds, U.K. January 25, 2010. http://www.eurekalert.org/pub_releases/2010-01/uol -ohh012210.php.

Kerr, Richard A. "Deep Chill Triggers Record Ozone Hole." *Science* 282 (October 16, 1998): 391.

Korhonen, H., et al. "Aerosol Climate Feedback Due to Decadal Increases in Southern Hemisphere Wind Speeds." *Geophysical Research Letters*, January 27, 2010 (online). doi: 10. 1029/2009GL041320.

"Ozone Hole Fans African Heat." *Nature* 502 (October 17, 2013): 275.

"Ozone Loss Reaches New Record." Environment News Service, October 2, 2006. http:// www.ens-newswire.com/ens/oct2006/2006-10-02-01.asp (no longer available).

Schiermeier, Quirin. "Poles Lose Out as Ozone Levels Begin to Recover." *Nature* 437 (September 8, 2005): 179.

Shindell, Drew T., David Rind, and Patrick Lonergan. "Increased Polar Stratospheric Ozone Losses and Delayed Eventual Recovery Owing to Increasing Greenhouse-Gas Concentrations." *Nature* 392 (April 9, 1998): 589–592.

"Watching the Ozone Hole Before and After the Montreal Protocol." NASA Earth Observatory. September 18, 2012. http://earthobservatory.nasa.gov/IOTD/view.php?id=79198 &src=eoa-iotd.

See also: Atmospheric Circulation; Temperatures, Global; Temperatures, Winter

OZONE DEPLETION AND WARMING, ARCTIC

By 2011, ozone depletion over the Arctic had become significant as intense cold in the stratosphere and record warmth at the surface (including record depletion of Arctic sea ice) began to replicate the pattern that had become well known over the Antarctic, scientists reported in the journal *Nature*.

Here we demonstrate that chemical ozone destruction over the Arctic in early 2011 was—for the first time in the observational record—comparable to that in the Antarctic ozone hole. Unusually long-lasting cold conditions in the Arctic lower stratosphere led to persistent enhancement in ozone-destroying forms of chlorine and to unprecedented ozone loss, which exceeded 80 percent over 18–20 kilometers altitude. Our results show that Arctic ozone holes are possible even with temperatures much milder than those in the Antarctic. (Manney et al. 2011)

At its maximum during February 2011, the area of depleted ozone hole reached into northwest Russia and Scandinavia and as far south as Mongolia. "The root cause is the residual products from the CFCs that were released throughout the 20th century. But they are very long-lived, and it will take a few decades for them to be cleansed from the atmosphere," said Michelle L. Santee, a planetary scientist at NASA's Jet Propulsion Laboratory and a joint author of the study (Barringer 2011).

Arctic Ozone Falls to Depleted Levels

During late March 2011, ozone levels in the Arctic fell to depleted levels (below 250 Dobson units) for the first time, with readings of 220–230 DU for a week. Levels under 250 DU were observed for a month. As Rolando R. Garcia observed in *Nature* (2011), "Temperature levels in the stratosphere fell to record lows for four months, below 194 Kelvin, the threshold for formation of polar stratospheric clouds, which provide platforms for chlorine monoxide, which destroys ozone. Record warmth at the surface occurred at the same time due at least in part to rising greenhouse gas levels."

Solar flares and frigid stratospheric temperatures during the winter of 2003–2004 provoked the worst depletion of ozone above the Arctic since records had been kept, according to a team of scientists reporting in the March 2, 2005, issue of *Geophysical Research Letters* (Randall et al. 2005). The team reported that levels of nitrous oxides as much as four times any previously observed, agitated by solar activity, combined with bitter cold (approximately −110°F) to drive the depletion of ozone to levels 60 percent below anything ever before observed (records reach to 1985). "I don't think we can be confident about whether or not we're seeing an ozone recovery or if we're attributing recovery to the correct causes," said Cora Randall, lead author of the report, who is an atmospheric scientist at the University of Colorado (Human and McGuire 2005).

The levels of protective ozone over most of Canada will not recover during the 21st century and probably will deteriorate during its second half, according to one scientific study. This study contradicted earlier, more optimistic forecasts that ozone levels around the world would begin to recover by midcentury thanks to a ban on synthetic chlorine compounds that destroy ozone. "The more we know, the more we realize we don't know," said Jack McConnell, an atmospheric science professor at York University (Calamai 2002). Lower ozone levels could mean jumps of up to 10 percent in some skin cancers, which now strike almost 60,000 Canadians a year. The research also found that ozone levels would be lowest in summer, the season of greatest danger. Ozone in the stratosphere—the layer 10 to 40 kilometers above Earth—screens out the ultraviolet rays linked to skin cancers.

In addition, scientists learned that as winter ends, ozone-depleted air tends to migrate southward over heavily populated areas of North America and Eurasia. "The largest ultraviolet increases from all of this are predicted to be in the mid-latitudes of the United States," said University of Colorado atmospheric scientist Brian Toon. "It affects us much more than the Antarctic [ozone hole]" (Borenstein 2000).

Permafrost Decay in One Lake

University of California–Santa Cruz Adina Paytan supervised a research team that studied methane concentrations in Toolik Lake, Alaska. Their findings published March 9, 2015, in the *Proceedings of the National Academy of Sciences* showed that seasonal thawing of soil's "active layer" of above permafrost adds "significantly" to concentrations of methane in the lake, influencing its emission to the atmosphere.

"Methane transport from the active layer to Toolik Lake can account for a large fraction of methane emissions from this lake," Paytan said. "This is important because warming in the Arctic may expand the active layer and increase the discharge, leading to increased emissions from Arctic lakes and driving additional global warming" (Stephens 2015).

The researchers assessed the significance of their work:

Methane, a greenhouse gas, contributes to global warming. We show that methane-rich water from the seasonally thawed active layer in the Arctic flows into Toolik Lake, Alaska. This may be an important previously unrecognized conduit for methane transport and emissions in Arctic lakes. The controls on methane input from the active layer are fundamentally different than those affecting methane production within lakes, and the response of these processes to climate and environmental change is also distinct. The accuracy of predictions of methane emissions and ultimately the extent of climate change that can be expected in the Arctic depend on a better understanding of methane dynamics in the region, including the controls over methane production and transport processes within the active layer. (Paytan et al. 2015)

Further Reading

Paytan, Adina, et al. "Methane Transport from the Active Layer to Lakes in the Arctic Using Toolik Lake, Alaska, as a Case Study." *Proceedings of the National Academy of Sciences* March 9, 2015. http://www.pnas.org/content/early/2015/03/05/14173 92112.abstract?sid=6bb09f2b-6806-46e4-981e-ade09ad31498.

Stephens, Tim. "Methane in Arctic Lake Traced to Groundwater from Seasonal Thawing." University of California Santa Cruz Newscenter, March 9, 2015. http://news .ucsc.edu/2015/03/arctic-methane.html.

"This depletion is not necessarily a big surprise," said Paul Newman, an atmospheric scientist and ozone expert at NASA's Goddard Space Flight Center. "The ozone layer remains vulnerable to large depletions because total stratospheric chlorine levels are still high, in spite of the regulation of ozone-depleting substances by the Montreal Protocol. Chlorine levels are declining slowly because ozone-depleting substances have extremely long lifetimes" ("Arctic Ozone" 2011).

Planetary Disaster Averted

A simulation assembled by NASA in 2008 indicated that if humankind had not united to curtail depletion of stratospheric ozone in the late 1980s, nearly two-thirds of Earth's ozone would have vanished worldwide, not just over the poles, by 2065. Severe ozone depletion would be a part of everyday life across the polar regions, and global warming would have been much worse than otherwise anticipated. Led by Goddard Institute for Space Sciences specialist Paul Newman, the team simulated "what might have been" if chlorofluorocarbons (CFCs) and similar chemicals were not banned by the Montreal Protocol ("Simulation" 2009). The simulation used a comprehensive model that included atmospheric chemical effects, wind changes, and radiation changes. The analysis was published online in the journal *Atmospheric Chemistry and Physics*.

"Ozone science and monitoring has improved over the past two decades, and we have moved to a phase where we need to be accountable," said Newman, who is cochair of the U.N. Environment Program's Scientific Assessment Panel to review the state of the ozone layer and the environmental impact of ozone regulation. "We are at the point where we have to ask: Were we right about ozone? Did the Montreal Protocol work? What kind of world was avoided by phasing out ozone-depleting substances?" ("Simulation" 2009).

"In the 2050s, according to this model, ozone levels in the stratosphere over the tropics would have collapsed to near zero in a process similar to the one that creates the Antarctic ozone hole. By the end of the model run in 2065, global ozone drops to 110 DU, a 67 percent drop from the 1970s. Year-round polar values hover between 50 and 100 DU (down from 300–500 in 1960). The intensity of UV radiation at Earth's surface doubles; at certain shorter wavelengths, intensity rises by as much as 10,000 times. Skin cancer-causing radiation soars" ("Simulation" 2009).

"Our world avoided calculation goes a little beyond what I thought would happen," said Goddard scientist and study coauthor Richard Stolarski, who was among the pioneers of atmospheric ozone chemistry in the 1970s. "The quantities may not be absolutely correct, but the basic results clearly indicate what could have happened to the atmosphere. And models sometimes show you something you weren't expecting, like the precipitous drop in the tropics" ("Simulation" 2009).

Paul J. Crutzen has asserted that problems with the stratospheric ozone layer could have been much worse if chemists had developed substances based on bromine, which is 100 times as dangerous for ozone when compared atom to atom with chlorine. "This brings up the nightmarish thought that if the chemical industry had developed organochlorine compounds instead of the CFCs—or, alternatively, if chlorine chemistry had behaved more like that of bromine—then without any preparedness, we would have faced a catastrophic ozone hole everywhere and in all seasons during the 1970s, probably before atmospheric chemists had developed the necessary knowledge to identify the problem" (Crutzen 2001, 10). "We have been extremely lucky" given the fact that no one seemed overly worried about this problem before 1974, wrote Crutzen. "That we should always be on our guard for the potential consequences of the release of new products into the environment . . . for many years to come" is critically important (Crutzen 2001, 10).

Crutzen emphasized the danger inherent in taking chances with Earth's climate system without understanding its chemistry. The history of atmospheric chemistry during the last few decades, he said, has been one of surprises. "There may be more of these things around the corner," he said (McFarling 2001).

Yukon Caskets Ride Melting Permafrost

Thawing permafrost has been causing Inuvialuit caskets, many more than 80 years old, to the surface on parts of Herschel Island, Yukon Territory. Animals, including caribou, have tampered with some of the caskets, while others have been looted of artifacts, presumably by people. According to DeNeen L. Brown of the *Washington Post*, "Graves are pushing up from the ground as the ice within the carpet of permafrost melts, churning the soil beneath it into a muddy soup, spitting up foreign contents, sending whole hill slopes sliding downward. On a far tip of this island an entire grave site one day got up and slipped into the sea" (Brown 2001). The grave sites are all that remains on the island of a once-thriving Inuvialuit community. The island now is mostly deserted, with only the few park rangers as summer residents. Tourists come in by plane to hike, camp, and watch birds.

As Brown further reported, "The older generation of Inuvialuit believes that anyone who touches the possessions of the dead after they are buried will be cursed. Some in the younger generation, many years removed from traditional life, are wrestling with both sides of the issue. They want to maintain tradition, but they also do not want to sit back while their ancestors' bones lie uncovered on melting ground" (Brown 2001).

Wayne Pollard, professor of geography at McGill University in Montreal, who began studying Herschel Island in 1988, said the island's landscapes are some of the world's most vulnerable to climate change. "If it were simply coastal erosion and the graves dropping into the ocean and disappearing, that wouldn't be a problem," he said. "Because as I understand it, the Inuvialuit are quite comfortable with the idea of bodies being returned to the environment, as part of the natural cycle that they accept in life" (Brown 2001). Most often, however, the graves do not reach the sea.

Further Reading

Brown, DeNeen L. "Waking the Dead, Rousing Taboo; In Northwest Canada, Thawing Permafrost Is Unearthing Ancestral Graves." Washington *Post*, October 17, 2001, A27.

Polar Stratospheric Clouds

Markus P. Rex and colleagues studied climatic conditions in the Arctic and found a surprisingly strong relationship between ozone loss and the number and density of polar stratospheric clouds (PSCs) that form in the stratosphere despite its extreme dryness. Results were reported in *Geophysical Research Letters* (Rex 2004). By the end of 2001, Michael Proffitt, the World Meteorological Organization's senior scientific officer, said, "The area with temperatures low enough for polar stratospheric clouds that initiate rapid ozone destruction to form during October is double that found during any earlier five-year period" (Tolbert and Toon 2001).

PSCs are not new, having been described as "nacreous clouds resembling giant abalone shells floating in the sky" (Tolbert and Toon 2001, 61). These clouds form 20 kilometers above the ground during winter and are sometimes called "mother-of-pearl clouds" because they shimmer. Some stratospheric clouds have been reported in Scandinavia for a century, and Edward Wilson noted them on Robert Falcon Scott's 1901 Antarctic expedition. Sometimes the clouds shine with green and orange shades at sunrise and sunset (Tolbert and Toon 2001, 61). Polar stratospheric clouds remained largely an atmospheric curiosity until the discovery of widespread ozone depletion over the Antarctic during the middle 1980s. Scientists surmised that the ozone loss was occurring in the only place where the stratosphere was cold enough to produce these clouds. The clouds form during springtime when sunshine is available.

In *Geophysical Research Letters*, Daniel Kirk-Davidoff and colleagues (2002) wrote that increasing coverage of polar stratospheric clouds "in a positive feedback loop" have been associated with dramatic polar warming at the surface during periods of high carbon dioxide levels in Earth's history, notably the Eocene Epoch (55 to 38 million years ago) and the Cretaceous Period (135 million to 65 million years ago). Land and surface ocean temperatures in the polar regions then are believed to have been much higher than at present, at a time when scientists' proxies indicate that tropical temperatures were similar to or only slightly higher than they currently are (Ball 2004).

Ozone Deplation: "Rocks" in the Stratosphere

As scientists probe the connections between surface warming and atmospheric cooling, they find more potentially dangerous complications. For example, a team of atmospheric scientists has discovered large particles inside stratospheric clouds over the Arctic that could further delay the healing of Earth's protective ozone layer. The team found large particles containing nitric acid that could delay the recovery and make the ozone layers over both poles more vulnerable to climate change, said atmospheric chemist David Fahey of the NOAA office in Boulder, Colorado.

Each winter in the stratosphere over the poles, water and nitric acid condense to form polar stratospheric clouds that unleash chlorine and bromine, which degrade ozone. Later in the winter, nitrogen compounds help shut down the destruction. Fahey's team found previously unknown nitric acid particles that remove nitrogen, allowing the destruction to continue. They nicknamed them "rocks" because they are hundreds of times bigger than other particles in the clouds.

These "rocks" form during the polar winter when temperatures in the stratosphere decline to below $-90°C$. If global warming forecasts become reality, the cooling of the stratosphere compelled by the retention of heat near the surface may cause more such particles to form, accelerating ozone depletion. "If it gets colder and you get more 'rocks,' the depletion period is going to last longer. The chlorine can continue to eat ozone," said Paul Newman, an atmospheric physicist at NASA's Goddard Space Flight Center in Maryland (Erickson 2001). "What he got is really outstanding," Newman said of the findings by Fahey and his team. "This mechanism that we now understand really will help us be able to more precisely predict what's going to happen in the future" (Erickson 2001).

Fahey led a team of 27 researchers that included scientists from NOAA in Boulder, the University of Colorado, the National Center for Atmospheric Research in Boulder, and the University of Denver. The scientists described their findings in the February 9, 2001, edition of *Science*. "It's a major puzzle piece in the process by which ozone comes to be destroyed," Fahey said of the discovery. It was made during a January 2000 flight over the Arctic in the ER-2, NASA's version of the U-2 spy plane (Erickson 2001). The discovery of the "rocks" occurred during a January 2000 research flight of a NASA high-flying ER-2 near the North Pole. A machine on the aircraft that was measuring nitrogen-containing gases "coughed out what looked like disastrous noise" (Johansen 2003, 92).

The "noise" turned out to be extremely large particles (compared to other Arctic cloud mass particles) that contained nitric acid (HNO_3), particles previously unknown to science. They averaged 3,000 times the size of other particles in the stratosphere. These polar stratospheric cloud particles (known in shorthand form as "PSC rocks") remove reactive nitrogen from the atmosphere that would otherwise "tie up chlorine and bromine in inactive, harmless forms through denitrification " (Johansen 2003, 92). The "rocks" also "provide surfaces where chlorine and bromine can be liberated from their inactive forms to enter their ozone-destroying forms" (Johansen 2003, 92). In addition, the size of the PSC "rocks" causes them to fall more quickly than other particles, removing even more nitrogen from the stratosphere. Given all these factors, the PSC "rocks" "have significant potential to denitrify the lower stratosphere" (Fahey et al. 2001, 1026).

Fahey et al. concluded:

Arctic ozone abundances will remain vulnerable to increased winter [and] spring loss in the coming decades as anthropogenic chlorine compounds are gradually removed from the atmosphere, particularly if rising concentrations of greenhouse gases induce cooling in the polar vortex and trends of increasing water vapor continue in the lower stratosphere. Both effects increase the extent of PSC formation and, thereby, denitrification and the lifetime of active chlorine. The role of denitrification in these future scenarios is likely quite important. (Fahey et al. 2001, 1030)

Fahey and his colleagues estimated that ozone depletion in the Arctic stratosphere may not reach its peak until 2070, even with a steady decline in chlorine levels.

Not everyone is as pessimistic as Fahey about the future of stratospheric ozone. The subject is a matter of some rather intense debate. Sherwood Rowland of the University of California–Irvine, who shared the 1995 Nobel Prize for Chemistry for his part in the discovery that CFCs destroy stratospheric ozone, said the effect of global warming on ozone depletion should be short-lived. "The [ozone depletion] story is approaching closure, and that's very satisfying," Rowland said (Schrope 2000).

Alan O'Neill, a climate modeler at England's University of Redding said that record-breaking ozone holes in 2000 were not surprising and that ozone holes should heal by 2050. He added, however, that "[higher] concentrations of greenhouse gases . . . could push that date back a few decades" (Schrope 2000). O'Neill believed that ozone losses would peak around 2005 and then decline. In 2016, his forecast appeared to have been correct, as ozone loss was declining. Shindell said that even though scientists are beginning to understand how global warming could delay ozone-shield recovery, the "agreement to limit production [of CFCs] has been an unqualified success. The science was listened to, the policy makers did something, and it actually worked" (Schrope 2000).

Arctic Ozone Depletion Rises

By 2015–2016, continued cooling in the stratosphere intensified by near-surface retention of heat by greenhouse gases continued to intensify ozone depletion over the Arctic. By February 2016, observers were expecting a record size and intensity ozone hole because Arctic spring sunlight speeds up ozone-destroying chemical reactions. "This winter [2015–2016] has been stunning," said Markus Rex, an atmospheric chemist at the Alfred Wegener Institute in Potsdam, Germany, in *Science* (Hand 2016). "By next week [February 14–21, 2016], about 25 percent of the Arctic's ozone will be destroyed," he said. "Conditions are primed," said Gloria Manney, an atmospheric scientist at NorthWest Research Associates in Socorro, New Mexico. "The last ingredient we need is sunlight" (Hand 2016). Pockets of depleted ozone have been slipping occasionally into populated areas of Western Europe and North America. Rising levels of ultraviolet radiation also could deplete phytoplankton that bloom in the Arctic Ocean during the spring. Continued increases in heat retention by rising greenhouse gas levels near the surface may provoke further stratospheric cooling in the future, aggravating Arctic ozone depletion.

Further Reading

"Arctic Ozone Loss." NASA's Earth Observatory March 30, 2011. http://earthobservatory.nasa.gov/IOTD/view.php?id=49874&src=eoa-iotd.

Ball, Philip. "Climate Change Set to Poke Holes in Ozone." *Nature* March 3, 2004. http://info.nature.com/cgi-bin24/DM/y/eOCB0BfHSK0Ch0JVV0AY.

Barringer, Felicity. "A Significant Ozone Hole Is Reported over the Arctic." *The New York Times*, October 3, 2011. http://www.nytimes.com/2011/10/04/science/earth/04ozone.html.

Borenstein, Seth. "Arctic Lost 60 percent of Ozone Layer; Global Warming Suspected." Knight-Ritter News Service, April 6, 2000 (LEXIS).

Calamai, Peter. "Alert over Shrinking Ozone Layer." *Toronto Star*, March 18, 2002, A8.

Crutzen, Paul J. "The Antarctic Ozone Hole, a Human-Caused Chemical Instability in the Stratosphere: What Should We Learn from It?" Pp. 1–11 in Lennart O. Bengtsson and Claus U. Hammer, eds., *Geosphere-Biosphere Interactions and Climate*. Cambridge, UK: Cambridge University Press, 2001.

Erickson, Jim. "Boulder Team Sees Obstacle to Saving Ozone Layer; 'Rocks' in Arctic Clouds Hold Harmful Chemicals." *Rocky Mountain News* (Denver), February 9, 2001, 37A.

Fahey, D. W., et al. "The Detection of Large HNO3-Containing Particles in the Winter Arctic Stratosphere." *Science* 291(February 9, 2001): 1026–1031.

Garcia, Rolando R. "Atmospheric Science: An Arctic Ozone Hole?" *Nature* 478 (October 27, 2011): 462–463.

Hand, Eric. "Record Ozone Hole May open over Arctic in the Spring." *Science*, February 10, 2016, 650. http://www.sciencemag.org/news/2016/02/record-ozone-hole-may-open-over-arctic-spring.

Human, Katy, and Kim McGuire. "Ozone Decline Stuns Scientists." *Denver Post*, March 2, 2005, A8.

Johansen, Bruce E. *The Dirty Dozen: Toxic Chemicals and the Earth's Future.* Westport, CT: Praeger, 2003.

Kirk-Davidoff, Daniel, Daniel P. Schrag, and James G. Anderson. "On the Feedback of Stratospheric Clouds on Polar Climate." *Geophysical Research Letters* 29(11) (2002): 14659–14663.

Manney, Gloria L., et al. "Unprecedented Arctic Ozone Loss in 2011." *Nature* 478 (October 27, 2011): 469–475.

McFarling, Usha Lee. "Fear Growing over a Sharp Climate Shift." *Los Angeles Times*, July 13, 2001, A1.

Randall, C. E., et al. "Stratospheric Effects of Energetic Particle Precipitation in 2003–2004." *Geophysical Research Letters* 32 (March 2, 2005): L05802. doi: 10.1029/2004GL022003.

Rex, Markus, et al. "Arctic Ozone Loss and Climate Change." *Geophysical Research Letters* 31 (March 10, 2004). http://www.eurekalert.org/pub_releases/2004-03/agu-ajh031004.php.

Schrope, Mark. "Successes in Fight to Save Ozone Layer Could Close Holes by 2050." *Nature* 408 (December 7, 2000): 627.

"Simulation Shows World without Ozone Layer." NASA Earth Observatory, March 18, 2009. http://earthobservatory.nasa.gov/Newsroom/view.php?id=37588&src=eoa-nnews (no longer available).

Tolbert, Margaret A., and Owen B. Toon. "Solving the PSC Mystery." *Science* 292 (April 6, 2001): 61–63.

See also: Atmospheric Circulation; Temperatures, Global; Temperatures, Winter

PALEOCENE–EOCENE THERMAL MAXIMUM

Scientists who seek clues to how Earth's geophysical system will react to large-scale, rapid injections of atmospheric carbon dioxide in the future sometimes turn to the past, to epochs during which the same type of rapid rises occurred. Increases in Earth's carbon dioxide and methane load from human industrialization using fossil fuels, however, have occurred much more quickly over the last two centuries than anything provoked by natural causes in the past, and many effects of the recent, continuing rise in emissions have yet to be realized because of thermal inertia.

Two periods of particular interests are the Pliocene Epoch, roughly 2 million to 3 million years ago, and the Paleocene–Eocene Thermal Maximum (PETM), 54 million to 56 million years ago. In each case, rises in atmospheric greenhouse gas levels were followed by significant increases in temperatures.

"Climate has always changed naturally, and this is not good news when contemplating a human-forced future," wrote Richard Alley in *Science* (2016, 151). During the PETM around 55.9 million years ago, for example, "a large, natural CO_2 release drove strong warming that caused amplifying feedbacks, dwarfing of large animals, ecosystem disruptions, soil degradation, water-cycle shifts, and other major changes." The time scale of those events was much slower than current human-provoked changes in greenhouse gas levels and temperatures. "The impacts of the latter may thus be even more severe," Alley said.

Paleoclimate also provides clues to warming and drought, as with recent studies of intense dryness that contributed to the demise of the classic Mayan civilization around 1000 CE. An international research team has produced the best climate record yet of Maya times: a subannual reconstruction of rainfall in the Maya heartland that extends back 2,000 years (Kennett et al. 2012; Pringle 2012). Kennett et al. wrote:

The role of climate change in the development and demise of Classic Maya civilization (300 to 1000 CE) remains controversial because of the absence of well-dated climate and archaeological sequences. We present a precisely dated sub-annual climate record for the past 2,000 years from Yok Balum Cave, Belize. From comparison of this record with historical events compiled from well-dated stone monuments, we propose that anomalously high rainfall favored unprecedented population expansion and the proliferation of political centers between 440 and 660 CE. This was followed by a drying trend between 660 and 1000 CE that triggered the balkanization of polities, increased warfare, and the asynchronous disintegration of polities, followed by population collapse in the context of an extended drought between 1020 and 1100 CE.

Past studies of sea level rise also are providing analogs to present conditions. One such episode about 400,000 years ago has been described by Jerry X. Mitrovica and Maureen E. Raymo in *Nature* (2012):

Contentious observations of Pleistocene shoreline features on the tectonically stable islands of Bermuda and the Bahamas have suggested that sea level about 400,000 years ago was more than 20 meters higher than it is today. Geochronologic and geomorphic evidence indicates that these features formed during interglacial marine isotope stage (MIS) 11, an unusually long interval of warmth during the ice age. Previous work has advanced two divergent hypotheses for these shoreline features: first, significant melting of the East Antarctic ice sheet, in addition to the collapse of the West Antarctic ice sheet and the Greenland ice sheet; or second, emplacement by a megatsunami during MIS 11.

Climate can change quite suddenly. For example, scientists have studied a sea-level rise known as meltwater pulse 1A (MWP-1A). During the waning years of the last ice age, approximately 14,500 years ago, the sea rose between 12 meters and 22 meters (as measured in Tahiti in the South Pacific) over some 300 years (a mere blink of the eye in geological time). The cause of this event, which has been estimated using coral samples, is thus far unknown; it has not yet been associated with a sudden rise in temperatures (Kopp 2012). Pierre Deschamps and colleagues wrote in *Nature* (2012):

> Past sea level records provide invaluable information about the response of ice sheets to climate forcing. Some such records suggest that the last deglaciation was punctuated by a dramatic period of sea level rise, of about 20 meters, in less than 500 years. Controversy about the amplitude and timing of this meltwater pulse (MWP-1A) has, however, led to uncertainty about the source of the meltwater and its temporal and causal relationships with the abrupt climate changes of the deglaciation. Here we show that MWP-1A started no earlier than 14,650 years ago and ended before 14,310 years ago, making it coeval with the Bølling warming. Our results, based on corals drilled offshore from Tahiti during Integrated Ocean Drilling Project Expedition 310, reveal that the increase in sea level at Tahiti was between 12 and 22 meters, with a most probable value between 14 and 18 meters, establishing a significant meltwater contribution from the Southern Hemisphere. This implies that the rate of eustatic sea level rise exceeded 40 millimeters per year during MWP-1A.

The Paleocene–Eocene Thermal Maximum in the Arctic

The first detailed analysis of the biological record from the seabed near the North Pole indicates that 55 million years ago the Arctic Ocean there was warm with a 74°F year-round average, much like the present-day ocean off a beach in Florida in late spring. The warming, which scientists call the *Palaeocene–Eocene Thermal Maximum*, was massive and sudden by natural standards. Scientists have been studying this episode as a possible analog to human-induced warming in out time. Quite suddenly, the Arctic's ice melted during a spike in greenhouse gas levels, as rapid changes in albedo (reflectivity) sped the change (Sluijs et al 2006, 610; Kintisch 2006).

"Something extra happens when you push the world into a warmer world, and we just don't understand what it is," said one lead author, Henk Brinkhuis, an expert on ancient Arctic ecology at the University of Utrecht in the Netherlands (Revkin 2006). According to these investigations, the Arctic Ocean warmed abruptly and robustly 55 million years ago during the Paleocene–Eocene Thermal Maximum. This warming may have been provoked by a large, sudden injection in to the atmosphere of methane and carbon dioxide. No one has yet found the "smoking gun," however. "The new research provides additional important evidence that greenhouse gas changes controlled much of climate history, which strengthens the argument that greenhouse gas changes are likely to control much of the climate future," said one such expert, Richard B. Alley, a geoscientist at Pennsylvania State University (Revkin 2006).

The PETM 55.8 million years ago presents an analog to the changes now taking place in the atmosphere and oceans with respect to pH change and rise in CO_2 level. Today, however, the change is occurring at least 10 times as quickly, according to paleoceanographer James Zachos of the University of California–Santa Cruz. Nature took several thousand years to vent 2,000 to 7,000 gigatons of carbon dioxide and methane (which oxidized to CO_2) from volcanoes, methane hydrates, or peat—or all of them. Human emissions of greenhouse gases are doing the same thing in a few centuries (Kerr 2010, 1500).

Warming-provoked acidification of the oceans has precedent in Earth's natural history, notably during the PETM 55 million to 56 million years ago when massive amounts of carbon dioxide oxidized from methane clathrates and surged into the atmosphere from the oceans, raising sea surface temperatures by 5°C in the tropics and around 9°C at high latitudes. An initial, rapid rise in temperatures over 1,000 years was followed by a slower increase during the next 30,000 years (Zachos et al. 2005, 1612).

Scientists have been studying the acidification of the oceans during this long-ago epoch as an analog to similar conditions anticipated in response to human-provoked increases in carbon dioxide. One study of the problem, published in *Nature*, pointedly concluded the following:

What, if any, implications might this have for the future? If combustion of the entire fossil fuel reservoir (about 4,500 gigatons of carbon) is assumed, the impacts on deep-sea pH [acidity] and biota will likely be similar to those in the Paleocene–Eocene Thermal Maximum. However, because the anthropogenic carbon input will occur within just 300 years, which is less than the mixing time of the ocean, the impacts on ocean surface pH and biota will probably be more severe. (Zachos et al. 2005, 1614)

Ocean pH is now lower (more acidic) than it has been in 20 million years, and it is heading lower, according to marine chemist Richard Feely of NOAA's Marine Environmental Laboratory in Seattle. Models by Feely and colleagues anticipate that ocean pH will decline from 8.2 before the Industrial Revolution to 7.8 by 2100, increasing acidity by around 150 percent (Kerr 2010, 1501).

Roughly 55 million years ago, the pH of seawater in the world's oceans may have dropped to a level similar to that expected in 2100. Ocean sediment from that time contains little carbonate "and no fossils of microorganisms with calcium carbonate shells, indicating that the seawater became too corrosive for calcifying algae such as deep-sea foraminifera, driving many to extinction. Today, acidification is progressing at least 10 times faster than it did 55 million years ago" (Schiermeier 2011; Zachos et al. 2005).

Global Warming in the Early Triassic

Global warming in the Early Triassic Epoch was so severe that equatorial latitudes were uninhabitable for many plants and animals. As Yadong Sun and colleagues wrote in *Science* (2012),

Global warming is widely regarded to have played a contributing role in numerous past biotic crises. Here, we show that the end-Permian mass extinction coincided with a rapid temperature rise to exceptionally high values in the Early Triassic that were inimical to life in equatorial latitudes and suppressed ecosystem recovery. This was manifested in the loss of calcareous algae, the near-absence of fish in equatorial Tethys, and the dominance of small taxa of invertebrates during the thermal maxima. High temperatures drove most Early Triassic plants and animals out of equatorial terrestrial ecosystems and probably were a major cause of the . . . crisis.

Researchers studying the PETM's sudden warming 56 million years ago have found that the size of mammals decreased as temperatures rose, sometimes dramatically. Witness *Sifrhippus*, the first genus of horses, whose species shrank to around 12 pounds (the size of a domestic cat) over 175,000 years of the Paleocene–Eocene Thermal Maximum as temperatures warmed 8 to 12°F in Wyoming's Bighorn Basin. Other mammals also may have shrunk during this warm period, a phenomenon common enough to have a scientific description—Bergmann's rule—which holds that "mammals of a given genus or species are smaller in hotter climates" (Gorman 2012).

Sifrhippus shrank 30 percent during the first 130,000 years of the PETM, with rapidly rising temperatures, but then grew 75 percent bigger during the next 45,000 years as temperatures cooled, ranging from eight to 15 pounds. "It seems to be natural selection," said Ross Secord, assistant professor of earth and atmospheric sciences at the University of Nebraska at Lincoln, a co-author of a paper in *Science*. He said that animals evolved to be smaller during warming because "smaller animals did better in that environment, perhaps because the smaller an animal is, the easier it is to shed excess heat" (Gorman 2012). "It's difficult to say that mammals are going to respond in the same way now," Dr. Secord said. "If I had to guess," he said, he thinks some will get smaller. And, he said, some studies have shown some birds to be getting smaller in response to warming (Gorman 2012).

Ross Secord and colleagues wrote in *Science* (2012) that body sizes of horses were driven by climate change in this period:

Body size plays a critical role in mammalian ecology and physiology. Previous research has shown that many mammals became smaller during the Paleocene–Eocene Thermal Maximum . . . , but the timing and magnitude of that change relative to climate change have been unclear. A high-resolution record of continental climate and equid body size change shows a directional size decrease of ~30% over the first [roughly] 130,000 years of the PETM, followed by [an approximately 76 percent] increase in the recovery phase of the PETM. These size changes are negatively correlated with temperature inferred from oxygen isotopes in mammal teeth and were probably driven by shifts in temperature and possibly high atmospheric CO_2 concentrations. These findings could be important for understanding mammalian evolutionary responses to future global warming. (Secord et al. 2012, 959)

Further Reading

Alley, Richard B. "A Heated Mirror for Future Climate." *Science* 352 (April 8, 2016): 151–152.

Deschamps, Pierre, et al. "Ice-Sheet Collapse and Sea-Level Rise at the Bølling Warming 14,600 Years Ago." *Nature* 483 (March 29, 2012): 559–564.

Gorman, James. "A Tiny Horse That Got Even Tinier as the Planet Heated Up." *The New York Times,* February 23, 2012. http://www.nytimes.com/2012/02/24/science/sifrhippus -the-first-horse-got-even-tinier-as-the-planet-heated-up.html.

Kennett, et al. "Development and Disintegration of Maya Political Systems in Response to Climate Change." *Science* 338 (November 9, 2012): 788–791.

Kerr, Richard A. "Ocean Acidification Unprecedented, Unsettling." *Science* 328 (June 18, 2010): 1500–1501.

Kintisch, Eli. "Hot Times for the Cretaceous Oceans." *Science* 311 (February 24, 2006): 1095.

Kopp, Robert E. "Paleoclimate: Tahitian Record Suggests Antarctic Collapse." *Nature* 483 (March 29, 2012): 549–550.

Mitrovica, Jerry X., and Maureen E. Raymo. "Collapse of Polar Ice Sheets during the Stage 11 Interglacial." *Nature* 483 (March 22, 2012): 453–456.

Pringle, Heather. "Did Pulses of Climate Change Drive the Rise and Fall of the Maya?" *Science* 338 (November 9, 2012): 730–731.

Revkin, Andrew C. "Studies Portray Tropical Arctic in Distant Past." *The New York Times,* June 1, 2006. http://www.nytimes.com/2006/06/01/science/earth/01climate.html.

Schiermeier, Quirin. "Earth's Acid Test." *Nature* 471 (March 9, 2011): 154–156. http://www .nature.com/news/2011/110309/full/471154a.html.

Secord, Ross, et al. "Evolution of the Earliest Horses Driven by Climate Change in the Paleocene–Eocene Thermal Maximum." *Science* 335 (February 24, 2012): 959–962. http://www.sciencemag.org/content/335/6071/959.

Sluijs, Appy, et al. "Subtropical Arctic Ocean Temperatures during the Palaeocene/Eocene Thermal Maximum." *Nature* 441 (June 1, 2006): 610–613.

Sun, Yadong, et al. "Lethally Hot Temperatures during the Early Triassic Greenhouse." *Science* 338 (October 19, 2012): 366–370.

Zachos, James C., et al. "Rapid Acidification of the Ocean during the Paleocene–Eocene Thermal Maximum." *Science* 308 (June 10, 2005): 1611–1615.

See also: Ice Melt, Antarctica; Pliocene Paleoclimate; Sea Ice, Arctic; Summer Ice, Arctic; Temperatures, Greenhouse Gas Levels and

PERMAFROST

Given the fact that one-third of Earth's carbon is stored in far northern latitudes (mainly in tundra and boreal forests) (Mack et al. 2004, 440), the speed at which ecosystem warming may release this carbon into the atmosphere is vitally important to forecasting global warming's pace and effects. The amount of carbon stored in Arctic ecosystems is two-thirds of the amount currently found in the atmosphere (Loya and Grogan 2004, 406). Its release into the atmosphere will depend on the pace of temperature rise—and the Arctic, according to several sources, has been the most rapidly warming regions of Earth. Roughly five times as much carbon is stored in frozen Arctic soils as has been emitted by all human activities since 1850.

Although emissions from permafrost currently account for less than 1 percent of global methane emissions, warming temperatures in high latitudes will augment human contributions in coming years.

"We know that permafrost soils are warming even faster than Arctic air temperatures—as much as 2.7 to 4.5 [°F or 1.5 to 2.5°C] in just the past 30 years," said Charles Miller, the principal investigator for the Carbon in Arctic Reservoirs Vulnerability Experiment. "What we do not understand well is the effect that is having on emissions of carbon dioxide and methane" ("Why Monitoring" 2016).

"A Slow-Motion Time Bomb"

As temperatures increase, methane and carbon dioxide escape from the soil. "It's kind of like a slow-motion time bomb," said Ted Schuur, a professor of ecosystem ecology at the University of Florida (Borenstein 2006). "The effects can be huge," said Katey Walter of the University of Alaska–Fairbanks. "It's coming out a lot and there's a lot more to come out" (Borenstein 2006; Walter et al. 2006, 71). "The higher the temperature gets, the more permafrost we melt, the more tendency it has to become a more vicious cycle," said Chris Field, director of global ecology at the Carnegie Institution of Washington (Borenstein 2006).

The large landmass of Siberia contains most of the permafrost that is capable of releasing methane and carbon dioxide. The amount of carbon trapped in a type of permafrost called *yedoma* may be 100 times the amount of carbon released into the air each year by the burning of fossil fuels (Borenstein 2006). Thus, human combustion of fossil fuels acts as a trigger by raising temperatures enough to release even more permafrost from natural sources.

As much as 90 percent of the Northern Hemisphere's permafrost (the top three meters, or 10 feet) could thaw by the end of the 21st century given business-as-usual increases in greenhouse gas emissions, according to projections by scientists affiliated with the National Center for Atmospheric Research (Lawrence and Slater 2005). Even with major reductions in emissions, Northern Hemisphere permafrost is projected to shrink from 4 million to around 1.5 million square miles by 2100.

Vladimir Romanovsky, who heads the University of Alaska–Fairbanks Permafrost Laboratory, has issued models that anticipate that one-third of Alaska's permafrost could thaw by midcentury and two-thirds by century's end. This rapid thawing is unprecedented and largely the result of fossil fuel emissions, said Kevin Schaefer of the National Snow and Ice Data Center in Boulder, Colorado ("Alaska Sinks" 2013). Schaefer said that the thaw is already adding carbon dioxide and methane to the atmosphere, and that permafrost decay "will skyrocket in the next 20 to 30 years. Once the emissions start, they can't be turned off" ("Alaska Sinks" 2013). In 2013, Fairbanks, Alaska, experienced a record number of high temperatures above 80°F as the pace of permafrost thawing accelerated.

Permafrost's "active" top layer thaws and then refreezes with the seasonal cycle each year, but a deeper layer that stores organic carbon remains frozen. With a warming climate, however, melting occurs farther down from the surface, releasing carbon dioxide and methane. The stores of organic carbon from decayed animals and plants contain roughly twice as much carbon dioxide as the atmosphere, so provoking its release by melting adds combustion gases from fossil fuels used by human beings. Temperature rise in and near the Arctic is accelerating the thaw: 3.4°F in Alaska, for example over the last 60 years, including 6°F in winter. Warming temperatures increase evaporation during the Arctic and near-Arctic summers, provoking increasing numbers of wildfires, which also add greenhouse gases to the atmosphere.

Scientists studying how methane emerges from thawing earth have found that it rises from lake bottoms "not as diffuse leaks, as many scientists had long assumed, but in a handful of scattered, vigorous plumes, some of them capable of putting out many quarts of gas per day" (Gillis 2011). In recent years, estimates of methane that may be released by warming earth have been rising. Frozen methane supports a surface almost as a concrete pillar supports a building, but thawing causes the ground to soften and then collapse, leading to swaying telephone poles, leaning buildings, and trees growing at angles in what has become known as the "drunken forests" of Alaska, all sinking into low-lying areas known as *thermokarst*. As snow thaws and melts on any given surface, darker ground absorbs more heat and speeds the melting of more permafrost.

As Arctic tundra becomes wetter, it emits more carbon dioxide, according to work released in 2009 by Walter Oechel, an ecologist at San Diego State University, principal investigator of a study on the subject. "It's a big deal," Oechel said (Witze 2009). "These are aspects that haven't been recognized before," reported Oechel's colleague Donatella Zona at the annual meeting of the Ecological Society of America in August 2009. According to Jeffrey Welker, an ecologist at the University of Alaska–Anchorage, "a deeper thaw means that older carbon—perhaps 1,000 years old rather than 100 years old—gets released from the soil. We're beginning to see evidence that deeper, older carbon will be entering the atmosphere in the future" (Witze 2009).

Thaw in Siberia

In August 2005, climate researchers who had recently returned from the area said that a large area of western Siberia was undergoing an unprecedented thaw that could dramatically increase the rate of global warming. Sergei Kirpotin at Tomsk State University in western Siberia and Judith Marquand at Oxford University reported that permafrost across a million square kilometers, an area as large as France and Germany combined, has started to melt (Sample 2005). What was until recently a barren expanse of frozen peat had turned, during the summer, "into a broken landscape of mud and lakes, some more than a kilometer across" (Sample 2005). In the winter, so much methane was bubbling up that it kept the surface from freezing.

Kirpotin told *The Guardian* (U.K.) that the situation was an "ecological landslide that is probably irreversible and is undoubtedly connected to climatic warming." He added that the thaw had probably begun three or four years earlier (Sample 2005). "When you start messing around with these natural systems, you can end up in situations where it's unstoppable. There are no brakes you can apply," said David Viner, a senior scientist at the Climatic Research Unit at the University of East Anglia. "This is a big deal because you can't put the permafrost back once it's gone. The causal effect is human activity and it will ramp up temperatures even more than our emissions are doing" (Sample 2005).

Euan Nisbet of London's Holloway College, who monitors methane emissions on an international scale, said that methane releases from western Siberia's peat bogs could account for as much as 100,000 tons a day—or more than all the human-generated emissions from the United States. Fred Pearce described the melting:

> Numberless lakes stretched to the horizon. From the air, they did not look like lakes that form naturally. They were generally circular, looking more like flooded potholes in a road. The lakes formed individually from small breaches in the permafrost. Wherever ice turned to water, a small pond formed. Then surrounding lumps of frozen peat would slump into the water and the pond would grow in an ever-widening circle, until mile after mile of frozen bog had melted into the mass of lakes. (Pearce 2007, 79–80)

Permafrost and Burning Peat Bogs across Russia

Some of the permafrost that covers as much as 65 percent of Russia has been melting. Scientists there expect the permafrost boundary to recede 150 miles northward during the next quarter-century. Already, the diamond-producing town of Mirny, in Yakutia, has evacuated a quarter of its residents because their houses melted into the previously frozen soil. Parts of the Trans-Siberian Railway's track has twisted and sunk because of melting permafrost, causing service delays of several days at a time.

By 2002, melting permafrost had damaged 300 apartment buildings in the Siberian cities of Norilsk and Yakutsk (Goldman 2002, 1494). "Assuming that the region [Siberia] continues to warm at the modest rate of 0.075°C per year," according to Lev Khrustalev, a geocryologist at Moscow State University. He estimated that every five-story structure built between 1950 and 1990 in Yakutsk, population 193,000, "could come crashing down [by 2030] unless steps are taken to strengthen them and preserve the permafrost" (Goldman 2002, 1494).

Russian climate experts described Siberia's problems with melting permafrost during an international conference on climate change in Moscow held in early fall 2003. Georgy Golitsyn, director of Moscow's Institute of Atmospheric Physics, said that by the end of the 21st century temperature increases in Siberia could be twice the worldwide rise at 1.4 to 5.8°C as anticipated by the Intergovernmental Panel on Climate Change (IPCC) (Meuvret 2003). "Extreme weather events might happen more frequently, [with] the melting of permafrost, which is already noticeable, and damage risks to buildings, roads and pipelines. Pipelines are always having some trouble," he said (Meuvret 2003).

"Here in Moscow and in European Russia, really cold episodes are becoming quite rare. And in Siberia, very heavy frosts have almost disappeared. Instead of

[−40 or −50°C] which were quite frequent, now these occur just occasionally and they usually experience [−30°C]," Golitsyn said (Meuvret 2003). The disastrous flooding of the Lena River basin in 2001 gave a foretaste of the problems to come, he noted. "The winter had been normal, but . . . the soil was frozen and then in May a heat wave came with temperatures up to 30 [°C] when the snow had not melted yet. . . . This is the type of catastrophe we might have more frequently" (Meuvret 2003).

Golitsyn's warning was supported by Michel Petit, who was a French representative to the IPCC until April 2002. Siberia, he said, is "one of the most sensitive regions to climate change on the planet" (Meuvret 2003). Although areas that could be farmed would probably expand in Siberia and maritime transport would be opened at seasons blocked until quite recently by ice, melting permafrost would be a "real catastrophe," turning large areas of Siberia into swamps. "Major industrial complexes, towns, and pipelines would subside. Carbon dioxide and methane would escape, and the greenhouse effect would become even more serious," said Gorgy Gruza, a Russian climate-change expert who opposed the Kyoto Protocol.

Lakes across the Siberian Arctic are shrinking and drying up, according to a comparison of satellite images taken of 10,000 large lakes over a 25-year period. Scientists found that 125 of the lakes disappeared completely and are now revegetated. Researchers at three U.S. universities described their research on lakes spread across over 200,000 square miles of Siberia. "Arctic warming has accelerated since the 1980s, driving an array of complex physical and ecological changes in the region," they said (Smith et. al. 2005). After three decades of rising soil and air temperatures in Siberia, "our analysis reveals a widespread decline in lake abundance and area, despite slight precipitation increases" (Smith et al. 2005). Why? Warming temperatures lead to thinning and eventual "breaching" of permafrost near lakes, greatly facilitating their drainage to the subsurface (Smith et al. 2005).

Russia also has been beset by burning peat bogs during warmer and drier summers. During September 2002, peat fires around Moscow produced so much smoke that motorists drove with headlights on at noon. The fires add carbon dioxide to the atmosphere, but they become especially intense because of hot, dry summers.

Further Reading

"Alaska Sinks as Climate Change Thaws Permafrost." *USA Today*, October 9, 2013. http://www.usatoday.com/search/Alaska%20Sinks%20as%20Climate%20Change%20Thaws%20Permafrost%20October%202013/.

Borenstein, Seth. "Scientists Find New Global Warming 'Time Bomb' Methane Bubbling Up from Permafrost." Associated Press, September 6, 2006 (LEXIS).

Gillis, Justin. "As Permafrost Thaws, Scientists Study the Risks." *The New York Times*, December 16, 2011. http://www.nytimes.com/2011/12/17/science/earth/warming-arctic-permafrost-fuels-climate-change-worries.html.

Goldman, Erica. "Even in the High Arctic, Nothing Is Permanent." *Science* 297 (August 30, 2002): 1493–1494.

Lawrence, D. M., and A. G. Slater. "A Projection of Severe Near-Surface Permafrost Degradation during the 21st Century." *Geophysical Research Letters* 32 (2005): 1–5.

Loya, Wendy M., and Paul Grogan. "Carbon Conundrum on the Tundra." *Nature* 431 (September 23, 2004): 406–407.

Mack, Michelle C., et al. "Ecosystem Carbon Storage in Arctic Tundra Reduced by Long-Term Nutrient Fertilization." *Nature* 432 (September 23, 2004): 440–443.

Meuvret, Odile. "Global Warming Could Turn Siberia into Disaster Zone: Expert." Agence France Presse, October 2, 2003.

Pearce, Fred. *With Speech and Violence: Why Scientists Fear Tipping Points in Climate Change.* Boston: Beacon Press, 2007.

Sample, Ian. "Warming Hits 'Tipping Point.'" *The Guardian* (U.K.), August 18, 2005, 1. http://www.guardian.co.uk/guardianweekly/story/0,12674,1550685,00.html.

Smith, L. C., et al. "Disappearing Arctic Lakes." *Science* 308 (June 3, 2005): 1429.

Walter, K. M., et al. "Methane Bubbling from Siberian Thaw Lakes as a Positive Feedback to Climate Warming." *Nature* 414 (September 7, 2006): 71–75. http://shadow.eas.gatech.edu/~kcobb/abrupt/walter06.pdf.

"Why Monitoring Emissions from Permafrost Matters." NASA Earth Observatory, April 12, 2016. http://earthobservatory.nasa.gov/IOTD/view.php?id=87850&src=eoa-iotd.

Witze, Alexandra. "Drowned Tundra Emits More Carbon." *Nature News* Published online, August 4, 2009. http://www.nature.com/news/2009/090804/full/news.2009.778.html. doi: 10.1038/news.2009.778.

See also: Carbon Cycle Feedbacks; Climate Change, Abrupt Nature of; Temperatures, Greenhouse Gas Levels and; Thermal Inertia

PLIOCENE PALEOCLIMATE

The Pliocene Epoch 3 million to 4 million years ago has received attention because atmospheric carbon dioxide levels at that time roughly parallel today's at around 400 parts per million (404 to 407 ppm as of 2016). However, today's rise from pre-industrial levels has occurred so quickly that thermal inertia has not caught up. Temperatures at that time were some 4 to 8°C higher worldwide, which are within some projections for the end of this century. At the same time, "surface temperatures in the polar regions were so much higher that continental glaciers were absent from the Northern Hemisphere, and sea level was about 25 meters higher" (Fedorov et al. 2006, 1485).

Allowing a century or two for thermal inertia in the seas to reach equilibrium, it may be a good analog to a warmer world into which we will be locked within 100 to 200 years. At that time, sea levels were 100 feet higher than today, without multiyear ice in the Northern Hemisphere and reduced ice in Antarctica. In such a world, Augusta (Georgia) and Richmond (Virginia) would be beachfront cities, and nearly all of the Florida peninsula would be underwater (Pollack 2009, 229–230).

One team of scientists has studied a sediment core from Lake El'gygytgyn in northeast Russia, provides a high-latitude climate record of the late Pliocene Epoch.

Understanding the evolution of Arctic polar climate from the protracted warmth of the middle Pliocene into the earliest glacial cycles in the Northern Hemisphere has been hindered by the lack of continuous, highly resolved

Arctic time series. Evidence from Lake El'gygytgyn, in northeast Arctic Russia, shows that 3.6 to 3.4 million years ago, summer temperatures were about 8°C warmer than today, when the partial pressure of CO_2 was about 400 parts per million. Multiproxy evidence suggests extreme warmth and polar amplification during the middle Pliocene, sudden stepped cooling events during the Pliocene–Pleistocene transition, and warmer than present Arctic summers until about 2.2 million years ago, after the onset of Northern Hemispheric glaciation. Our data are consistent with sea level records and other proxies indicating that Arctic cooling was insufficient to support large-scale ice sheets until the early Pleistocene. (Brigham-Grette 2013, 1421)

Deer and Forests on Ellesmere Island

According to the July 2010 issue of *Geology*, work by U.S., Dutch, and Canadian scientists found fossil evidence that during the Pliocene Epoch 2.6 to 5.3 million years ago, Ellesmere Island in the far northern Canadian Arctic was home to small deer and other mammals living in pine forests even as the atmospheric carbon dioxide level was 400 parts per million.

A Mid-Cretaceous Supergreenhouse?

Seeking an analog to a world dominated by severe global warming, Paul A. Wilson and Richard D. Norris investigated the climate of the mid-Cretaceous Period, "a time of unusually warm polar temperatures, repeated reef-drownings in the tropics, and a series of oceanic anoxic events. . . . with maximum sea-surface temperatures 3 to 5 [°C] warmer than today" (Wilson and Norris 2001, 425). This was a period with considerable greenhouse forcing not unlike what the end of the 21st century may resemble, except that the forcing was natural, not anthropogenic. Writing in *Geophysical Research Letters*, Orsi et al. (2001) found no evidence of a recent reduction in generation of Antarctic bottom water (see also Chin 2001, 575). They did not rule out the possibility, however, that global warming could affect formation rates of North Atlantic deep water or Antarctic bottom water in the future.

The possibility that water temperatures rose to 15°C or perhaps higher in the Arctic during the Cretaceous Period also has been explored by Hugh C. Jenkyns and colleagues in *Nature* (Jenkyns et al. 2004). At the time, atmospheric carbon dioxide levels may have been three to six times today's levels, creating a "supergreenhouse" caused mainly by volcanic outgasing. Today's climate models have had a difficult time accommodating such warmth in the Arctic without raising projected temperatures at lower latitudes to levels that would have been intolerable for animals and plants living at the time.

Further Reading

Chin, Gilbert. "No Deepwater Slowdown?" *Science* 293 (July 27, 2001): 575.

Jenkyns, Hugh C., et al. "High Temperatures in the Late Cretaceous Arctic Ocean." *Nature* 432 (December 16, 2004): 888–892.

Orsi, Alejandro H., et al. "Cooling and Ventilating the Abyssal Ocean." *Geophysical Research Letters* 2(15) (August 1, 2001): 2923–2926.

Wilson, Paul A., and Richard D. Norris. "Warm Tropical Ocean Surface and Global Anoxia during the Mid-Cretaceous Period." *Nature* 412 (July 26, 2001): 425–429.

"Our findings indicate that CO_2 levels of approximately 400 parts per million are sufficient to produce mean annual temperatures in the high Arctic of approximately zero degrees Celsius (32°F)," wrote lead author Ashley Ballantyne of the University of Colorado–Boulder. "As temperatures approach zero degrees Celsius, it becomes exceedingly difficult to maintain permanent sea and glacial ice in the Arctic. Thus current levels of CO_2 in the atmosphere of approximately 390 parts per million may be approaching a tipping point for irreversible ice-free conditions in the Arctic," Ballantyne concluded ("Ancient Fossils" 2010).

"Our findings are somewhat disconcerting regarding the temperatures and greenhouse gas levels during the Pliocene," said Jaelyn Eberle, curator of fossil vertebrates at the University of Colorado Museum of Natural History and an associate professor in the geological sciences department. "We already are seeing evidence of both mammals and birds moving northward as the climate warms, and I can't help but wonder if the Arctic is headed toward conditions similar to those that existed during the Pliocene" ("Ancient Fossils" 2010). At that time, according to this research, "[the] island was inhabited by fish, frogs, and mammals now extinct, including tiny deer, ancient relatives of the black bear, three-toed horses, small beavers, rabbits, badgers, and shrews" ("Ancient Fossils" 2010).

Sea Levels 25 Meters Higher?

In the decade of the early 2000s, James Hansen, then director of NASA's Goddard Institute for Space Studies, estimated that a sea level rise of 6 meters (the Greenland ice cap, the West Antarctic ice sheet, or half of each) would displace 235 million people, based on 2000 population figures, including 11 million in the United States, 93 million in China and Taiwan, 46 million in India and Sri Lanka, 24 million in Bangladesh, 23 million in Indonesia and Malaysia, 12 million in Japan, and 26 million in Western Europe.

Raise the sea level 25 meters (Pliocene level), and the number of displaced people rises to 700 million, including 47 million in the United States, 224 million in China and Taiwan, 146 million in India and Sri Lanka, 109 million in Bangladesh, 72 million in Indonesia and Malaysia, 39 million in Japan, and 66 million in Western Europe (Hansen 2005, 2006). A 25-meter sea level rise would drown most of Florida and sizable U.S. cities on the East and Gulf Coasts.

According Hansen, "If emissions of greenhouse gases continue to increase as in the 'business-as-usual' scenario, the rate of isotherm movement will double in this century to at least 70 miles per decade. If we continue on this path, a large fraction of the species on Earth will go extinct. . . . For all foreseeable human generations, [Earth] will be a far more desolate planet than the one in which civilization developed and flourished over the past several thousand years" (Hansen 2006, 12).

The tropics extended much farther toward the during the early Pliocene Period, roughly 4 million years ago, so the temperature gradient between the equator and subtropics was much less than in our time. This period is now being used as an analog to a globally warmed Earth a century from now. The fact that the Arctic is warming far more rapidly than the tropics may be a clue. Chris M. Brierley and colleagues wrote in *Science*, "Corroborating evidence indicates that the Pacific temperature contrast between the equator and 32° N has evolved from about 2°C 4 million years ago to about 8°C today. The meridional warm pool expansion evidently had enormous impacts on the Pliocene climate, including a slowdown of the atmospheric Hadley circulation and El Niño–like conditions in the equatorial region" (Brierley et al. 2009, 1714). The Pliocene Epoch was not only warm enough to melt much of the world's ice but also able to trigger a semipermanent El Niño–like condition in the Pacific Ocean, with warm water spanning the Pacific Ocean (Kerr 2005, 1456).

A Day at the Beach in the Pliocene

Scientists are looking for shells on beaches along the U.S. East Coast, from near Washington, D.C., to Orlando, Florida. These are not ordinary shells on today's beaches but the surf line of the Pliocene. The shells are fossilized, and they washed up the last time the carbon dioxide level in the atmosphere was 400 parts per million (ppm), the level we are now breathing. The project, funded by the National Science Foundation, is called "Pliomax."

The shoreline during the Pliocene between Washington, D.C., and Orlando roughly followed today's "fall line," where rivers become too narrow and rocky to navigate most commercial ships, often between 60 and 100 miles inland. Temperatures in the Pliocene were roughly equal to those projected for the end of the 21st century. A century or two after that, plus another century to melt ice as thermal inertia in the ocean catches up with surface temperatures, and anywhere from 60 to 100 miles of the U.S. East Coast may be under water. This equation originally assumed that atmospheric carbon dioxide would stop dead in its tracks at 400 ppm, which it did not.

Within 300 years of "business as usual," today's greenhouse gas levels (which will be higher by then, speeding this process) will have melted much of Greenland's ice, as well as the West Antarctic ice sheet. Portions of the larger East Antarctic ice sheet will be eroding. Mountain glaciers and the Arctic Ocean's ice cap will exist only in history books and dusty copies of the *National Geographic*.

Given today's CO_2 levels, major sea level rise is not a question of if but when. Most people think 300 years is a long time, but it is a blink of the eye in geologic time, much less than the time period required by the natural causes that raised the ocean to the levels that washed up shells near present-day Washington, D.C. Unless the levels of greenhouse gases (mainly CO_2 and methane) stop rising and then decline to no higher than 350 ppm, this script has already been written. Every pump of the gas pedal today makes it more likely. We do not know the exact timetable, because land itself also slowly rises and falls. We do know the seas will rise and that storms will cause short-term inundation above the average. We also know that most of the land that along the Eastern Seaboard is subsiding—that is, slowly sinking. The height at which some of the fossilized shells have been found indicate that the level of sea level rise is not a question with a single linear answer. In South Africa, some shells have been found 64 feet above modern shorelines. The most commonly cited figure for Pliocene sea level is around 80 feet above today's level.

"I wish I could take people that question the significance of sea level rise out in the field with me," Maureen E. Raymo told *The New York Times*. Raymo works at the Lamont–Doherty Earth Observatory, a unit of Columbia University just outside New York City. "Because you just walk them up 30 or 40 feet in elevation above today's sea level and show them a fossil beach, with shells the size of a fist eroding out, and they can look at it with their own eyes and say, 'Wow, you didn't just make that up'" (Gillis 2013).

"Absolutely, unequivocally, nature has changed before," Richard B. Alley, a leading climate scientist at Pennsylvania State University told the *Times*. "But it looks like we're going to do something bigger and faster than nature ever has." "I can merely tell you that every time in recent Earth history where we've had these kinds of temperatures for any protracted period of time, two polar ice sheets have catastrophically collapsed," said Jerry X. Mitrovica, an earth physicist at Harvard University (Gillis 2013).

Further Reading

"Ancient Fossils Show Arctic Now Near Climate Tipping Point." Environment News Service, June 29, 2010. http://www.ens-newswire.com/ens/jun2010/2010-06-29-01.html.

Brierley, Chris M., et al. "Greatly Expanded Tropical Warm Pool and Weakened Hadley Circulation in the Early Pliocene." *Science* 323 (March 27, 2009): 1714–1718.

Brigham-Grette, Julie, et al. "Pliocene Warmth, Polar Amplification, and Stepped Pleistocene Cooling Recorded in NE Arctic Russia." *Science* 345 (June 21, 2013): 1421–1427.

Fedorov, A.V., et al. "The Pliocene Paradox (Mechanisms for a Permanent El Niño)." *Science* (June 9, 2006): 1485–1489.

Gillis, Justin. "How High Could the Tide Go?" *The New York Times*, January 21, 2013. http://www.nytimes.com/2013/01/22/science/earth/seeking-clues-about-sea-level-from-fossil-beaches.html.

Hansen, James E. "Is There Still Time to Avoid 'Dangerous Anthropogenic Interference' with Global Climate? A Tribute to Charles David Keeling." Paper delivered to the American Geophysical Union, San Francisco, December 6, 2005. http://www.columbia.edu/~jeh1/keeling_talk_and_slides.pdf.

Hansen, James E. "The Threat to the Planet." *New York Review of Books,* July 2006, 12–16.

Kerr, Richard A. "Looking Back for the World's Climatic Future." *Science* 312 (June 9, 2005): 1456–1457.

Pollack, Henry. *A World without Ice.* London: Avery/Penguin, 2009.

See also: Carbon Cycle Feedbacks; Climate Change, Abrupt Nature of; Ice Melt, Antarctica; Sea Ice, Arctic; Temperatures, Greenhouse Gas Levels and; Thermal Inertia

SEA ICE, ARCTIC

In 1906, Robert Peary first surveyed 3,900 square miles of ice shelves, of which 90 percent have now broken up, said Luke Copland, the director of the University of Ottawa's Laboratory for Cryospheric Research. "The quick pace of these changes right now is what stands out," he said (Revkin 2008). In 1907, when Peary traversed the northern coast of Ellesmere Island on a dog sled, he described a "glacial fringe" along most of the area. At the time, Peary was describing a continuous ice shelf covering 8,900 square kilometers (some 3,500 square miles). Half a century later, a large part of that ice had disintegrated. By July 2008, only five isolated ice shelves remained: Serson, Petersen, Milne, Ward Hunt, and Markham. (The Ayles ice shelf had broken off in 2005.) These were the last remaining ice shelves in Canada.

Rapid ice melt in the Arctic was evident to some observers by the 1980s. By 1989, the Arctic ice was reported to be thinning quickly. The Scott Polar Institute of the United Kingdom reported that ice in an area north of Greenland had thinned from an average of 6.7 meters in 1976 to 4.5 meters in 1987. At about the same time, a British Antarctic Survey base reported unusual losses of ice as well (Kelly 1990, 83).

"Our ice is getting thinner and going out. This year, we had very thin ice and it went out real fast," said Inuit elder Alexander Akeya. "It's getting more windy, and we have to hunt farther out than we used to before" (Braem 1997). Elder Jimmy Toolie had a similar observation: "Today it has changed, the ice moves all the time, and goes away quicker. We used to have it to mid-June. This year it went out in mid-May. Even in the whaling time, it seemed it went away so quick," Toolie said (Braem 1997). Another Inuit, Joe Noongwook, agreed: "This is true. We used to have huge icebergs come first before the young ice formed. We used to have thick ice from here to the [Russian] mainland to the north, but now there is thin ice. There is no hardpack ice" (Braem 1997). Several other Inuit residents said the ice has been thinning, especially since the late 1970s.

Every winter since 1912, when an iceberg claimed the *Titanic* in the North Atlantic, the U.S. Coast Guard has issued warnings for an average of 500 icebergs that float south of 48° north latitude, calving from Greenland's ice sheet. These icebergs threaten the shipping lanes of "Iceberg Alley," the Grand Banks southeast of Newfoundland. During the El Niño winter of 1998–1999, for the first time in 85 years, the Coast Guard did not issue a single iceberg alert for that area. "The lack of ice is remarkable," said Steve Sielbeck, commander of the International Ice Patrol (Wuethrich 1999).

Before 2000, 40 percent to 50 percent of Arctic sea ice remained frozen for several years; by 2008, that ratio was down to 10 percent, according to a team of University of Colorado—Boulder scientists led by Charles Fowler. "Ice extent is an important measure of the health of the Arctic, but it only gives us a two-dimensional view of the ice cover," said Walter Meier, research scientist at the center and the University of Colorado–Boulder. "Thickness is important, especially in the winter, because it is the best overall indicator of the health of the ice cover. As the ice cover in the Arctic grows thinner, it grows more vulnerable to melting in the summer" ("Satellites Show" 2009).

"We're seeing more melting of multiyear ice in the summer," said Stroeve. "We may soon reach a threshold beyond which the sea ice can no longer recover." "We have already witnessed major losses in sea ice, but our research suggests that the decrease over the next few decades could be far more dramatic than anything that has happened so far," said Holland. "These changes are surprisingly rapid" ("Arctic Sea Ice Melt" 2006).

As levels of atmospheric greenhouse gases increase, Arctic sea ice will continue to decline, said Serreze. "While the Arctic is losing a great deal of ice in the summer months, it now seems that it also is regenerating less ice in the winter," said Serreze. "With this increasing vulnerability, a kick to the system just from natural climate fluctuations could send it into a tailspin" ("Arctic Sea Ice Melt" 2006). In the late 1980s and early 1990s, shifting wind patterns from the North Atlantic Oscillation flushed much of the thick sea ice out of the Arctic Ocean and into the North Atlantic where it drifted south and eventually melted, he said.

In addition, beginning in the mid-1990s, scientists observed pulses of relatively warm water from the North Atlantic entering the Arctic Ocean, further speeding ice melt there. According to Serreze, "This is another one of those potential kicks to the system that could evoke rapid ice decline and send the Arctic into a new state" ("Arctic Sea Ice Melt" 2006). "As the ice retreats, the ocean transports more heat to the Arctic and the open water absorbs more sunlight, further accelerating the rate of warming and leading to the loss of more ice," explained scientist Marika Holland of the National Center for Atmospheric Research. "This is a positive feedback loop with dramatic implications for the entire Arctic region" ("Arctic Sea Ice Melt" 2006).

Winter Arctic sea ice thinned between 2004 and 2008, with thinner, single-season ice replacing thicker, older ice on more than half the ice cap as the dominant type of ice for the first time since records have been kept in 1979, according to data from NASA's ICESat spacecraft.

At the annual American Geophysical Union meeting in San Francisco during December 2007, scientists reported that water temperatures near Alaska and Russia were as much as to 9°F above average. A study by scientists at the University of Washington said that the sun's heat made the greatest contribution to the record melting of the Arctic ice cap at the end of summer 2007. Sunlight added twice as much heat to the water as was typical before 2000. Relatively warm water entering the Arctic Ocean from the Atlantic and the Pacific Oceans was a less important factor, said Michael Steele, an oceanographer at the University of Washington. Energy from the warmer water delays the expansion of ice growth as it warms the air. Wieslaw Maslowski of the Naval Postgraduate School in Monterey, California, foresaw blue Arctic Ocean summers by 2013 (Revkin 2007).

In addition to a decrease in the extent of Arctic sea ice, by 2007 large areas averaged only a meter thick, half what they were in 2001, according to measurements taken by an international team of scientists aboard the research ship *Polarstern*. "The ice cover in the North Polar Sea is dwindling, the ocean and the atmosphere are becoming steadily warmer, the ocean currents are changing," said chief scientist Ursula Schauer from the Alfred Wegener Institute for Polar and Marine Research, then aboard the *Polarstern* in the Arctic Ocean ("Arctic Ocean Ice" 2007).

The speed of ice breakup can sometimes be astonishing. For example, the 100-foot-thick Ayles shelf of floating ice roughly 25 miles square that had extended into the Arctic Ocean from the north coast of Ellesmere Island in the Canadian Arctic for roughly 3,000 years detached during the summer of 2005 because of wind and waves in warming water. "The breakup was observed by Laurie Weir of the Canadian Ice Service in satellite images of Ellesmere and surrounding ice on and after Aug. 13, 2005. In less than an hour, a broad crack opened and the ice shelf was on its way out to sea" (Revkin 2008).

The breakup of Ellesmere glaciers continued in August 2008 when a large piece of the 4,500-year-old Markham ice shelf—nearly as large as Manhattan Island—separated and floated into the sea. The glacier's disintegration was sudden and unexpected by scientists, taking a matter of a day or two during a cloudy spell that hid the event from satellites.

Air temperatures above large areas of the Arctic Ocean from January 2006 to August 2006 were approximately 2 to 7°F warmer than the long-term average during the previous 50 years ("Arctic Sea Ice Melt" 2006). Temperatures continued an irregular rise after that. High winter temperatures contributed to limited ice growth, and much of the ice that did form was thinner than normal. "Unusually high temperatures through most of July then fostered rapid melt," Serreze explained. Cooler weather in August slowed the melt, he said, and storm conditions led to wind patterns that helped spread existing ice over a larger area ("Arctic Sea Ice Melt" 2006). A *polynya*, an area of persistent open water surrounded by sea ice, formed north of Alaska. According to one report, "Calculations show[ed] that in early September, the polynya was the size of the state of Indiana, a size never before seen in the Arctic. Unusual wind patterns and an influx of warmer ocean waters may have caused the polyny" ("Arctic Sea Ice Melt" 2006).

Temperatures in the Arctic continued to rise into 2008, reaching an all-time record high of more than 9°F (5°C) above average during the fall of 2007. Temperatures remained almost as high a year later, according to the annual "Arctic Report Card," the product of 46 scientists from 10 countries, edited by Jackie Richter-Menge, a climate expert at the Cold Regions Research and Engineering Laboratory in Hanover, New Hampshire. Although the Arctic has been warming on average, effects vary regionally. The Bering Sea, for example, is in a cooling spell, as an unusually severe winter during 2007–2008 added mass to many of Alaska's glaciers (Boyd 2008). Greenland's ice cap shrank by 88 square miles (220 square kilometers) as a result of an unusually warm spring and summer, according to Jaxon Box of the Byrd Polar Research Center in Columbus, Ohio.

Record Low Maxima in 2015 and 2016

Year by year, in an irregular fashion, the Arctic is becoming warmer and greener. The maximum extent of Arctic sea ice reached a record low on February 25, 2015, at 14.54 million square kilometers (5.61 million square miles), according to the National Snow and Ice Data Center. The year's maximum also was reached 15 days earlier than the 1981 to 2010 average of March 12. A record-low sea ice maximum extent does not always lead to a record-low summertime minimum extent, NASA commented. "The winter maximum gives you a head start, but the minimum is so much more dependent on what happens in the summer," said Walt Meier, a sea ice scientist at NASA's Goddard Space Flight Center. "Scientifically, the yearly maximum is not as interesting as the minimum because it is highly influenced by weather. We're looking at the loss of thin, seasonal ice that is going to melt in the summer anyway, and it won't become part of the permanent ice cover. With the summertime minimum, when the extent decreases, it's because we're losing the thick ice component, and that is a better indicator of warming temperatures" ("Annual Peak" 2015).

Warming ocean waters (which absorb more of the sun's heat than ice and snow) may be playing an important role in the ice shelves' retreat, along with warming air. Air temperatures in the area have risen about 5°C. in 50 years. "As sea ice retreats, the exposed water in front of the shelf warms, and as the ice in the fjord behind the shelf retreats, that water is also warmed. It's possible that warmer water is circulating under the shelf and causing basal melting" said Ted Scambos, lead scientist at the National Snow and Ice Data Center (Scott 2008; Jeffries 2002).

When the Arctic's sea ice reached its annual maximum extent on March 24, 2016, it was at 14.52 million square kilometers (5.607 million square miles). This was less area covered than any other similar point in the satellite record (since 1979), continuing a steady erosion. The 13 smallest maxima have all occurred in the past 13 years, according to the National Snow and Ice Data Center (NSIDC) and NASA. The low sea ice maximum coincided with record high air temperatures in December, January, and February in the Arctic and around the planet. According to Walt Meier, a sea ice scientist at NASA's Goddard Space Flight Center, the atmospheric warmth contributed to the low ice extent because air temperatures along the edges of the ice pack—where sea ice is thinnest—were as much as 6°C (10°F) above average. Wind patterns in the Arctic were also unfavorable; southerly winds in January and February brought warm air and also prevented ice cover from moving toward lower latitudes ("Annual Peak" 2016).

Further Reading

"Annual Peak of Arctic Sea Ice Hit a Record Low." NASA Earth Observatory, April 8, 2016. http://earthobservatory.nasa.gov/IOTD/view.php?id=87831&src=eoa-iotd.

"Arctic Ocean Ice Thinner by Half in Six Years." Environment News Service, September 14, 2007. http://www.ens-newswire.com/ens/sep2007/2007-09-14-03.asp (no longer available).

"Arctic Sea Ice Melt Accelerating." Environment News Service, October 4, 2006. http://www.ens-newswire.com/ens/oct2006/2006-10-04-02.asp (no longer available).

Boyd, Robert S. "Arctic Temperatures Hit Record High." Knight Ridder Washington Bureau, October 16, 2008 (LEXIS).

Braem, Nicole M. "Greenpeace Activists Visit Yup'ik and Inupiat Villages in Alaska to Gather Information about Global Warming." *Nunatsiaq News*, August 8, 1997. http://www.nunanet.com/~nunat/week/70808.html (no longer available).

Jeffries, M. O. "Ellesmere Island Ice Shelves and Ice Islands." *Satellite Image Atlas of Glaciers of the World: Glaciers of North America—Glaciers of Canada.* J147-J164. March 7, 2002 (approved for publication). https://pubs.usgs.gov/pp/p1386j/iceshelves/iceshelves-hires.pdf.

Kelly, Mick. "Halting Global Warming." Pp. 83–112 in Jeremy Leggett, ed., *Global Warming: The Greenpeace Report.* New York: Oxford University Press, 1990.

Revkin, Andrew C. "Arctic Update: Resilient Bears, Shrinking Ice." *The New York Times*, December 13, 2007. http://dotearth.blogs.nytimes.com/2007/12/12/arctic-update-resilient-bears-vanishing-ice/index.html.

Revkin, Andrew C. "Arctic Ocean Ice Retreats Less Than Last Year." *The New York Times*, December 30, 2008. http://www.nytimes.com/2008/09/17/science/earth/17ice.html.

"Satellites Show Arctic Literally on Thin Ice." NASA Earth Observatory. April 7, 2009. http://earthobservatory.nasa.gov/Newsroom/view.php?id=37803&src=eoa-nnews (no longer available).

Scott, Michon. "Rapid Retreat: Ice Shelf Loss Along Canada's Ellesmere Coast." NASA Earth Observatory, September 9, 2008. http://earthobservatory.nasa.gov/Study/Ellesmere.

Wuethrich, Bernice. "Lack of Icebergs Another Sign of Global Warming?" *Science* 285 (July 2, 1999): 37.

See also: Adaptation, Animals, Lizards, and Warming Habitats; Animal Life, Arctic; Arctic Hunters; Atmospheric Circulation; Climate Change, Abrupt Nature of; El Niño and La Niña; Ice Melt, Arctic; Inuit; Permafrost; Summer Ice, Arctic; Temperatures, Global; Temperatures, Greenhouse Gas Levels and

SOLAR INFLUENCE

During the Little Ice Age, roughly 1600 to 1800, New Yorkers occasionally walked from Manhattan Island to Staten Island across the frozen harbor. Changes in the sun's activity may have had a role in this centuries-long cold episode. The cold spell that doomed Viking settlements in Greenland from 1420 to 1570 also occurred during a decline of the sun's magnetic activity, as measured by the prevalence of sun spots. In 2006, astronomers said that the sun was about to enter another quiet period. Others asked whether this cycle might stall some of Earth's warming influences from greenhouse gases. A few climate contrarians even heralded a new ice age, regardless of rapidly rising greenhouse gas levels in the atmosphere (Haigh 2001, 2109).

How Powerful Is the Sun's Influence on Climate?

Calculations by scientists in Germany largely agreed with James E. Hansen's assertion that even a "grand solar minimum" similar to the Maunder Minimum (1645–1715) that caused notable cooling before the Industrial Age would be overwhelmed by increasing greenhouse gas emissions. Georg Feulner and Stefan Rahmstorf of the Potsdam Institute for Climate Impact Research used a global climate model to estimate the effect of a Maunder-type minimum by 2100. Their best estimate was that even the deepest of solar minima would offset only 0.3°C of warming from other sources. Results were published in *Geophysical Research Letters* (Feulner and Rahmstorf 2010).

The sun's output unquestionably plays a role in temperature changes on Earth's surface. From the standpoint of humanity's impact on temperatures, however, the sun's role has declined as anthropogenic emissions of greenhouse gases have risen, illustrating just how strong the human influence has become. As James Lawrence Powell explained,

> Not only does the sun provide virtually all of Earth's surface energy, the sun's activity and global temperatures have correlated in the past. From 1880 until about 1975, solar activity and global temperatures tracked closely. But then the correlation stopped: solar activity has stayed about the same during the period in which global temperatures rose sharply. The year 2009 had the lowest solar activity in 50 years but was the second warmest year on record. (Powell 2011, 134)

Illustrating the power of human effects on climate vis à vis solar forcing, global temperatures in 2010 reached a record high on the instrumental record despite the fact that solar irradiance also was at a cyclical low, at maximum cooling effect—the same cooling effect that some climate contrarians were using to anticipate a cooling cycle (Hansen et al. 2010). The deniers have never appreciated the relative power of human-generated greenhouse gas emissions on Earth's environment.

Humankind's use of fossil fuels is only one influence on climate, although it grows relatively more powerful as the atmosphere's overload of greenhouse gases increases. A team writing in *Nature* (Solanki et al. 2004) found that the level of sunspot activity during the last two-thirds of the 20th century was "exceptional," the highest in roughly 8,000 years. Although the sun's cycles have had long-term effects on climate, these authors asserted that "solar variability is unlikely to have been the dominant cause of the strong warming during the past three decades" (Solanki et al. 2004, 1084).

The level of greenhouse gases in the atmosphere influences surface temperature but not in a simple one-on-one relationship. As Michael E. Mann, Raymond S. Bradley, and Michael K. Hughes point out in *Nature*, the level of greenhouse gases in the atmosphere is one part in a cacophony of influences on a given climate at any particular time and place. In addition to the possibility of warming from greenhouse gases during the past century, there is evidence that both solar irradiance and explosive volcanism have played important roles in forcing climate variations over the past several centuries (Mann et al. 1998, 779).

James E. Hansen, retired director of the Goddard Institute for Space Studies, described a role for the sun in climate change that is overshadowed by Earth-bound forcings such as greenhouse gas levels (in the long term) and La Niña and El Niño and Southern Oscillation cycles, which produce pronounced short-term changes:

> This cyclic solar variability yields a climate forcing change of about 0.3 W/m_2 [watts per square meter] between solar maxima and solar minima. [Although solar irradiance of an area perpendicular to the solar beam is about 1366 W/m_2, the absorption of solar energy averaged over day and night and the Earth's surface is about 240 W/m_2.] Several analyses have extracted empirical global temperature variations of amplitude about 0.1°C associated with the 10–11 year solar cycle, a magnitude consistent with climate model simulations, but this signal is difficult to disentangle from other causes of global temperature change including unforced chaotic fluctuations. The solar minimum forcing is thus about 0.15 W/m_2 relative to the mean solar forcing. For comparison, the human-made GHG climate forcing is now increasing at a rate of about 0.3 W/m_2 per decade. (Goddard Institute for Space Studies 2007)

Hansen and colleagues wrote that "If the sun were to remain 'stuck' in its present minimum for several decades, as has been suggested in analogy to the solar Maunder Minimum of the seventeenth century, that negative forcing would be balanced by a five-year increase of greenhouse gases. Thus, in the current era of rapidly increasing greenhouse gases, such solar variations cannot have a substantial impact on long-term global warming trends is about to get underway" ("GISS 2007" 2008).

In a paper published online by *Science* (www.scinceexpress.org), paleoceanographer Gerard Bond of the Lamont–Doherty Earth Observatory in Palisades, New York, and colleagues reported that the climate of the northern North Atlantic has warmed and cooled nine times in the past 12,000 years in step with the waxing and waning of the sun. "It really looks like the sun has mattered to climate," said Richard Alley. "The Bond et al. data are sufficiently convincing that [solar variability] is now the leading hypothesis" to explain the roughly 1,500-year oscillation of climate seen since the last ice age, including the Little Ice Age of the 17th century, said Alley (Kerr 2001, 1431).

This cycle is now in a rising mode and "could also add to the greenhouse warming of the next few centuries," according to a report by Richard Kerr in *Science* (Kerr 2001, 1431). According to Alley (cited by Kerr), solar variations can "gain leverage on the atmosphere" by changing circulation patterns in the stratosphere, which then affects the lower atmosphere and, finally, ocean circulation, where they affect such climate drivers as the rate at which "deep water" forms in polar regions (Kerr 2001, 1432).

The Sun in Climatic Context

Using lake sediments from southwestern Alaska, Feng Sheng Hu and colleagues have found indications that "small variations in solar irradiance induced cyclic changes in northern high-latitude environments" (Hu et al. 2003, 1890). These changes on century-long time scales were detected in Holocene climate, "similar between the subpolar regions of the North Atlantic and North Pacific, possibly because of sun-ocean-climate linkages" (Hu et al. 2003, 1890).

British Atlantic Survey research suggests that in years to come the sun will exert less impact on climate. According to Mark Clilverd, who led this study, "This work is speculative and relies on the idea that the sun shows regular cycles of activity on timescales of 10 to 10,000 years, and that its heat output and activity are related. . . . We believe the work is well-grounded and the effect of solar activity on Earth's environmental system will not increase in the way it has during the [20th] century" (Radowitz 2003). The reduction in solar activity may be temporary, however. Clilverd said he expected solar activity to return to 20th-century levels by 2200.

A study by Swiss and German scientists also suggested that increasing radiation from the sun may be responsible for part of the recent warming. Sami Solanki, director of the Max Planck Institute for Solar System Research in Gottingen, Germany, and research leader, said. "The sun has been at its strongest over the past 60 years and may now be affecting global temperatures. The sun is in a changed state. It is brighter than it was a few hundred years ago and this brightening started relatively recently—in the last 100 to 150 years" (Leidig and Nikkhah 2004). Solanki said that the brighter sun and higher levels of greenhouse gases have contributed to recent sharp rises in Earth's temperature. Determining the exact proportion of the warming caused by changes in solar activity is impossible with present knowledge, he said.

Solanki's research team measured several hundred years of records kept on sunspots, which are believed to have intensified the sun's energy output. Sunspot activity has been increasing for the past 1,000 years, they found, in a record that roughly conforms to temperature records. Along with other research (cited previously), these researchers found that the increase in sunspot activity was especially notable during the 20th century. Gareth Jones, an English climate researcher, said that Solanki's findings were inconclusive because the study had not incorporated other potential climate-change factors. "The sun's radiance may well have an impact on climate change but it needs to be looked at in conjunction with other factors such as greenhouse gases, sulphate aerosols, and volcano activity," he said (Leidig and Nikkhah 2004).

Further Reading

Feulner, G., and S. Rahmstorf. "On the Effect of a New Grand Minimum of Solar Activity on the Future Climate on Earth." *Geophysical Research Letters* 37 (2010): L05707. doi: 10.1029/2010GL042710.

"GISS 2007 Temperature Analysis." Goddard Institute for Space Studies. January 15, 2008. http://www.columbia.edu/~jeh1/mailings/20080114_GISTEMP.pdf.

Goddard Institute for Space Studies. "GISS Surface Temperature Analysis." Washington, DC: National Aeronautics and Space Administration, 2007.

Haigh, Joanna D. "Climate Variability and the Influence of the Sun." *Science* 294 (December 7, 2001): 2109–2111.

Hansen, James E., et al. "Global Surface Temperature Change." *Reviews of Geophysics*, 2010. http://onlinelibrary.wiley.com/doi/10.1029/2010RG000345/abstract.

Hu, Feng Sheng, et al. "Cyclic Variation and Solar Forcing of Holocene Climate in the Alaskan Subarctic." *Science* 301 (September 26, 2003): 1890–1893.

Kerr, Richard A. "A Variable Sun Paces Millennial Climate." *Science* 294 (November 16, 2001): 1431–1432.

Leidig, Michael, and Roya Nikkhah. "The Truth about Global Warming—It's the Sun That's to Blame." *Sunday Telegraph* (London), July 18, 2004, 5.

Mann, Michael E., Raymond S. Bradley, and Michael K. Hughes. "Global-Scale Temperature Patterns and Climate Forcing over the Past Six Centuries." *Nature* 392 (April 23, 1998): 779–787.

Powell, James Lawrence. *The Inquisition of Climate Science*. New York: Columbia University Press, 2011.

Radowitz, John von. "Calmer Sun Could Counteract Global Warming." Press Association, October 5, 2003 (LEXIS).

Solanki, S. K., et al. "Unusual Activity of the Sun during Recent Decades Compared to the Previous 11,000 Years." *Nature* 431 (October 28, 2004): 1084–1086.

See also: Atmospheric Circulation; Temperatures, Global

SUMMER ICE, ARCTIC

Global warming is being felt most intensely in the Arctic and coastal regions of Antarctica, where a world based on ice and snow has been melting away. Arctic sea ice cover shrank more dramatically between 2000 and 2015 than at any time since detailed records have been kept. A report produced by 250 scientists under the auspices of the Arctic Council found that Arctic sea ice was half as thick in 2003 as it was 30 years earlier. Erosion of ice cover continued after that until at least 2015. If current rates of melting continue, there may be no summer ice in the Arctic ocean by as early as 2020.

Pal Prestrud, vice chairman of the steering committee for the aforementioned report for the Arctic Council, said that, "Climate change is not just about the future; it is happening now. The Arctic is warming at twice the global rate" (Harvey 2004). Since that report, ice loss has accelerated, most notably during 2007 and 2008, giving rise to forecasts of an ice-free Arctic Ocean in summer between 2020 and 2040.

Melting of the Arctic ice cap could lead to summertime ice-free ocean conditions not seen in the area in a million years, a group of scientists wrote during 2005 in *Eos*, the transactions journal of the American Geophysical Union. Jonathan Overpeck of the University of Arizona and chair of the National Science Foundation's Arctic System Science Committee and colleagues wrote that the Arctic during the 21st century is moving beyond the glacial and interglacial cycle that has characterized the last million years. "At the present rate of change," they wrote, "a summer ice-free Arctic ocean within a century is a real possibility . . . a state [that is] driven largely by feedback-enhanced global climate warming . . . [into] a 'super interglacial' state" (Overpeck et al. 2005, 309).

The eighth annual "Arctic Report Card" issued in 2014 by the National Oceanic and Atmospheric Administration (NOAA) said that Arctic ice continued to melt because darkening surfaces were absorbing more sunlight and heat. Spring snow cover was at a record low in spring 2014 across Europe and Asia. Greenland's reflectivity was lower in August of that year than at any other time since records have been kept. The report was compiled by 63 scientists from 13 countries.

"We can't expect records every year. It need not be spectacular for the Arctic to continue to be changing," said report lead editor Martin Jeffries, an Arctic scientist for the Office of Naval Research (Borenstein 2014). The Arctic's drop in reflectivity is crucial because "it plays a role like a thermostat in regulating global climate," Jeffries told an interviewer. The replacement of bright areas, even temporarily, with dark heat-absorbing dark areas "has global implications" (Borenstein 2014). Between 2000 and 2014, NASA calculates that solar radiation absorbed in the Arctic during summer increased 5 percent because of changes in albedo (reflectivity) (Chang 2014).

"The large area of the Arctic Ocean promises to become much warmer," said Igor Polyakov, principal investigator at Nansen and Amundsen Basins Observational Systems and a research professor at the International Arctic Research Center. Instruments first detected a surge of abnormally warm water approximately 150 to 800 meters below the surface during February 2004 on the continental slope of the Laptev Sea north of Siberia ("Warm Water" 2006).

Ice-Melt Records Set Year after Year

During September 2007, the Arctic ice cap shrank to its smallest extent since satellite records first started being kept in 1979—1.59 million square miles versus the previous record low of 2.04 square miles in 2005, more than a 20-percent loss of ice cover, an area the size of Texas and California combined. By 2007, the Arctic ice cap had lost 43 percent of its extent since 1979 (Kerr 2007, 33).

Arctic Hockey on Artificial Ice

By the winter of 2002–2003, a warming trend was forcing hockey players in Canada's far north to seek rinks with artificial ice. As Canada's *Financial Post* reported, "Officials in the Arctic say global warming has cut hockey season in half in the past two decades and may hinder the future of development of northern hockey stars such as Jordin Tootoo" ("Ice a Scarce Commodity" 2003). According to the report, hockey rinks in northern communities were raising funds so they could install cooling plants to create artificial ice because of the reduced length of time during which natural ice was available. In Rankin Inlet, Tootoo's hometown on Hudson Bay in Nunavut, a community of 2,400 residents installed artificial ice during the summer of 2003 after eight years of lobbying for the funds ("Ice a Scarce Commodity" 2003).

Hockey season on natural ice, which ran from September until May in the 1970s, often now begins around Christmas and ends in March, according to Jim MacDonald, president of Rankin Inlet Minor Hockey. "It's giving us about three months of hockey. Once we finally get going, it's time to stop. At the beginning of our season, we're playing teams that have already been on the ice for two or three months," MacDonald said ("Ice a Scarce Commodity" 2003.) According to Tom Thompson, president of Hockey Nunavut, there are around two dozen natural ice rinks in tiny communities throughout the territory, but only two with artificial ice. Both are in the capital, Iqaluit ("Ice a Scarce Commodity" 2003).

"In my lifetime I will not be surprised if we see a year where Hudson Bay doesn't freeze over completely," said Jay Anderson of Environment Canada. "It's very dramatic. Yesterday [January 6, 2003], an alert was broadcast over the Rankin Inlet radio station warning that ice on rivers around the town is unsafe. The temperature hovered around [−12°C]. It's usually [−37°C] there at this time of year" ("Ice a Scarce Commodity" 2003).

Further Reading

"Ice a Scarce Commodity on Arctic Rinks: Global Warming Blamed for Shortened Hockey Season." *Financial Post* (Canada), January 7, 2003, A3.

In 2008, the Arctic ice cap did not shrink quite as much as in 2007, but it was still 33 percent below the average since 1979. At summer's end, the ice cap was 1.74 million square miles, according to the National Snow and Ice Data Center (NSIDC) in Boulder, Colorado. The year 2008 also was notable because two potential shipping routes, the Northwest Passage off Canada and the Northern Sea Route adjacent to Russia, opened at the same time. Walter Meier, a research scientist at the center, said that small variations from one year to the next were less significant than the long-term trajectory, which remained toward progressively more open water. "It's hard to see the summer ice coming back in any substantial way," he said (Revkin 2008).

During 2008, NASA's ICESat satellite measured the largest annual decline in "perennial" ice (that frozen more than one year) on record. The proportion of the Arctic covered with this stable type of ice had declined from more than 50 percent during the mid-1980s to less than 30 percent in February 2008. Josefino Comiso of NASA's Goddard Space Flight Center said that Arctic Ocean temperatures appear to be rising quickly because less of the water is covered by ice, which reflects sunlight and keeps water temperatures lower (Kaufman 2008). Flying over the Arctic, one might perceive the sea ice cover as broad, Meier said, but that apparent breadth hides the fact that the ice is so thin. "It's a facade, like a Hollywood set," he said. "There's no building behind it" (Kaufman 2008).

On July 22, 2008, a new wave of ice shelf disintegration began, and these ice shelves lost another 214 square kilometers (83 square miles) by late August.

The Ward Hunt ice shelf lost a total of 42 square kilometers (16 square miles). The Serson ice shelf lost 122 square kilometers (47 square miles), 60 percent of its previous area. The Markham ice shelf, with a total area of 50 square kilometers (19 square miles), completely broke away from the Ellesmere coast (Scott 2008; Copland et al. 2007).

Arctic sea ice cover continued to thin during 2009. Winter ice early in 2009 continued to freeze less thickly and in a more fragile way than in previous years. "We know that short-lived carbon forcers like methane, black carbon, and tropospheric ozone contribute significantly to the warming of the Arctic," said Secretary of State Hillary Clinton said. "And because they are short-lived, they also give us an opportunity to make rapid progress if we work to limit them" (Eilperin and Sheridan 2009). Satellite data released by NASA and the NSIDC showed that the maximum extent of the 2008–2009 winter sea ice cover was the fifth-lowest since measurements began 30 years previously. All five low maxima had been during the previous five years. Multiyear ice shrank to les than 10 percent of the total, its lowest ever. "We're seeing an ice cover that's younger and that's thinner as we head into summer," Walt Meier, a scientist at the NSIDC, said in a telephone news conference. "It's been a pretty sharp decline" (Eilperin and Sheridan 2009).

After slight increases in 2009, Arctic ice resumed its decline in 2010 as temperatures rose to 7.2°F above average in northern Canada. Winter snow accumulations in the Arctic were the lowest since records began in 1966, The Petermann glacier in Greenland lost an area equal to several times the size of Manhattan Island. The summer Arctic ice minimum in 2010 was the third smallest in the satellite era. At its smallest extent, on September 10, 4.76 million square kilometers (1.84 million square miles) of Arctic Ocean was covered with ice, more than in 2007 and 2008 but less than in every other year since 1979.

On September 19, 2010, Arctic sea ice reached a minimum of 1.78 million square miles, or 4.6 million square kilometers, the second lowest since satellite records began in 1979, according to NASA. The National Snow and Ice Data Center called it the third lowest ice extent on record. About six months later, in March 7, 2011, the sea ice reached its second lowest maximum on the satellite record, at 5.62 million square miles, or 14.56 million square kilometers. The new ice had formed with nearly none older than four years that once had dominate much of the Arctic Ocean ("Arctic Sea Ice" 2011).

> The now-clearly-accelerating decline of summer ice—punctuated by exceptional losses in 2007 and now in 2012—has persuaded everyone that summer Arctic sea ice will be a goner far sooner than the end of the century, as current models predict. So the full knock-on effects of an ice-free Arctic Ocean— from the loss of polar bear habitat to possible increases of weather extremes at mid-latitudes—could be here in many people's lifetimes. How far wrong the models might be, however, is still very much in dispute. (Kerr 2012, 1591)

In 2011, the NSIDC reported that Arctic sea ice reached its lowest extent recorded in January since satellite records had begun in 1979. Areas in Hudson Bay, Hudson Strait, and Davis Strait, which usually freeze by late November, did not completely freeze until mid-January 2011. In the meantime, displaced Arctic air pushed southward over the United States and combined with air masses from the Gulf of Mexico to produce winter storms of historic proportions. One of these in the first week of February 2011 enveloped most of the United States east of the Rockies ("Record Low" 2011). In September 2011, sea ice covering the Arctic Ocean declined to the second-lowest extent on record. Satellite data from NASA and the National Snow and Ice Data Center showed that the summertime ice cover narrowly avoided a new record low ("2011 Arctic" 2011).

Melt season in 2011 brought higher-than-average summer temperatures but not the unusual weather conditions that contributed to the extreme melt of 2007, the record low. "Atmospheric and oceanic conditions were not as conducive to ice loss this year, but the melt still neared 2007 levels," said Walt Meier of the NSIDC. "This probably reflects loss of multiyear ice in the Beaufort and Chukchi Seas, as well as other factors that are making the ice more vulnerable" ("2011 Arctic" 2011).

"The low sea ice level in 2011 fits the pattern of decline over the past three decades," said Joey Comiso of NASA's Goddard Space Flight Center. Since 1979, September Arctic sea ice extent has declined by 12 percent per decade. "The sea ice is not only declining; the pace of the decline is becoming more drastic," he noted. "The older, thicker ice is declining faster than the rest, making for a more vulnerable perennial ice cover" ("2011 Arctic" 2011).

In August and September 2012, sea ice covered less of the Arctic Ocean than at any other time since at least 1979 when the first reliable satellite measurements began. In mid-September, Arctic sea ice had declined to 3.41 million square kilometers (1.32 million square miles), much less than the previous record of 4.17 million square kilometers (1.61 million square miles) in 2007. Melting sea ice feeds itself. With less ice to reflect sunlight, the dark sea surface absorbs more energy. Less ice survives year to year, and the new ice is thinner and melts faster.

A few years previously, scientific forecasts had anticipated an ice-free Arctic Ocean in summer by the middle of the 21st century, but after several years of rapid ice melt those projected dates were being moved up to as early as 2020. "It's an example of how uncertainty is not our friend when it comes to climate-change risk," said Michael E. Mann, a climate scientist at Pennsylvania State University. "In this case, the models were almost certainly too conservative in the changes they were projecting, probably because of important missing physics" (Gillis 2012).

By 2012, nearly all of Greenland's ice was melting in the summer, and some scientists were anticipating an ice-free Arctic Ocean for part of the summer within a decade or two. "Things are happening much faster than what any scientific model predicted," said Dr. Morten Rasch, who runs the Greenland Ecosystem Monitoring program at Aarhus University in Denmark (Rosenthal 2012).

In mid-September 2012, sea-ice extent in the Arctic Ocean declined to 3.41 million square kilometers (1.32 million square miles), significantly below the previous record of 4.17 million square kilometers (1.61 million square miles) set in 2007.

As NASA's Earth Observatory reported, "The summer of 2012 witnessed periods of extremely rapid sea ice melt" ("Visualizing" 2012). "In June, sea ice lost about 170,000 to 175,000 square kilometers per day, but only for a few days," explained scientist Walt Meier of the National Snow and Ice Data Center. "Sea ice melt usually slows down in August to about 60,000 to 70,000 square kilometers a day. But this year, we saw melt rates upward of 100,000 to 150,000 square kilometers a day, and those high melt rates persisted through the first half of the month" ("Visualizing" 2012).

Summer 2012's low 3.41 million square miles of water-laced ice was 18 percent below the 2007 previous record (an area as large as Texas) and half of the area measured by the first satellite photos in 1979. Between 1979 and 2001, ice declined in the Arctic on an average of 6.5 percent a year. In the next 11 years, the average rate of decline doubled. "The rules are changing," said Mark Serreze of the National Snow and Ice Data Center. "The ice is so thin, it just can't survive the summer" (Kerr 2012).

The Arctic's summer of 2013 was cooler than 2012, when ice extent hit a record low, but David Kennedy of the National Oceanic and Atmospheric Administration (NOAA), said that this sort of year-to-year variation in some parts of the Arctic did little to "to offset the long-term trend of the last 30 years" (Rice 2013). The annual "Arctic Report Card" issued by NOAA noted that although the extent of summer sea ice did not match 2012, it was the sixth lowest since satellite records began in 1979. Signs of warming were evident, however: Fairbanks, Alaska, had a record 36 days above 80°F. Sea surface temperatures in the Barents and Kara Seas were as much as 7°F above long-term averages.

Arctic sea ice had a relatively average year in 2015, reaching its annual minimum on September with the fourth lowest coverage since space-based observations began in 1978. "We haven't seen any major weather event or persistent weather pattern in the Arctic this summer that helped push the extent lower, as often happens," said Walt Meier, a sea ice scientist at NASA's Goddard Space Flight Center ("Arctic Sea Ice Reaches" 2015). By 2015, the 10 lowest minimum ice extents on the satellite record had occurred during the previous 11 years, illustrating the steady erosion of the Arctic ice cap.

Further Reading

"Annual Peak of Arctic Sea Ice Is Far Below the Norm." NASA Earth Observatory, March 21, 2015. http://earthobservatory.nasa.gov/IOTD/view.php?id=85562&src=eoa-iotd.

"Arctic Sea Ice." NASA Earth Observatory. May 3, 2011. http://earthobservatory.nasa.gov/IOTD/view.php?id=50365&src=eoa-iotd.

"Arctic Sea Ice Reaches Annual Low." NASA Earth Observatory, September 16, 2015. http://earthobservatory.nasa.gov/IOTD/view.php?id=86607&src=eoa-iotd.

Borenstein, Seth. "Report: Arctic Loses Snow, Ice; Absorbs More Heat." *Houston Chronicle*, December 20, 2014. http://www.houstonchronicle.com/news/politics/article/Report-Arctic-loses-snow-ice-absorbs-more-heat-5971210.php#/0.

Chang, Kenneth. "Snow Is Down and Heat Is Up in the Arctic, Report Says." *The New York Times*, December 18, 2014. http://www.nytimes.com/2014/12/18/science/snow-is-down-and-heat-is-up-in-the-arctic-report-says.html.

Copland, L., D. R. Mueller, and L. Weir. "Rapid Loss of the Avies Ice Shelf, Ellesmere Island, Canada." *Geophysical Research Letters* 34 (2007): L21501.

Eilperin, Juliet, and Mary Beth Sheridan. "New Data Show Rapid Arctic Ice Decline." *Washington Post*, April 7, 2009, A3. http://www.washingtonpost.com/wp-dyn/content/article/2009/04/06/AR2009040601634_pf.html.

Gillis, Justin. "Satellites Show Sea Ice in Arctic Is at a Record Low." *The New York Times*, August 27, 2012. http://www.nytimes.com/2012/08/28/science/earth/sea-ice-in-arctic -measured-at-record-low.html.

Harvey, Fiona. "Arctic May Have No Ice in Summer by 2070, Warns Climate Change Report." *Financial Times* (London), November 2, 2004, 1.

Kaufman, Marc. "Perennial Arctic Ice Cover Diminishing, Officials Say." *Washington Post*, March 19, 2008, A3. http://www.washingtonpost.com/wp-dyn/content/article/2008/03 /18/AR2008031802903_pf.html.

Kerr, Richard A. "Is Battered Arctic Sea Ice Down for the Count?" *Science* 318 (October 5, 2007): 33–34.

Kerr, Richard A. "Ice-Free Arctic Sea May Be Years, Not Decades, Away." *Science* 337 (September 28, 2012): 1591.

Overpeck, J. T., et al. "Arctic System on Trajectory to New, Seasonally Ice-Free State." *Eos* 86(34) (August 23, 2005): 309, 312.

"Record Low Arctic Sea Ice Extent for January." NASA Earth Observatory, February 8, 2011. http://earthobservatory.nasa.gov/IOTD/view.php?id=49132&src=eoa-iotd.

Revkin, Andrew C. "Arctic Ocean Ice Retreats Less Than Last Year." *The New York Times*, December 30, 2008. http://www.nytimes.com/2008/09/17/science/earth/17ice.html.

Rice, Doyle, "Arctic Gets a Break in 2013, but Warming Still Ongoing." *USA Today*, December 13, 2013, 3A.

Rosenthal, Elisabeth. "Race Is On as Ice Melt Reveals Arctic Treasures." *The New York Times*, September 18, 2012. http://www.nytimes.com/2012/09/19/science/earth/arctic-resources -exposed-by-warming-set-off-competition.html.

Scott, Michon. "Rapid Retreat: Ice Shelf Loss along Canada's Ellesmere Coast." NASA Earth Observatory, September 9, 2008. http://earthobservatory.nasa.gov/Study/Ellesmere.

"2011 Arctic Sea Ice." NASA Earth Observatory. October 4, 2011. http://earthobservatory .nasa.gov/IOTD/view.php?id=53108&src=eoa-iotd. Accessed October 5, 2011.

"Visualizing the 2012 Sea Ice Minimum." NASA Earth Observatory, September 27, 2012. http://earthobservatory.nasa.gov/IOTD/view.php?id=79256&src=eoa-iotd.

"Warm Water Surging into Arctic Ocean." Environment News Service, September 27, 2006. http://www.ens-newswire.com/ens/sep2006/2006-09-27-01.asp (no longer available).

See also: Adaptation, Animals, Lizards, and Warming Habitats; Animal Life, Arctic; Arctic Hunters; Atmospheric Circulation; Climate Change, Abrupt Nature of; El Niño and La Niña; Ice Melt, Arctic; Inuit; Permafrost; Sea Ice, Arctic; Temperatures, Global; Temperatures, Greenhouse Gas Levels and

TEMPERATURES, GLOBAL

In August 2010, meteorologists from 189 countries under the aegis of World Meteorological Organization (WMO) said, "Several diverse extreme weather events are occurring concurrently around the world, giving rise to an unprecedented loss of human life and property. They include the record heat wave and wildfires in the Russian Federation, monsoonal flooding in Pakistan, rain-induced landslides in China, and calving of a large iceberg from the Greenland ice sheet" ("Extreme Weather" 2010).

"These should be added to the extensive list of extreme weather-related events, such as droughts and fires in Australia and a record number of high-temperature days in the eastern United States of America, as well as other events that occurred earlier in the year," said the WMO, a specialized agency of the United Nations ("Extreme Weather" 2010). "While a longer time range is required to establish whether an individual event is attributable to climate change, the sequence of current events matches IPCC [Intergovernmental Panel on Climate Change] projections," the WMO said ("Extreme Weather" 2010).

The NASA Goddard Institute for Space Studies (GISS) keeps tabs on 6,300 stations worldwide (on land and sea) and issues maps digesting these data that are publicly available on the Internet. "When you look at all the red in that map, there is no doubt that April through September was unusually warm in most of the world, but it's the decadal trend that is more significant," said Gavin Schmidt, GISS director. "Earth has experienced rapid warming in the last few decades, and the most recent decade was the warmest of all. What has happened so far in 2014 extends this ongoing trend. But in the context of climate change, it does not make sense to try to derive much meaning from a single month—or, for that matter, even a single year" ("Rising Temperatures" 2014). Temperatures during the record year of 2014 were warmer over the oceans than land compared to averages.

Then 2015 became the hottest year in recorded history by far, breaking the record set just the year before. In fact, the only areas that were colder than average in 2015 were parts of the oceans off Greenland and Antarctica where rapidly melting ice is cascading into the oceans, chilling the air above. In geophysical terms, the margin of the new record was astonishing—0.23°F (0.23°C) according to NASA, and 0.29°F (0.16°C) as measured by the National Oceanic and Atmospheric Administration (NOAA). New global highs or lows often are measured in few hundredths of a degree. A strong El Niño was an important factor, but so was long-term planetary warming caused by human emissions of greenhouse gases. "The whole system is warming up, relentlessly," said Gerald A. Meehl, a scientist at the National Center for Atmospheric Research in Boulder, Colorado (Gillis 2016).

The worldwide rise in temperatures is not a simple, linear reaction to greenhouse gas emissions. For example, rapid warming in the Arctic may be related to heat waves in the midlatitudes because it inhibits atmospheric circulation by reducing the temperature gradient between the midlatitudes and the poles. High-pressure systems in midlatitudes then move more slowly and build summer heat at the surface (Coumou et al. 2015).

Scotland: Climate Changes

More than 12,000 forts, castles, and other archaeological sites that stand along Scotland's shorelines are being eroded into the sea "as rising tides and relentless storms, brought about by global warming, eat away at the coastline" (Grant 2001). Among the sites most at risk of eroding into the sea is

Scrabster Castle, built in the 12th century by the bishop of Caithness, near Thurso. This structure is referenced in the Vikings' Orkney Inga saga, which dates to 1196.

Dunbar Castle, where Mary, Queen of Scots, sought refuge after the murder of her second husband, also has been threatened by the sea. Tom Dawson, a researcher at St. Andrews University's School of History, said, "Parts of the coastline are receding up to one meter a year" (Grant 2001). A third-century fort on Orkney, the Borch of Borwick, also has been crumbling into the sea "with pottery and human skeletons falling from ruins into the water" (Grant 2001).

In another sign that Scotland's climate has warmed, hops—the dried flowers that give beer its bitter flavor—have invaded the Scottish countryside. Previously confined to the southern British Isles and the southern English Midlands, hops have now been sighted in large patches in at least a dozen locations along roadways near Pathhead, Midlothian. Botanics specialist Brian Moffat was quoted in Glasgow's *The Herald* as saying, "Finding one hop plant here is usually a little bit of a freak event; finding so many is really extraordinary" ("Changes in Climate" 2000).

Elsewhere in Scotland, two of the area's five ski resorts were put up for sale early in 2004 after they suffered large financial losses. The owners of Glencoe and Glenshee put the resorts on the market after deciding that they could no longer afford to keep them open. Mild winters and lack of snow have left the winter sports industry in Scotland in poor financial shape. With the pace of global warming accelerating, some in Scotland expected that the country's skiing industry would cease to exist within 20 years.

Fred Last, president of the Royal Caledonian Horticultural Society, who has recorded changes in plants for a quarter century, said he had observed significant changes in Scotland. Based upon his continued observations of 800 species, he said: "There is good evidence that some plants are flowering earlier than before but the patterns of flowering have changed. Some flowering plants are showing about three weeks earlier, gardeners will need to think about their herbaceous borders to fill in any possible gaps that may occur" ("Climate Change" 2002). "If all the predictions are right, 2050 could turn out to be an excellent year for Scottish wine," quipped *The Herald* (Crilly 2002).

Further Reading

"Changes in Climate Bring Hops Northward." *The Herald* (Glasgow), September 29, 2000, 13.

"Climate Change in the Back Garden." *The Scotsman* (Edinburgh), November 20, 2002. http://www.scotsman.com/news/climate-change-in-the-back-garden-1-629292.

Crilly, Rob. "2050 to Be Good Year for Scottish Wine; Global Warming Will Bring Grapes North." *The Herald* (Glasgow), November 20, 2002, 11.

Grant, Christine. "Swelling Seas Eating Away at Country's Monuments." *The Scotsman* (Edinburgh), December 24, 2001, 5.

With the percentage of the world experiencing extreme heat soaring from 0.2 percent to as much as 13 percent within a decadal time scale, James E. Hansen and colleagues concluded that such changes could not be natural: "The main thing is just to look at the statistics and see that the change is too large to be natural. . . . It confirms people's suspicions that things are happening" to the climate. "It's just going to get worse" (Gillis 2012).

Hansen wrote of the 2012 study:

This is not a climate model or a prediction but actual observations of weather events and temperatures that have happened. Our analysis shows that it is no longer enough to say that global warming will increase the likelihood of extreme weather and to repeat the caveat that no individual weather event can be directly linked to climate change. To the contrary, our analysis shows that, for the extreme hot weather of the recent past, there is virtually no explanation other than climate change. (Hansen 2012)

Hansen included here the European heat wave of 2003 that killed more than 50,000 people, the Russian heat wave of 2010 that destroyed several million acres of crops, as well as the extremely hot summer during 2012 in the United States. That event drove water temperatures in some streams to nearly 100°F and killed uncounted numbers of fish. "These weather events are not simply an example of what climate change could bring. They are caused by climate change. The odds that natural variability created these extremes are minuscule, vanishingly small," Hansen wrote (2012).

Hansen was one of the first scientists to identify threats from climate change and advocate action to counter carbon dioxide emissions. He testified before Congress in 1988 that global warming had begun, and he has been arrested several times during protests of coal mining that uses mountain-top removal and the proposed Keystone XL pipeline. In 2008, a government watchdog agency, responding to a complaint from Hansen, said political appointees in NASA's public-affairs office from 2004 through early 2006 "reduced, marginalized or mischaracterized" information on climate-change science for political advantage (Drajem 2012).

Temperatures at a Critical Level

According to NASA scientists, the world's temperatures in 2006 reached levels never before seen in thousands of years. James Hansen, then director of NASA's Goddard Institute for Space Studies, joined other scientists in concluding that because of a rapid warming trend during the past 30 years Earth is now reaching and passing through the warmest levels in the current interglacial period, which has lasted nearly 12,000 years. This evidence implies that we are getting close to dangerous levels of human-made pollution, said Hansen.

Notable Warming in the Coldest Places

A taste of what is to come in Greenland was provided in late 2004 in a report in *Nature* describing how Greenland's Jakobshavn glacier's advance toward the sea had accelerated significantly since 1997. The speed of the glacier nearly doubled between 1997 and 2004 to nearly 50 cubic kilometers a year. This single glacier was responsible for around 4 percent of worldwide sea level rise during the 20th century (Joughin et al. 2004, 608). The glacier's speed continued to increase after that.

Ocean warming has been most notable, relatively speaking, where the sea is usually the coldest, including waters adjacent to Antarctica. A study using seven decades of temperature data indicated that mid-depth water around Antarctica was warming nearly twice as quickly as the oceans' worldwide average (Cowen 2002). Sarah T. Gille's analysis found that the mid-depth water had warmed 0.17°C since 1950 (Gille 2002, 1275). Gille, of the University of California–San Diego, said, "We thought the ocean between 700 and 1,100 meters (2,300 and 3,600 feet) was pretty well insulated from what's happening at the surface. But these results suggest that the mid-depth Southern Ocean is responding and warming more rapidly than global ocean temperatures [generally]" (Cowen 2002).

Gille noted that the Southern Ocean "is a very climatically sensitive region" (Cowen 2002). It is at one end of the conveyor-belt circulation that carries heat toward the poles in upper-level currents and returns cold water back toward the equator at great depths—a key part of the system that maintains Earth's current climate. Any change in Antarctic waters could directly affect circulations in the Atlantic, Indian, and Pacific Oceans (Cowen 2002). Ocean circulation already may have been affected. Gille, who reported her findings in *Science*, said that her study suggests that this current has moved closer to the polar continent as the mid-depth water has warmed (Cowen 2002).

Further Reading

Cowen, Robert C. "Into the Cold? Slowing Ocean Circulation Could Presage Dramatic—and Chilly—Climate Change." *Christian Science Monitor*, September 26, 2002, 14.

Gille, Sarah T. "Warming of the Southern Ocean since the 1950s." *Science* 295 (February 15, 2002): 1275–1277.

Joughin, I., W. Abdalati, and M. Fahnestock. "Large Fluctuations in Speed on Greenland's Jakobshavn Isbrae Glacier." *Nature* 432 (December 2, 2004): 608–610.

"The most important result found by these researchers," according to a NASA statement, "is that the warming in recent decades has brought global temperature to a level within about one degree Celsius (1.8°F) of the maximum temperature of the past million years" (NASA Study 2006). According to James Hansen,

That means that further global warming of 1 [°C] defines a critical level. If warming is kept less than that, effects of global warming may be relatively manageable. During the warmest interglacial periods the Earth was reasonably similar to today. But if further global warming reaches 2 or 3 [°C], we will likely see changes that make Earth a different planet than the one we know. The last time it was that warm was in the middle Pliocene, about three million years ago, when sea level was estimated to have been about 25 meters (80 feet) higher than today. ("NASA Study" 2006)

Although no single temperature reading "proves" that Earth's atmosphere is undergoing a sustained warming spell because of human activities, several hundred readings over many years establish a fund of evidence that indicates a trend. Averages as well as daily temperature reports indicate just such a trend. According to an analysis by the Goddard Institute for Space Studies,

The year 2007 tied for second warmest in the period of instrumental data [to that time], behind the record warmth of 2005. The year 2007 tied 1998, in a remarkably different climatic context. First, temperatures in 1998 leapt 0.2°C above the prior record, a very steep increase, aided by the most intense El Niño conditions in a century. Nine years later, a La Niña condition was in sway, which usually cools world temperatures (2008, also a La Niña year, was slightly cooler worldwide).

Where Is the "Tipping Point"?

How high will temperatures go if greenhouse gas levels continue to increase as they have been? Under business-as-usual greenhouse gas emissions, by 2100 New York City may feel like Las Vegas does today (with additional humidity), and San Francisco will have a climate comparable to that of today's New Orleans. By the same date, Boston will have average temperatures similar to Memphis, Tennessee, today, according to a report, "The Cost of Climate Change." The report, released on May 23, 2008, by researchers at Tufts University (and commissioned by the Natural Resources Defense Council) forecast that temperatures will rise 13°F on average in the United States and 18°F in Alaska ("Global Warming Sticker Shock" 2008).

Researchers have reported that Earth's ancient stores of peat are gasifying into the atmosphere at an accelerating rate that is adding significantly to the atmosphere's overload of greenhouse gases. Given the fact that one-third of Earth's carbon is stored in far northern latitudes (mainly in tundra and boreal forests), the speed by which warming of the ecosystem releases this carbon dioxide to the atmosphere is vitally important to forecasts of global warming's speed and effects. The amount of carbon stored in Arctic ecosystems also makes up two-thirds of the amount currently found in the atmosphere. Its release into the atmosphere will depend on the pace of temperature rise—and the Arctic, according to several sources, has been the most rapidly warming region on Earth.

Further Reading

"Global Warming Sticker Shock." Environment News Service, May 23, 2008. http://www.ens-newswire.com/ens/may2008/2008-05-23-01.asp (no longer available).

The unusual warmth in 2007 was also noteworthy because it occurred when solar irradiance was at a minimum in the recent 10- to 11-year cycle. Thus, the 2007 averages, according to GISS, "continue the strong warming trend of the past 30 years that has been confidently attributed to the effect of increasing human-made greenhouse gases" ("GISS 2007" 2008).

The GISS summary rebutted assertions by contrarians that warming has been stopped in its tracks:

Because both of these natural effects were in their cool phases in 2007, the unusual warmth of 2007 is all the more notable. It is apparent that there is no letup in the steep global warming trend of the past 30 years. "Global warming stopped in 1998" has become a recent mantra of those who wish to deny the reality of human-caused global warming. The continued rapid increase of the five-year running mean temperature exposes this assertion as nonsense. In reality, global temperature jumped two standard deviations above the trend line in 1998 because the "El Niño of the century" coincided with the calendar year, but there has been no lessening of the underlying warming trend. ("GISS 2007" 2008)

The year 2012 was the ninth warmest year on the instrumental record (since 1880), according to NASA's Goddard Institute for Space Studies, cooperating with the U.S. National Climatic Data Center, the Japanese Meteorological Agency, and the British Met Office. The average temperature was 14.6°C (58.3°F), which was 0.55°C (1.0°F) warmer than the mid-20th century, which is used as a base period. By 2012, the average worldwide temperature had increased 0.8°C (1.4°F) since record keeping began, most it within the previous four decades. In the United States, 34,008 record daily highs were set, as were 6,664 record lows.

"One more year of numbers isn't in itself significant," GISS climatologist Gavin Schmidt said. "What matters is [that] this decade is warmer than the last decade, and that decade was warmer than the decade before. The planet is warming. The reason it's warming is because we are pumping increasing amounts of carbon dioxide into the atmosphere" ("Long-Term" 2013). In 2012, the United States experienced its warmest year on record with heat waves in March and summer. July 2012 was the hottest more than 1,400 months on the instrumental record since 1895, exceeding July 1936 at the peak of the Dust Bowl drought. In 2013 and 2014, however, the country's temperatures fell as the world average rose.

Worldwide temperatures reached records in the oceans before they did the same on land. During June 2009, the average temperature of the world's ocean surface rose to a record high since records began in 1880, according to NOAA's National Climatic Data Center in Asheville, North Carolina. The previous record had been set in 2005. Worldwide temperatures (land and ocean) during the same month were the second-warmest in the record at 1.12°F (0.62°C) above the 20th-century average of 59.9°F (15.5°C). Ocean surface temperature for June 2009 was the warmest on record: 1.06°F (0.59°C) above the 20th-century average of 61.5°F (16.4°C) ("Warmest June" 2009). The land surface temperature was 1.26°F (0.70°C) above the 20th-century average of 55.9°F (13.3°C) and ranked as the sixth-warmest June on record.

"It's Like Farming in Hell"

During the record-warm year of 2012, high corn prices induced farmers in the United States to sow as much as possible. A freakish heat wave, however (three 90°F days in late March and early April in Omaha, plus others in the 80s), led them to plant early. By June and July, searing heat and a sudden, intense drought kept much of the corn from pollinating. Prices soared (little help for farmers with no corn to sell), and everyone had a preview of what the world will be like on a more regular basis in a few decades as thermal inertia begins to reflect its 40- to 50-year lag time. "You couldn't choreograph worse weather conditions for pollination," said Fred Below, a crop biologist at the University of Illinois at Urbana–Champaign. "It's like farming in Hell" (Kolbert 2012, 19).

Even as wildfires devastated hundreds of homes in the U.S. West (and ravaged Siberia as well), Kuwait City experienced a night July 27 during which the temperature did not fall below 100°F. Even so, the U.S. presidential campaign carried on with little mention of climate change. "The familiar disconnect continues," wrote Elizabeth Kolbert in *The New Yorker.* "There's no discussion of what could be done to avert the worst effects of climate change, even as the insanity of doing nothing becomes increasingly obvious" (2012, 20).

In July 2012, scientists reported that thunderstorms were growing so large that their moisture was spilling into the stratosphere, eroding the protective ozone layer, a product of a hydrological cycle jacked up by warming. At ground level, a huge line of thunderstorms reached hurricane intensity and knocked out power to tens of millions of homes across the U.S. Midwest and East Coast, causing people to suffer several days of intense heat without air conditioning. "This is certainly what I and other climate scientists have been warning about. This is what global warming looks like," said Jonathan Overpeck, professor of geosciences and atmospheric sciences at the University of Arizona (Kolbert 2012, 20). "Behind this summer's heat are greenhouse gases emitted decades ago," wrote Kolbert (2012, 20). "Before many effects of today's emissions are felt, it will be time for the Summer Olympics of 2048."

Tropical Zones Expand as Earth Warms

A study published December 2, 2007, in the scientific journal *Nature Geoscience* (an affiliate of the London-based *Nature*) indicated that the tropics have been expanding more quickly than climate models had anticipated. Independent teams using four different meteorological measurements found that the tropical atmospheric belt had grown by anywhere between 2° and 4.8° in latitude since 1979. That translates to a total north and south expansion of 140 to 330 miles. Dian Seidel, a research meteorologist with NOAA in Silver Spring, Maryland, and lead author of the study, said, "They are big changes. . . . It's a little puzzling" (Borenstein 2007). Seidel said that global warming may be coupling with stratospheric ozone depletion and more frequent El Niño conditions to produce expansion of tropical zones.

"Several lines of evidence show that over the past few decades the tropical belt has expanded," Seidel and colleagues wrote.

> This expansion has potentially important implications for subtropical societies and may lead to profound changes in the global climate system. Most importantly, poleward movement of large-scale atmospheric circulation systems, such as jet streams and storm tracks, could result in shifts in precipitation patterns affecting natural ecosystems, agriculture, and water resources. . . . The observed recent rate of expansion is greater than climate model projections of expansion over the 21st century, which suggests that there is still much to be learned about this aspect of global climate change. (Seidel et al. 2007)

Climate scientists Andrew Weaver of the University of Victoria, British Columbia, and Richard Somerville of the Scripps Institution of Oceanography in San Diego said that computer models have often underestimated the effects of global warming. "Every time you look at what the world is doing it's always far more dramatic than what climate models predict," Weaver said (Borenstein 2007).

Dry areas on the edge of the tropics (where weather is governed by descending air in Hadley cells) such as the U.S. Southwest, parts of the Mediterranean. and southern Australia could become drier as this effect intensifies. "You're not expanding the tropical jungles, what you're expanding is the area of desertification," Weaver said (Borenstein 2007).

Further Reading

Borenstein, Seth. "Earth's Tropics Belt Expands, May Mean Drier Weather for U.S. Southwest, Mediterranean." Associated Press, December 2, 2007 (LEXIS).

Seidel, Dian, et al. "Widening of the Tropical Belt in a Changing Climate." *Nature Geoscience*, December 2, 2007. http://www.nature.com/ngeo/journal/v1/n1/full/ngeo .2007.38.html. doi: 10.1038/ngeo.2007.38.

The Northern Hemisphere is warming more quickly than the Southern Hemisphere because of the uneven distribution of land and sea around the globe. According to a report in *Nature*,

Land masses warm more quickly than the ocean—in part because more heat is required to raise the temperature of water than that of land, and sea-surface evaporation produces some cooling. Yangyang Xu and Veerabhadran Ramanathan of the Scripps Institute of Oceanography in La Jolla [California] confirmed this relationship with climate simulations for warming periods (1910–1940 and 1975–2005) as well as a cooling period (1940–1975). For the future, the same model forecasts intense droughts in western North America, parts of Africa, the Amazon Valley, and Australia. ("Lopsided" 2012)

Further Reading

Coumou, Dim, Jascha Lehmann, and Johanna Beckmann. "The Weakening Summer Circulation in the Northern Hemisphere Mid-Latitudes." *Science* 348 (April 17, 2015): 324–327.

Drajem. Mark. "Recent Heat Waves Caused by Global Warming, Hansen Says." Bloomberg Business News. August 06, 2012. http://www.businessweek.com/printer/articles/301255 ?type=bloomberg (no longer available).

"Extreme Weather Events Signal Global Warming to World's Meteorologists." Environment News Service, August 17, 2010. http://www.ens-newswire.com/ens/aug2010/2010-08-17 -01.html.

Gillis, Justin. "Study Finds More of Earth Is Hotter and Says Global Warming Is at Work." *The New York Times*, August 6, 2012. http://www.nytimes.com/2012/08/07/science/earth /extreme-heat-is-covering-more-of-the-earth-a-study-says.html.

Gillis, Justin. "2015 Was Hottest Year in Historical Record, Scientists Say." *The New York Times*, January 20, 2016. http://www.nytimes.com/2016/01/21/science/earth/2015 -hottest-year-global-warming.html.

"GISS 2007 Temperature Analysis." Goddard Institute for Space Studies. January 15, 2008. http://www.columbia.edu/~jeh1/mailings/20080114_GISTEMP.pdf.

Hansen, James E., Makiko Sato, and Reto Riuedy. "Perception of Climate Change." *Proceedings of the National Academy of Sciences*, August 6, 2012. http://www.pnas.org/content /early/2012/07/30/1205276109.abstract. doi: 10.1073/pnas.1205276109.

Hansen, James E. "Climate Change Is Here—and Worse Than We Thought." *Washington Post*, August 3, 2012. http://www.washingtonpost.com/opinions/climate-change-is-here— and-worse-than-we-thought/2012/08/03/6ae604c2-dd90-11e1-8e43-4a3c4375504a _print.html.

Kolbert, Elizabeth. "The Big Heat." *The New Yorker,* July 23, 2012, 19–20.

"Long-Term Global Warming Trend Continues." NASA Earth Observatory, January 16, 2013. http://earthobservatory.nasa.gov/IOTD/view.php?id=80167&src=eoa-iotd.

"Lopsided Warming North to South." *Nature* 487 (July 5, 2012): 9.

"NASA Study Finds World Warmth Edging Ancient Levels." NASA, September 25, 2006. http://www.giss.nasa.gov/research/news/20060925.

"Rising Temperatures: A Month Versus a Decade." NASA Earth Observatory, October 24, 2014. http://earthobservatory.nasa.gov/IOTD/view.php?id=84589&src=eoa-iotd.

"Warmest June on Record for Global Ocean Surface Temperature." Environment News Service, July 221, 2009. http://www.ens-newswire.com/ens/jul2009/2009-07-21-096.asp (no longer available).

See also: Atmospheric Circulation; Climate Change, Abrupt Nature of; Drought, Worldwide; El Niño and La Niña; Temperatures, Greenhouse Gas Levels and; Temperatures, Winter; Thermal Inertia

TEMPERATURES, GREENHOUSE GAS LEVELS AND

The levels of carbon dioxide, methane, and other greenhouse gases in the atmosphere are important because—allowing for achievement of thermal equilibrium over time—they exercise an important role over temperature and thus Earth's habitability. The exploitation of fossil fuels for human comfort, employment, and profit plays a critical role in greenhouse gas levels because their use removes organic materials that have been locked in Earth and ejects them into the atmosphere.

Shale Oil, Tar Sands, and Greenhouse Gas Levels

For many years, there were occasional concerns that the world's industrial infrastructure would consume all of Earth's available oil and natural gas (coal reserves have long been known to be plentiful). The case for "peak oil" dissolved around 2000 when rising prices and oil company ingenuity combined to provide access to vast new reserves of oil extracted from shale (via hydrologic fracturing, also known as "fracking") and tar-laden sand deposits in such places as Alberta, Canada. So much oil and gas are ultimately available that if extractive industries remove (and consumers burn) anything close to most of it, carbon dioxide and methane levels will rise so high within a few hundred years that the much of planet will become uninhabitable as temperatures rise. The new technologies also are highly water intensive as well, often in areas such as the U.S. Midwest that are prone to drought. The *hydraulic* in "hydraulic fracturing" is all about water.

Anticipation of such a future is the major reason why the climate consequences of the Keystone XL pipeline became a cardinal environmental issue after the year 2010, as NASA climate scientist James Hansen (and many hundreds of other people) willingly offered themselves for arrest at the White House in opposition to it. Farmers, ranchers, and Native American peoples were concerned about oil spills as well, but opposition was largely symbolic because oil companies were building a rail network to carry their oil from fracking and tar sands, with a much higher risk of spills than pipelines. (A conservative-dominated government in Canada also was promoting pipelines east and westward from Alberta, until it was replaced by liberals in 2015.)

By 2015, the United States was approaching oil production from fracking that might make it the world's largest producer, with the top producing state becoming North Dakota because of its shale oil in the Bakken Formation, an area that has been turned into a massive oilfield within a few years with application of new oil-drilling technology. At the same time, a glut of oil and natural gas cut the benchmark price of crude in half (from about \$110 to less than \$50 a barrel) within eight months in 2014 and 2015—and then to less than \$30 early in 2016—undermining the boom. Oil from shale and tar sands must be mined and refined to produce usable products at a cost of \$50 to \$85 a barrel, whereas oil that simply squirts out of the ground (in Saudi Arabia, for example) can be shipped to refiners for \$10 to \$15 a barrel. If the price of oil remains relatively low, fracking and mining of tar sands may cease to be profitable, causing supplies to tighten and prices to rise.

Bakken oil shale and the Canadian tar sands introduce huge new sources of fossil fuels. The carbon dioxide level in the atmosphere passed 400 parts per million as fracking and tar sands production boomed. That level is already close to that of the Pliocene Epoch some 2 million to 3 million years ago when temperatures and sea levels were much higher than today. Given thermal inertia in the air (a lag of about 50 years) and the ocean (roughly a century and a half), the atmospheric levels of greenhouse gases are at levels that could replicate those conditions.

Earth "Another Planet"?

When James Hansen was arrested, he knew that if tar sands and oil shale are fully exploited, the level of carbon dioxide in the atmosphere will rise so high that, given the eventual effects of thermal inertia, humanity will have plenty of fossil fuels long after its water is polluted and the air is too hot during the growing season for most plants to pollinate. To fully exploit these resources will be, in Hansen's words, "game over" for the climate as Earth becomes "another planet."

The stakes involved in using this new technology were sketched by Edwin Dobb in the March 2013 issue of the *National Geographic*:

> The implications are already reverberating far beyond North Dakota. Bakken-like shale formations occur across the United States, indeed across the world. The extraction technology refined in the Bakken is in effect a skeleton key that can be used to open other fossil fuel treasure chests. . . . The technology . . . coupled with shifts in the marketplace that favor exploiting deposits that are harder and more expensive to tap has convinced some experts that the carbon-based economy can continue much longer than they had imagined. Billionaire oilman and Bakken pioneer Harold Hamm argues that the assumption we are running out of oil and gas is false. America, in his view, needs a national policy based on abundance, one that doesn't favor developing renewable sources of energy. (Dobb 2013, 37)

Renewable energy such as wind and solar are being developed to decrease reliance on fossil fuels, especially coal. At the same time, increasing consumption of oil, gas, and coal in rapidly developing countries with large populations (most notably India and China, which together contain almost half of Earth's people) are keeping upward pressure on greenhouse gas levels. Reports by the U.N. Intergovernmental Panel on Climate Change anticipate that decades of stalling on worldwide greenhouse-gas limits raise the probability of economic disruption that will require emergency responses such as seeding the atmosphere with sulfur to block incoming solar radiation. According to a report in *The New York Times*, "Governments of the world [are] still spending far more money to subsidize fossil fuels than to accelerate the shift to cleaner energy, thus encouraging continued investment in projects like coal-burning power plants that pose a long-term climate risk" (Gillis 2014).

Greenhouse gas emissions have fallen slightly in recent years in some of the wealthiest countries (in parts of Europe and the United States, for example). But this is "somewhat of an illusion [because] growth of international trade means many of the goods consumed in wealthy countries are now made abroad—so that those countries have, in effect, outsourced their greenhouse gas emissions to countries like China" (Gillis 2014). In other words, China and other countries now host industries that emit greenhouse gases when manufacturing goods that are sold to the United States and other wealthy countries.

Climate models try to project how high temperatures will rise over time given various forecasts of greenhouse gas levels. Early indications are that the models may err on the low side. Climate scientists agree Earth will be hotter by the end of the century, but their simulations do not agree on how much. Now a 2012 study has suggested that gloomier predictions may be closer to the mark. "Warming is likely to be on the high side of the projections," said John Fasullo of the National Center for Atmospheric Research in Boulder, Colorado, coauthor of a report based on satellite measurements of the atmosphere (Vastag 2012). His study projects a rise of roughly 8°F by the end of the 21st century, nearly double the widely accepted target of 3.6°F "as the threshold for triggering catastrophic consequences" (Vastag 2012).

Catastrophe at What Level of Emissions?

A key question in climate change—*the* key issue, according to some—is how do we determine which given level of carbon dioxide and other greenhouse gases in the atmosphere will cause a specific rise or fall in temperature after other "forcings" have been factored in? "For example, how much will global temperatures rise, on average, if the carbon dioxide level doubles, all other factors being equal (which they rarely are)? Humankind has been altering the Earth's climate for several thousand years; cultivation of rice, for example, raises methane levels; biomass burning increases the amount of carbon dioxide in the air" (White 2004, 1610). In effect, human interference with the carbon cycle began in a small way the first time a human being cooked over an open fire.

Writing in *Journal of Climate*, David W. Lea noted that several models have produced varying answers to this question. To sharpen the debate, Lea turned to paleoclimatic data describing both tropical sea surface temperatures and ice cores taken from Antarctica. Using proxy records from the eastern equatorial Pacific Ocean and Vostok (Antarctic) ice cores, Lea arrived at an estimate of "a tropical climate sensitivity of 4.4 to 5.6 [°C] (error estimated at plus or minus 1.0 [°C]) for a doubling of atmospheric CO_2 concentration" (Lea 2004, 2170). According to Lea, this suggests "that the equilibrium response of tropical climate to atmospheric CO_2 changes is likely to be similar to the upper end of available global predictions from global models" (Lea 2004, 2170).

Considerable debate over basic questions takes place in the world of climate modeling. For example, how much will the temperature at Earth's surface increase if carbon dioxide levels double? Models have produced a range of 2.0 to 4.6°C, more than a 100-percent variation, much of it related to differences in estimating how cloudiness influences what scientists call *equilibrium climate sensitivity*. Early in 2016, a scientific team said it had sharpened the focus by accounting more accurately for the role of cloudiness and—at 5.0 to 5.3°C—sharply raising the amount of potential warming at a given temperature level.

Although this study raised the probability that warming may accelerate even with small increases in CO_2 levels, some scientists believe it may have gone too far. Kevin Trenberth, a distinguished senior scientist at the climate and global dynamics laboratory of the National Center for Atmospheric Research, said the new paper "is fine as a first step, but it is not the last step, and much more is needed to establish how clouds change as the climate changes." Trenberth suggested that the model from Yale University's Ivy Tran and colleagues "has too strong a climate sensitivity rather than too small because of clouds" (Schwartz 2016; Tan et al. 2016).

Temperatures Cycle with CO_2 Levels—or Not

During most of Earth's history, temperatures have risen and fallen roughly in tandem with carbon dioxide levels. To add another mystery to the equation, however, this has not *always* been the case. Jan Veizer and colleagues (2000) presented a temperature record using the oxygen isotopic composition of tropical marine fossils that describe two periods with mismatches between carbon dioxide levels and temperatures: the late Ordovician Period (440 million years ago) and the Jurassic and early Cretaceous Periods (120 to 220 million years ago). The findings of Veizer and colleagues indicated that very occasionally rising levels of carbon dioxide were accompanied by a *drop* in average temperatures.

For example, during the Jurassic, 145 million to 208 million years ago, carbon dioxide may have been as much as 10 times as high as current levels; temperature averages, however, were cooler. Veizer believes that the sun may be a major driver of climate change, disrupting the relationship of temperature and CO_2 (Veizer et al. 2000, 698). This conclusion is far from settled, however. Some scientists question the climate proxies used in these studies, asserting that they may be producing inaccurate data (Kump 2000, 651; Pearson et al. 2001, 481).

As Lee R. Kump commented in *Nature*, "Either the CO_2 proxies are flawed, or our understanding of the relationship between CO_2 and climate . . . needs rethinking" (Kump 2000, 651). He believes that a "[lack] of close correspondence between climate change and proxy indicators of atmospheric CO_2 may force us to reevaluate the proxies, rather than disallow the notion that substantially increased atmospheric CO_2 will indeed lead to marked warming in the future" (Kump 2000, 652).

Climate modelers have long been perplexed by an apparent discrepancy between "climate models with increased levels of carbon dioxide [that] predict that global warming causes heating in the tropics [compared to] . . . investigations of ancient climates based on paleodata [that] have generally indicated cool tropical temperatures during supposed greenhouse episodes" (Pearson et al. 2001, 481). During the late Cretaceous Period (970 million years ago) and Eocene Epoch (50 million years ago), for example, the poles were believed to have been ice free, and tropical sea-surface temperatures are estimated to have been between 15 and 23°C, based on oxygen isotope paleothermometry of surface-dwelling planktonic forminifer shells, "which provide proxy information on ocean temperatures" (Kump 2001, 470). These data indicate an ice-free Earth in which tropical sea surface temperatures were cooler than today in most areas—a highly unusual circumstance by present-day standards. Either Earth was strikingly different at that time or the proxies being used to estimate temperatures are flawed—or both.

Debating Proxies

Writing in *Nature* in 2001, Paul N. Pearson and colleagues questioned the validity of most such data "on the grounds of poor preservation and diagenetic alteration" (Pearson et al. 2001, 481). They presented new data "from exceptionally well-preserved foraminifera shells extracted from impermeable clay-rich sediments" that indicate temperatures of at least 28 to 32°C (compared to 25 to 27°C today) in the tropics during these periods. (Foraminifera shells are valuable in paleoclimatic research because they can be dated.)

Such estimates would indicate warming in the tropics consistent with the rest of Earth during periods of high atmospheric carbon dioxide. These new data indicate that "for those parts of the late Cretaceous [Period] and Eocene [Epoch] that we have sampled, tropical temperatures were at least as warm as today, and probably several degrees warmer . . . [allowing] us to dispose of the 'cool tropic paradox' that has bedeviled the study of past warm climates" (Pearson et al. 2001, 486).

This study also confirms the relationship between past episodes of elevated atmospheric carbon dioxide and global warming. These results fit well with existing knowledge that temperature-sensitive organisms such as corals were often displaced northward when the atmosphere's carbon dioxide levels were relatively high (Kump 2001, 471). Such episodes are being used as predictors of conditions in the future when "fossil fuel burning is likely to drive atmospheric CO_2 to perhaps six times the preindustrial level" (Kump 2001, 470).

In 2011, James Hansen and Makiko Sato wrote, "Earth is poised to experience strong amplifying polar feedbacks in response to moderate global warming," so even allowing a 2°C warming (3.6°F) is no cure, but "a prescription for disaster." They continued:

Ice sheet disintegration is nonlinear, spurred by amplifying feedbacks. We suggest that ice sheet mass loss, if warming continues unabated, will be characterized better by a doubling time for mass loss rate than by a linear trend.

Satellite gravity data, though too brief to be conclusive, are consistent with a doubling time of 10 years or less, implying the possibility of multimeter sea level rise this century. Observed accelerating ice sheet mass loss supports our conclusion that Earth's temperature now exceeds the mean Holocene value. Rapid reduction of fossil fuel emissions is required for humanity to succeed in preserving a planet resembling the one on which civilization developed. (Hansen and Sato 2011)

Further Reading

Dobb, Edwin. "The New Oil Landscape." *National Geographic*, March 2013, 28–59.

Gillis, Justin. "U.N. Says Lag in Confronting Climate Woes Will Be Costly." *The New York Times*, January 16, 2014. http://www.nytimes.com/2014/01/17/science/earth/un-says-lag-in-confronting-climate-woes-will-be-costly.html.

Hansen, James E., and Makiko Sato. "Paleoclimate Implications for Human-Made Climate Change." In A. Berger, F. Mesinger, and D Šijački, eds., *Climate Change at the Eve of the Second Decade of the Century: Inferences from Paleoclimate and Regional Aspects: Proceedings of Milutin Milankovitch 130th Anniversary Symposium*, July 2011. New York: Springer.

Kump, Lee R. "What Drives Climate?" *Nature* 408 (December 7, 2000): 651–652.

Kump, Lee R. "Chill Taken Out of the Tropics." *Nature* 413 (October 4, 2001): 470–471.

Lea, David W. "The 100,000-Year Cycle in Tropical SST, Greenhouse Forcing, and Climate Sensitivity." *Journal of Climate* 17(11) (June 1, 2004): 2170–2179. https://courses.seas.harvard.edu/climate/seminars/pdfs/Lea_JofClimate2004.pdf.

Pearson, Paul N., et al. "Warm Tropical Sea-Surface Temperatures in the Late Cretaceous and Eocene Epochs." *Nature* 413 (October 4, 2001): 481–487.

Schwartz, John. "Climate Models May Overstate Clouds' Cooling Power, Research Says." *The New York Times*, April 8, 2016. http://www.nytimes.com/2016/04/08/science/climate-models-may-overstate-clouds-cooling-power-research-says.html.

Tan, Ivy, Trude Storelvmo, and Mark D. Zelinka. "Observational Constraints on Mixed-Phase Clouds Imply Higher Climate Sensitivity." *Science* 352 (April 8, 2016): 224–227.

Vastag, Brian. "Warmer Still: Extreme Climate Predictions Appear Most Accurate, Report Says." *Washington Post*, November 8, 2012. http://www.washingtonpost.com/national/health-science/warmer-still-extreme-climate-predictions-appear-most-accurate-study-says/2012/11/08/ebd075c6-29c7-11e2-96b6-8e6a7524553f_print.html.

Veizer, Jan, Yves Godderis, and Louis M. Francois. "Evidence for Decoupling of Atmospheric CO_2 and Global Climate during the Phanerozoic Eon." *Nature* 408 (December 7, 2000): 698–701.

White, James C. "Do I Hear a Million?" *Science* 304 (June 11, 2004): 1609–1610.

See also: Atmospheric Circulation; Carbon Cycle Feedbacks; Climate Change, Abrupt Nature of; Drought, Worldwide; El Niño and La Niña; Thermal Inertia

TEMPERATURES, WINTER

Although summer heat gets most of the headlines, winters usually produce the largest proportional increases in temperatures under the influence of rising greenhouse gases. On March 16, 2015, for example, a high of 88°F hit Omaha, the highest calendar winter reading (for that early in the year) since records began in

1871. At street level, many people regarded this as a salubrious day after a usually bitter Nebraska winter. Climatically, however, it presents reason for alarm—one more bit of evidence that Earth is warming rapidly.

On a basis basis, surges of unusual late fall and winter warmth have been highly dramatic. In Anchorage, Alaska, 20 days in November 2002 saw low temperatures above the average high. The city ended the month with no snow on the ground, a most unusual event. During October and November, Anchorage had less than 10 percent of its average snowfall. At the same time, the U.S. Northeast and mid-Atlantic region experienced repeated snowfalls. More snow was recorded that winter in Richmond, Virginia, than in Anchorage. A snow drought also afflicted Fairbanks. For the first time in recorded history, Fairbanks during the winter of 2002–2003, did not record a single nighttime temperature below −40°F.

During the week before and after Christmas 2015, a tongue of warm air from the tropics ran up the U.S. East Coast and set all-time highs for the month, one of which was 82°F at Norfolk, Virginia. A suburb of Montreal hit 70°F at 10 a.m. The unusual warmth also spawned several tornadoes that would have been unusual for their ferocity even in springtime; they killed more than 20 people from Indiana to

The White Christmas: Soon to Be Just a Memory?

Statistics provided by researchers at the Oak Ridge National Laboratory in Tennessee who examined weather records of 16 cities, mainly in the northern United States after 1960, indicated that the number of white Christmases declined between the 1960s and 1990s. In Chicago, for example, the number of white Christmases (defined as at least one inch of snow on the ground) dropped from seven in the 1960s to two during the 1990s. In New York, the number declined from five in the 1960s to one during the 1990s, and Detroit had just three white Christmases during the 1990s compared to nine in the 1960s ("Are White" 2001). The snowfall analysis was performed by Dale Kaiser, a meteorologist with the Carbon Dioxide Information Analysis Center at Oak Ridge (which is run by the Department of Energy), and Kevin Birdwell, a meteorologist in the lab's Computational Science and Engineering Division.

There are exceptions. In 2009, for example, 63 percent of the U.S. land area had snow on the ground on Christmas day, including a notable blizzard that left 12 to 20 inches from Omaha to Sioux City, Iowa, and points east and north.

Further Reading

"Are White Christmases Just a Memory?" Environment News Service, December 21, 2001. http://ens-news.com/ens/dec2001/2001L-12-21-09.html (no longer available).

Texas, Arkansas, and Mississippi. In New York City, the average temperature in December 2015 was 51.6°F, 14.1°F above the average of 37.5. Daffodil bulbs rose in suburban gardens along with asters, toad lilies, chrysanthemums, winter jasmines, rhododendrons, and azaleas, among others.

The stormy Christmas week of 2015 combined the ferocity of tornadoes reminiscent of May with blizzards that rivaled the worst of midwinter as tropical air from the south combined with an Arctic blast. At least 56 people died in weather-related events. The ferocious week climaxed December 27 with EF3 and EF4 tornadoes that killed a dozen people in the Dallas, Texas, area, devastating the suburbs of Garland and Rowlett, the most severe tornadoes to hit that area in 27 years at *any* season.

In the climatic context, the tornadic Christmas of 2015 was highly unseasonal and abnormal, perhaps a window on a future that will be characterized by severe warming. A dramatic contrast between East Coast warmth and cold air over the northern Plains (Omaha had a foot of snow during this period) unleashed record rains as well as tornadoes from Texas north and eastward across the southern and midwestern sections of the United States, along with the tornadoes. By the end of December 2015, the Mississippi River was experiencing its second-highest crest in recorded history. St. Louis reported widespread floods in early winter of a type heretofore associated with April or May.

The same storm that provoked extreme warmth in the U.S. East and tornadoes in late December drove temperatures to 50°F above average and above the freezing point at the North Pole. In terms of barometric pressure, the storm was one of the five most intense on record in that area. Heat from the south was conveyed to the pole by sustained south winds of 70 miles an hour, gusting to 100 (Fritz 2015).

A Record of Unusual Warmth

The 21st century climatic record abounds with references to usual warmth. November and early December 2001 were exceptionally warm, especially in the northern and eastern sections of the United States. John Kahionhes Fadden, Mohawk culture bearer, said on December 2 that there was no snow at his house high in the Adirondacks of far upstate New York—extremely unusual for that date. Buffalo, New York, reported no snow in November for the first time since records have been kept, and Marquette, Michigan, reported a record high average of 39.9°F, 10.4°F above average. (Weather is still variable, of course. Two years later, Fadden reported copious snow and temperatures as low as −36°F.)

Ski Slopes to Swimming Pools

As snow becomes scarcer in Europe's Alps, resorts are transforming ski resorts into spas and shopping centers, advertising them as escapes from sweltering lowland cities. In Davos, Switzerland, the InterContinental Resort opened in late 2013. The owners intended for the facility to become a huge spa with luxury hotel rooms, residential apartments, shops, and conference rooms. "A lot of people are telling us: You guys are doing fine because you're far above the critical height line where ski areas will have a problem," said Armin Egger, former director of Davos Tourism. "But we know if about 40 percent of skiing areas in the European Alps will be gone in 50, 100 years, then we will have a problem as well" (Williams 2007). Davos earns money hosting United Nations conferences on climate change.

Atop Klein ("Little") Matterhorn, the highest point in the Alps that tourists can reach on cable cars at 13,120 feet above sea level, the featured attraction has become indoor swimming pools. According to an account in *The New York Times*, the new Aqua Dome near the Austrian ski resort of Solden is "hard to miss [because of] the three enormous concrete bowls that resemble outdoor birdbaths. Each contains a different soaking experience: one is a supersize whirlpool tub, the second has a battery of massage jets, and the third is filled with saltwater and has piped-in underwater music. Inside, a dome-topped spa has two more pools and a waterfall" (Williams 2007).

Further Reading

Williams, Gisela. "Resorts Prepare for a Future without Skis." *The New York Times*, December 2, 2007. http://www.nytimes.com/2007/12/02/travel/02skiglobal.html.

Omaha's November 2001 average temperature in was 49.5°F—10.4°F above average, an all-time record high (dating from 1871) that exceeded the previous record set two years earlier by a remarkable 2.4°F. Some trees were budding in late fall. The local newspaper contained no speculation about global warming in its page-one article on the unusual warmth. The headline on the piece merely bid readers to "enjoy" the warm spell (Gaarder 2001, A1).

In Omaha on December 5, 2001, the temperature was 65°F at 6 a.m., a record daily high before the sun came up, the weather feeling more like August. The warm, humid early morning was followed by thunder, lightning, and a torrential downpour, portending passage of a summer-style cold front. The December 6, 2001, edition of the *Omaha World-Herald* published a page-one story headlined "Warm Weather a Mixed Blessing" that concentrated on flagging sales of snowmobiles in a season that had been utterly bereft of snow to that date.

The unusual weather was not restricted to Omaha, of course. Josh Rubin commented in the *Toronto Star*, "A rose by any other name might smell as sweet. But blooming in Toronto, in December?" (Rubin 2001). "Roses are blooming, bulbs are sprouting, golfers are still hitting the links, and climatologists are puzzled" (Rubin 2001). As John Manning, manager of White Rose Nursery in East York, told Rubin, "People are coming in and asking what to do about their roses still blooming, and wondering if it's safe. Some of their trees are starting to bud. It's pretty weird for this time of year" (Rubin 2001). At Stouffville's Rolling Hills Golf Club, manager John

Finlayson said the phone had been ringing off the hook with golfers requesting tee times. "It's pretty much been the busiest it's been all year," said Finlayson, who kept the course open well past the usual end of the season. "We'd normally close down some time in November, and it wouldn't usually be so busy late in the season," he Finlayson (Rubin 2001).

At about the same time, *Boston Globe* columnist Derrick Z. Jackson wrote,

On Thanksgiving morning here in the cradle of the new North, I went to my backyard to clip a full bunch of cilantro for my guacamole. Earlier that week, I snipped four nice-sized eggplants from the vine, as well as my last jalapeno peppers to smoke into the chipotles that would go into the guacamole. More fresh cilantro is still poking up out of the ground. It is not soaring to reach the summer's sun, but its greens are still vibrant and the taste still a pungent surprise. (Jackson 2001)

According to Jackson, the "vigorous multitudes who ran and biked along the Charles River in tank tops, the smiling families who picked up Christmas trees in T-shirts, and the shoppers who walked in their shorts as if this were Tampa would rather have this than blustery zero-degree wind chill" (Jackson 2001). To the north, Canada by late 2001 was in the midst of its 17th straight warmer than usual season.

Maple Syrup Production Wanes in New England

New England's maple trees require cold weather to produce the sugary sap that becomes maple syrup. In warmer winters, the trees yield considerably less sap and thus affect production of the syrup, which typically requires 40 gallons of sap to make a single gallon of syrup. An analysis of syrup production between 1920 and 2000 indicated a decline in every New England state except Maine (Donn 2002). At the same time, titmice, red belly woodpeckers, northern cardinals, and mockingbirds are being observed more often at bird feeders in Vermont. All of these birds have migrated from more southerly latitudes as temperatures have warmed.

University of New Hampshire forester Rock Barrett, who supervised the survey, said that pervasive warming already may have doomed New England's maple syrup industry. "I think the sugar maple industry is on its way out, and there isn't much you can do about that," he said. Even in 2002, however, roughly one in four Vermont trees was a sugar maple. Vermonters made almost 60 percent of New England's 850,000 gallons of syrup that year, according to federal farm data (Donn 2002).

A newspaper account described the endangered industry:

Every year, as winter begins to melt away, Vermont's sugarhouses come back to life. They puff thick, white smoke from stainless steel evaporators that boil sap down to syrup. Punctuated by vacuum gauges, sap-carrying plastic tubing—a technology that is replacing suspended buckets—snakes through thick woods to collection tubs. Vermont kids, as always, freeze maple treats in the snow. (Donn 2002)

The winter of 2009–2010 was known across much of the United States for its plentiful snowfalls, but in New England above-normal temperatures continued a series of several seasons in which the maple syrup industry suffered from lack of sap induced by unusual warmth. Many farms had planned public exhibitions of sap tapping in March 2010 but found many trees nearly dry ("Mild Weather" 2010). Maine had its third warmest winter on the instrumental record.

Further Reading

Donn, Jeff. "New England's Brilliant Autumn Sugar Maples—and Their Syrup—Threatened by Warmth." Associated Press, September 23, 2002 (LEXIS).
"Mild Weather Zaps Sap of Maple-Syrup Farmers; Some New England Producers Report Shortened Seasons." *USA Today*, March 26, 2010, 4A.

Weather is never a one-way street, however. To be climatically fair, the next winter in eastern Canada (as well as the eastern United States) was colder and snowier than usual, following a global warming tease of several seasons with unusual warmth. In 2003–2004, the northeastern quadrant of the United States and adjacent Canada experienced one of its coldest winters on record. Boston (at 20.7°F) and New York City (at 24.8°F) had their coldest months (both January, in this case) in 70 years. At the same time, the rest of the United States was warmer than usual. Early in 2005, the Boston area was hit by a monumental blizzard that blasted easy certainties about global warming with as much as three feet of snow.

Winter was so late in Moscow during the winter of 2006–2007 that even a chance of snow flurries was newsworthy. Peter Finn wrote in the *Washington Post*:

Scattered flurries teased Moscow . . . with the promise of a real winter, the birthright of a city whose people take pride in trudging through snow and in ice fishing and cross-country skiing in white countryside beyond the outer beltway. The winter of 2006 has yet to arrive, however, and Muscovites are deeply discombobulated. "I want snow. I want the New Year's feeling," said Viktoria Makhovskaya, a street vendor who sells gloves and mittens. "This is a disgusting winter. I don't like it at all." (Finn 2006)

In late December 2006, bears at the Moscow Zoo, who usually begin to hibernate in early November, finally fell asleep after weeks of insomnia because of record mild temperatures. The season was Moscow's warmest on record.

With unusual warmth in Siberia, Canada, northern Asia, and Europe, the world's land areas were 3.4°F warmer than an average during the following January, according to the U.S. National Climatic Data Center in Asheville, North Carolina. That did not just nudge past the old record set in 2002—it broke that mark by 0.81°F, a significantly large increase.

The Met Office, Britain's national weather service, and the University of East Anglia released data at about the same time indicating that 2006 was the warmest year in Britain since 1659, when record keeping began (Finn 2006). Swiss trees put out leaves in midwinter as meadows around low-altitude ski resorts turned green. "We are currently experiencing the warmest period in the Alpine region in 1,300 years," said Reinhard Boehm, climatologist at Austria's Central Institute for Meteorology.

Malcolm Ritter of the Associated Press sketched the national weather scene:

Let's put it this way: People played golf this winter in Maine. In shorts. Buttercups have been blooming in Montana. In Ohio, an ice-free Lake Erie allowed an early start to seasonal ferry service. And the sap started running early in Vermont. While January plunged much of Europe and Russia into the deep freeze, it appeared to be remarkably mild across the United States. Federal scientists haven't calculated yet whether it ranks as the warmest January on record nationwide, but "it's certainly going to be right up there," said Michael Halpert, a meteorologist at the National Oceanic and Atmospheric Administration's Climate Prediction Center. (Ritter 2006)

Winters: Flowers in January

With temperatures approaching 73°F in Washington, D.C., during the first week of January 2007, some scientists ventured the opinion that the year might become the warmest in the instrumental record with the aid of a brief El Niño episode. As Joel Achenbach of the *Washington Post* wrote,

Never has good weather felt so bad. Never have flowers inspired so much fear. Never has the warm caress of a sunbeam seemed so ominous. The weather is sublime, it's glorious, it's *the end of the world*. January is the new March. The daffodils are busting out everywhere. It's porch weather. Put on a T-shirt and shorts, fire up the grill. . . . Everyone out for volleyball! The normal high for this time of year is 43 degrees; yesterday's high at Reagan National was a record-breaking 73. And yet it's all a guilty pleasure. Weather is both a physical and a psychological phenomenon. Meteorology, meet eschatology. We've read the articles, we've seen the Gore movie, we've calculated our carbon footprint, and we're just not intellectually capable anymore of fully enjoying warm winter weather. . . . Greenland melting faster than the Wicked Witch of the West. Just ain't right. Ain't natural. Cherry blossoms during the NFL playoffs? *Run for your lives.* (Achenbach 2008)

In 2006, the twin cities of Minneapolis and St. Paul had their warmest January in 160 years. Ice sculptures at the St. Paul Winter Carnival disintegrated and melted into puddles of water nearly as quickly as they were carved. Several annual ice-fishing contests in Minnesota were canceled or moved because of thin ice. Temperatures in Bismarck, North Dakota, stayed above zero the entire month, a record. The Nonesuch River Golf Club in Scarborough, Maine, hosted 250 players on January 21 and had to turn away 200 more (Ritter 2006).

Further Reading

Achenbach, Joel. "Global Warming Did It! Well, Maybe Not." *Washington Post*, August 3, 2008, B1. http://www.washingtonpost.com/wp-dyn/content/article/2008/08/01/AR20 08080103014_pf.html.

Finn, Peter. "In Balmy Europe, Feverish Choruses of 'Let It Snow.'" *Washington Post*, December 20, 2006, A1. http://www.washingtonpost.com/wp-dyn/content/article/2006/12/19 /AR2006121901681_pf.html.

Fritz, Angela. "Freak Storm Pushes North Pole 50 Degrees Above Normal to Melting Point." *Washington Post*, December 30, 2015, n.p.

Gaarder, Nancy. "State Enjoying 'Exceptional' Warmth." *Omaha World-Herald*, December 4, 2001, A1, A2.

Jackson, Derrick Z. "Sweltering in a Winter Wonderland." *Boston Globe*, December 5, 2001, A23.

Ritter, Malcolm. "This Is Winter? Much of Nation Basked in Warm January," Associated Press, February 3, 2006 (LEXIS).

Rubin, Josh. "Toronto's Blooming Warm; Gardens, Golfers Spring to Life as Record High Nears." *Toronto Star*, December 5, 2001, B2.

See also: Atmospheric Circulation; Climate Change, Abrupt Nature of; Drought, Worldwide; El Niño and La Niña; Temperatures, Global; Thermal Inertia

THERMAL INERTIA

The concept of thermal inertia describes the process by which Earth's atmosphere reaches an equilibrium to balance incoming and outgoing radiation; in the long run, this determines surface temperatures within variations imposed by other aspects of weather and climate. With greenhouse gas levels increasing, outgoing radiation is impeded, resulting in a radiative imbalance. Determining climatic equilibrium involves scientific formulas that measure the pace at which the warming we feel in the atmosphere catches up with the "forcing" of greenhouse gases. The observed level of temperatures in the air is probably some 50 years behind actual emission levels; in the oceans, equilibrium is reached much more slowly (100 to 150 years), given their great thermal inertia.

Venus, Climatic Precedents

Planetary ecosystems are forever evolving, a fact that may bring cold comfort to students of the greenhouse effect who cast their eyes on Venus, where catastrophic greenhouse-effect warming has raised surface temperatures to a searing 850°F (464°C), which is hot enough to melt lead.

The planet Venus is known today to have one of the most hellish environments anywhere in the solar system, an example of what can happen to a planet in the grip of a runaway greenhouse effect. A scientific case has been made that not long ago by the standards of geologic time the planet's atmosphere far more closely resembled Earth's. The geophysical career of Venus has long been a subject of inquiry among leading scientists of global warming such as James E. Hansen, who began his professional life as a student of that planet. The climatic transect of Venus may be a cautionary tale for our Earth.

Some contemporary theories argue that Venus may have once had a climate more like that of today's Earth, "complete with giant rivers, deep oceans, and teeming with life" (Leake 2002). Two British scientists believe they have found evidence that rivers the size of the Amazon once flowed for thousands of miles across the Venusian landscape before emptying into liquid water seas. According to a report by Jonathan Leake in the *Sunday Times*, these scientists used radar images from a NASA probe to trace the river systems, deltas, and other features they say could have been created only by moving water (Leake 2002).

No one on Earth knew much about Venusian topography or climate until 1990, the year NASA's Magellan probe used radar to penetrate clouds and map the planet's surface. Magellan's images showed that the surface had been carved by large riverlike channels that scientists then thought were caused by volcanic lava flows. Jones and colleagues reanalyzed the same images using latest computer technology and found they were too long to have been created by lava (Leake 2002).

Adrian Jones, a planetary scientist at University College of London who carried out the research, said the findings suggested life on Venus could have evolved on a parallel with Earth's. "If the climate and temperature were right for water to flow, then they would have been right for life, too. It suggests life could once have existed there" (Leake 2002). Studies compiled by David Grinspoon of the Southwest Research Institute at Boulder, Colorado, suggest that Venus may have been habitable for as long as 2 billion years before an accelerated greenhouse effect dried its oceans.

Roughly 20 exploration missions to Venus have returned enough data to construct an image of Venus today as a hellishly hot place, "its skies dominated by clouds of sulfuric acid, poisoned further by hundreds of huge volcanoes that belch lava and gases into an atmosphere lashed by constant hurricane winds" (Leake 2002). Venus differs from Earth in one important respect, however: it has no tectonic plates that permit stresses to express themselves piecemeal via earthquakes and volcanic eruptions. Venus may have had a shallow ocean for as long as 2 billion years, but the last liquid water may have evaporated as recently as 715 million years ago.

Something (perhaps a surge of volcanic eruptions long contained by the lack of tectonics) triggered runaway global warming that radically changed the Venusian climate and eventually boiled away the oceans (Leake 2002). According to this research, warming on Venus may have been accelerated by heat released into its atmosphere by billions of tons of carbon dioxide from rocks and possibly vegetation. Today, Venus' atmosphere is 95 percent carbon dioxide.

Further Reading

Leake, Jonathan. "Fiery Venus Used to Be Our Green Twin." *Sunday Times* (London), December 15, 2002, 11.

Feedback loops involve climatic forcings (influences) that compound each other and add to the speed and intensity of global warming. The release of carbon dioxide and methane from melting permafrost is an example of a feedback loop; changes in albedo from white snow to heat-absorbing dark surfaces (forests or oceans) are another. In the meantime, around the Arctic Circle, permafrost melts, adding still more carbon dioxide and methane and accelerating the natural process that feeds on itself. To these natural processes add the trigger of increasing human emissions.

At a tipping point, feedbacks take control and propel a climatic forcing such as the effect of increasing levels of greenhouse gases beyond a point where human control ("mitigation") is possible. A tipping point may occur when a small additional forcing causes large climate change. At a tipping point, feedbacks force temperatures to rise long after their provocation—human greenhouse gas generation—has been removed. This process is accelerated by thermal inertia. In the case of carbon dioxide, these effects continue and intensify for centuries.

Mars: Climate-Change Precedents

Astronomers and science-fiction writers long have considered Mars a possible twin of Earth, with flowing water and possibly life. Although today's Martian climate probably would not support any kind of life (it is too dry, probably too cold, and has an atmosphere that is more than 90 percent carbon dioxide), recent probes of the planet's soil raise a possibility that at one time Mars had a climate much more similar to Earth's, with a warmer and richer atmosphere and copious liquid water that flowed on the surface, at least occasionally. If so, what happened, and does the change harbor any lessons for Earth's changing climate?

The surface topography of Mars suggests that water once flowed there. Mars Rovers have also found soil chemistry that suggests the presence of water. The change may have begun 4 billion years ago. "I think the short story is the atmosphere went away and the oceans froze but are still there, locked up in subsurface ice," said Chris McKay, an astrobiologist and Mars expert at NASA's Ames Research Center (Overbye 2014). Radiation from the sun ("solar wind") may have played a role.

Some astronomers believe that long ago Mars had a denser atmosphere and just enough carbon dioxide (as Earth has now) to keep water from freezing most of the time. Others theorize that asteroid impacts changed the planet's axial tilt and may have created conditions conducive to water flow for relatively short periods. Perhaps it had (and gradually lost) a magnetic field similar to Earth's, which led the atmosphere to slowly escape.

Dennis Overbye described one theory in *The New York Times* (2014): "The impact that created the huge crater called the Hellas basin, for example, would have hurled vast amounts of vaporized rock into the sky—leading to decades or centuries of hot rain and flash floods, said Brian Toon of the University of Colorado. It might have been followed perhaps by a lingering era of nearly freezing temperatures as clouds left over from the steam bath produced a mild greenhouse effect" (2014).

The upshot is that no one yet knows why Martian climate changed. The takeaway from Mars (and Venus) is that planetary climate changes over time, a cautionary tale for everyone on Earth as human emissions of greenhouse gases alter our atmosphere.

Further Reading

Overbye, Dennis. "Looking to Mars to Help Understand Changing Climates." *The New York Times*, December 8, 2014. http://www.nytimes.com/2014/12/09/science/looking-to-mars-to-help-understand-changing-climates.html.

The amount of time that temperatures and sea levels may continue to rise even after greenhouse gas levels stabilize is not yet known with any degree of certainty, however. As one scientific observer noted, "The rise in mean global SAT [surface average temperature] and the world ocean level (due to thermal expansion) may . . . continue for several centuries after the stabilization of the carbon dioxide concentration [in the atmosphere], due to the gigantic thermal inertia of the oceans." The same observer also asserts, "The response of ice sheets to earlier climate changes may continue for several centuries after the climate stabilizes" (Kondratyev et al. 2004, 45).

Portending a Hot Future

Thus, given the workings of thermal inertia, climatic equilibrium, feedback loops, and tipping points, somewhere in the middle of this century, when our children are grandparents, the planetary emergency of which we are now just tasting the first course will be a dominant theme in everyone's life, unless there is drastic reduction in greenhouse gas levels today. Within a few decades, these geophysical mechanisms will provoke acceleration of the warming rate, portending a hot and miserable future for coming generations. Thermal inertia explains why so many scientists find the problem so urgent now. The real problem with global warming is not how warm it is today. Today's carbon emissions do not give us tomorrow's temperature. The real debate is not over how much the oceans may rise from melting ice by the end of this century (one to four feet, perhaps) but how much melting will be "in the pipeline" by that time for the next century and beyond.

Because of the great thermal inertia of the ocean, probably only half or so of the eventual warming from gases already in the air has been realized. Earth has warmed 2°F since widespread use of fossil fuels began 250 years ago, but there is another one degree already in the pipeline. Moreover, there is surely more warming in the pipeline.

The Earth's Changing Radiative Balance

Over the last 400,000 years, Earth has usually been nearly in energy balance. The amount of radiation from the sun has equaled that radiating back into space. Thus, calculated temperature change was simply a product of the forcing and equilibrium (i.e., long-term) climate sensitivity. This assumption is no longer valid in modern times because climate forcings are changing so quickly, especially in the past few decades, that the climate system has not come to equilibrium with today's climate forcings. Thus, the observed temperature rise in the past century is only part of that expected for the current levels of greenhouse-gases. The additional amount of global warming in the pipeline is probably 0.5°C (1°F) (Hansen 2006, 10).

"If warming approaches the range of 2 to 3 [°C], large-scale disastrous climate impacts for humans as well as for other inhabitants of the planet will be virtually certain," said James E. Hansen, former director of the Goddard Institute for Space Studies. "An important point to note is that the tripwire between keeping global warming less than 1°C, as opposed to having a warming that approaches the range 2–3°C, may depend upon a relatively small difference in human-made direct forcings." Hansen has written. "The reason for that conclusion is that keeping global warming less than 1°C requires having both a moderate limit on CO_2 and a reduction of non-CO_2 forcings. However, if the warming becomes greater than 1°C, it may be come impractical to achieve and maintain the presumed reductions of the non CO_2 forcings, in part because positive [greenhouse gas] feedbacks from tundra or other terrestrial sources may become significant in the warmer climate" (Hansen 2006, 15–16).

Sir John Houghton, one of the world's leading experts on global warming, told *The Independent* (London), "We are getting almost to the point of irreversible meltdown, and will pass it soon if we are not careful" (Lean 2004, 8). A report by the National Academy of Sciences prepared during 2001 at the request of the presidential administration of George W. Bush estimated that in coming years only 40 percent of potential warming will be caused directly by increasing levels of greenhouse gases. The other 60 percent may be caused by feedback mechanisms. "Together," according to this report, "these two feedbacks amplify the simulated climate response to the greenhouse gas forcing by a factor of 2.5" (Victor 2004, 146). The same summary pointed out that this factor is an estimate and that different climate models forecast varying ranges of feedback responses, producing differing estimates of future warming.

Further Reading

Hansen, James E. "Declaration of James E. Hansen." *Green Mountain Chrysler-Plymouth-Dodge-Jeep, et al., Plaintiffs v. Thomas W. Torti, Secretary of the Vermont Agency of Natural Resources, et al., Defendants.* Case Nos. 2:05-CV-302 and 2:05-CV-304, Consolidated. United States District Court for the District of Vermont. August 14, 2006. http://www.giss.nasa.gov/~dcain/recent_papers_proofs/vermont_14aug2 0061_textwfigs.pdf (no longer available).

Lean, Geoffrey. "Global Warming Will Redraw Map of the World." *The Independent* (London), November 7, 2004, 8.

Victor, David G. *Climate Change: Debating America's Policy Options.* New York: Council on Foreign Relations, 2004.

Earth's rising temperature is gradually raising the planet's sea level both through thermal expansion of the oceans and the melting of glaciers and ice sheets. Scientists are particularly concerned by the melting of the Greenland and West Antarctic ice, which has accelerated sharply in recent years. If these ice sheets, which are a mile thick in some places, were to melt entirely, the sea level would rise by several meters. Even a one-meter (39-inch) sea level rise would inundate vast areas of low-lying coastal land, including many of the rice-growing river deltas and floodplains of India, Thailand, Viet Nam, Indonesia, and China.

Too Little, Too Late?

Diplomats and climate scientists have been gathering for more than 20 years to debate and negotiate a worldwide pact to stall the rise in greenhouse gas emissions amid a rising chorus of warnings that the results may be too little, too late. The

United States and China reached a first-time agreement in 2014, including a joint announcement that the United States will cut its emissions as much as 28 percent by 2025, and China's emissions will peak by 2030. In India, which became the world's third-largest source of greenhouse gases by 2014, the government does not expect that emissions will peak until at least 2040. Action to reduce emissions is all "on speculation"—in the future—as the levels of carbon dioxide and methane continue to rise. And as long as these levels rise, humankind is losing its battle against warming.

Thermal inertia may allow large bodies of water to hold heat long after greenhouse gas levels reach peak levels. For example, the Amundsen Sea embayment in West Antarctica, one of Earth's most quickly eroding masses of ice, accelerated by nearly three times within two decades, from 1992 to 2013. According to recent research, these glaciers "are hemorrhaging ice faster than any other part of Antarctica and are the most significant Antarctic contributors to sea level rise" (Sutterley et al. 2014; "West Antarctica" 2014). "The mass loss of these glaciers is increasing at an amazing rate," said scientist Isabella Velicogna, who took part in the study and works at the University of California–Irvine and NASA's Jet Propulsion Laboratory. Tyler Sutterley, lead author of an article on the Amundsen ice melt, and a UC Irvine doctoral candidate, calculated that the Amundsen lost as much ice every two years on average as Mount Everest weighs.

Earth's Planetary Energy Imbalance

James Hansen and several other scientists have published temperature readings from the deep ocean that trace a clear warming trend indicative of the planet's thermal inertia and energy imbalance—the difference between the amount of heat absorbed by Earth and the amount radiated out into space. This thermal imbalance (0.85 watts plus or minus 0.15 per square meter) is evidence of a steadily warming world, raising the odds of a catastrophic sudden change marked by rising seas and melting ice caps (Hansen et al. 2005, 1431).

Hansen and colleagues concluded that the unusual magnitude of the warming trend could not be explained by natural variability but instead fit precisely with theories suggesting that human activity is the dominant "forcing agent." "This energy imbalance is the 'smoking gun' that we have been looking for," Hansen said in a prepared summary of the study, which was published in the journal *Science*. "The magnitude of the imbalance agrees with what we calculated using known climate forcing agents, which are dominated by increasing human-made greenhouse gases. There can no longer be substantial doubt that human-made gases are the cause of most observed warming" (Hall 2005).

By Hansen's estimate, 25 to 50 years are required for Earth's surface temperature to reach 60 percent of its equilibrium response—a key concept when diplomacy and policy usually respond to experience rather than expectations of climate change. Regarding those expectations, 0.85 plus or minus 0.15 watts per square meter of excess heat absorbed by Earth may not seem like much, but it adds up when multiplied by many square meters over many years.

Earth's planetary energy imbalance did not exceed more than a few tenths of a watt per square meter before the 1960s, according to Hansen and colleagues. Since then, the amount of excess heat absorbed by the atmosphere has grown steadily, except for years after large volcanic eruptions, to a level much above historical averages. Much of this excess heat ends up in the oceans, where it helps to melt ice in the Arctic and Antarctic. In the measured tones with which scientists express themselves, accelerating ice melt could "create the possibility of a climate system in which large sea-level change is practically impossible to avoid" (Hansen et al. 2005, 1434). Hansen and colleagues wrote that continuing the present energy imbalance could lead to a climate that is "out of control" (Hansen et al. 2005, 1434).

Further Reading

Hall, Carl T. "Ocean Tells the Story: Earth Is Heating Up; Human Activity, Not Variables in Nature, Cited as Culprit." *San Francisco Chronicle*, April 29, 2005, A1.

Hansen, James E., et al. "Earth's Energy Imbalance: Confirmation and Implications." *Science* 308 (June 3, 2005): 1431–1435.

Jeff K. Ridley and Helene T. Hewitt, both of whom work at the United Kingdom's Met Office Hadley Center in Exeter, found that ice melt on or near Antarctica may be irreversible, even if carbon dioxide levels stabilize or fall. Their simulation tested ice melt at preindustrial CO_2 levels and found that it did not reform. The problem was thermal inertia in the ocean around Antarctica, which kept ocean water relatively warm and continued to inhibit formation of new ice.

They wrote in *Geophysical Research Letters*:

We find evidence that ocean processes during global warming may result in irreversible changes to the Antarctic sea ice, whereas the Arctic sea ice changes appear to be reversible. Increased forcing gives rise to strong heat uptake in the Southern Ocean, and existing pathways provide an increased transport of heat to the Weddell Sea. As atmospheric concentrations of CO_2 are returned to preindustrial levels, the Antarctic ice extent at first recovers, but a rapid change in the position of the ocean front in the South Atlantic maintains the heat transport into the Weddell Sea. A cooling surface initiates deep convection, accessing the stored heat, resulting in a substantial loss of sea ice, which has not recovered after a further 150 years at preindustrial CO_2. (Ridley and Hewitt 2014)

Paleoclimatic Precedents

According to a theory first advanced by Anthony Hallam and Paul Wignall (1997), the volcanic eruptions 251 million years ago provoked several climatic feedback loops that accelerated global warming of some 6°C. Michael Benton elaborated on this idea in "What Caused the Biggest Catastrophe of all Time?," a chapter of his book *When Life Nearly Died: The Greatest Mass Extinction of All Time* (2003). Benton sketched how the warming (which was accompanied by anoxia) may have fed on itself:

> The end-Permian runaway greenhouse may have been simple. Release of carbon dioxide from the eruption of the Siberian Traps [volcanoes] led to a rise in global temperatures of 6 [°C] or so. Cool polar regions became warm and frozen tundra became unfrozen. The melting might have penetrated to the frozen gas hydrate reservoirs located around the polar oceans, and massive volumes of methane may have burst to the surface of the oceans in huge bubbles. This further input of carbon into the atmosphere caused more warming, which could have melted further gas hydrate reservoirs. So the process went on, running faster and faster. The natural systems that normally reduce carbon dioxide levels could not operate, and eventually the system spiraled out of control, with the biggest crash in the history of life. (Benton 2003, 276–277)

And What about the Far Future?

Most scientific projections of global warming's effects on our atmosphere, our climate, and our lives—the latest report of the Intergovernmental Panel on Climate Change (IPCC) is an example—do not extend past the end of the present century. The projections do not end because global warming will end but because consensus science does not have the tools to forecast any further. It is not a matter of humanity, having had enough, and being able to set a thermostat and reverse things. In fact, thermal inertia and various other feedbacks virtually guarantee a rapid amplification of warming into the next century and beyond if fossil fuels continue to be burned at anything like today's rate.

The lag in temperatures caused by thermal inertia and natural variability have allowed climate-change deniers to argue that geophysical facts do not matter, even as the atmospheric carbon dioxide level passes 400 parts per million, the level during parts of the Pliocene Epoch, when oceans were 100 feet higher. Where is the warming, they ask? Unfortunately, they ignore the role of thermal inertia and climate equilibrium, feedbacks, and tipping points.

Geophysics does matter. At some point—exactly when we do not yet really know—if greenhouse gas emissions are not sharply curtailed, then feedbacks may compound each other rapidly enough to accelerate changes so quickly that remedy by the hand of humanity will be impossible. The IPCC roughly calculates that time as around 2040 to 2050.

By that time, the climate deniers will have been discredited, but it will be too late. We will be on our way to a world where the air in most places will eventually become simply too hot to sustain human (or most other) life. "It is not an exaggeration to suggest," wrote James E. Hansen in an e-mail post on September 26, 2013, "based on best available scientific evidence, that burning all fossil fuels could result in the planet being not only ice-free but human-free" (Merchant 2013).

Further Reading

Benton, Michael J. *When Life Nearly Died: The Greatest Mass Extinction of All Time*. London: Thames & Hudson, 2003.

Hallam, Anthony, and Paul Wignall. *Mass Extinctions and Their Aftermath*. Oxford, UK: Oxford University Press, 1997.

Kondratyev, Kirill, Vladimir F. Krapivin, and Costas A. Varotsos. *Global Carbon Cycle and Climate Change*. Berlin, Germany: Springer/Praxis, 2004.

Merchant, Brian. "The Nation's Top Climate Scientist Predicts an 'Ice-Free, Human-Free' Planet." Motherboard, April 17, 2013. http://motherboard.vice.com/blog/the-nations-top -climate-scientist-predicts-an-ice-free-human-free-planet (no longer available).

Ridley, Jeff K., and Helene T. Hewitt. "A Mechanism for Lack of Sea Ice Reversibility in the Southern Ocean." *Geophysical Research Letters*, December 1, 2014. doi: 10.1002 /2014GL062167.

Sutterley Tyler C., et al. "Mass Loss of the Amundsen Sea Embayment of West Antarctica from Four Independent Techniques." *Geophysical Research Letters*, December 5, 2014. doi: 10.1002/2014GL061940.

"West Antarctic Glacier Loss Appears Unstoppable, UCI-NASA Study Finds." UCI News. University of California–Irvine. December 2, 2014. http://news.uci.edu/press-releases /west-antarctic-melt-rate-has-tripled-uc-irvine-nasa/West.

See also: Carbon Cycle Feedbacks; Climate Change, Abrupt Nature of; Pliocene Paleoclimate; Temperatures, Global; Temperatures, Greenhouse Gas Levels and

Index

Bold indicates volume numbers.